Graduate Texts in Mathematics 114

А. Т. ФОМЕНКО

Neal Koblitz

# A Course in Number Theory and Cryptography

Second Edition

Springer

Neal Koblitz
Department of Mathematics
University of Washington
Seattle, WA 98195
USA



Mathematics Subject Classification (2000): 11-01, 11T71

With 5 Illustrations.

Library of Congress Cataloging-in-Publication Data
Koblitz, Neal, 1948–
A Course in number theory and cryptography / Neal Koblitz. — 2nd ed.
p. cm. — (Graduate texts in mathematics ; 114)
Includes bibliographical references and index.
ISBN 0-387-94293-9 (New York : acid-free). — ISBN 3-540-94293-9
(Berlin : acid-free
1. Number theory. 2. Cryptography. I. Title. II. Series.
QA169.M33 1998
512′.7—dc20 94-11613

ISBN 0-387-94293-9 Printed on acid-free paper.



12 11 10 9 8

springer.com

# Foreword

> ...both Gauss and lesser mathematicians may be justified in rejoicing that there is one science [number theory] at any rate, and that their own, whose very remoteness from ordinary human activities should keep it gentle and clean.
>
> — G. H. Hardy, *A Mathematician's Apology*, 1940

G. H. Hardy would have been surprised and probably displeased with the increasing interest in number theory for application to "ordinary human activities" such as information transmission (error-correcting codes) and cryptography (secret codes). Less than a half-century after Hardy wrote the words quoted above, it is no longer inconceivable (though it hasn't happened yet) that the N.S.A. (the agency for U.S. government work on cryptography) will demand prior review and clearance before publication of theoretical research papers on certain types of number theory.

In part it is the dramatic increase in computer power and sophistication that has influenced some of the questions being studied by number theorists, giving rise to a new branch of the subject, called "computational number theory."

This book presumes almost no background in algebra or number theory. Its purpose is to introduce the reader to arithmetic topics, both ancient and very modern, which have been at the center of interest in applications, especially in cryptography. For this reason we take an algorithmic approach, emphasizing estimates of the efficiency of the techniques that arise from the theory. A special feature of our treatment is the inclusion (Chapter VI) of some very recent applications of the theory of elliptic curves. Elliptic curves have for a long time formed a central topic in several branches of theoretical mathematics; now the arithmetic of elliptic curves has turned out to have potential practical applications as well.

Extensive exercises have been included in all of the chapters in order to enable someone who is studying the material outside of a formal course structure to solidify her/his understanding.

The first two chapters provide a general background. A student who has had no previous exposure to algebra (field extensions, finite fields) or elementary number theory (congruences) will find the exposition rather condensed, and should consult more leisurely textbooks for details. On the other hand, someone with more mathematical background would probably want to skim through the first two chapters, perhaps trying some of the less familiar exercises.

Depending on the students' background, it should be possible to cover most of the first five chapters in a semester. Alternately, if the book is used in a sequel to a one-semester course in elementary number theory, then Chapters III–VI would fill out a second–semester course.

The dependence relation of the chapters is as follows (if one overlooks some inessential references to earlier chapters in Chapters V and VI):

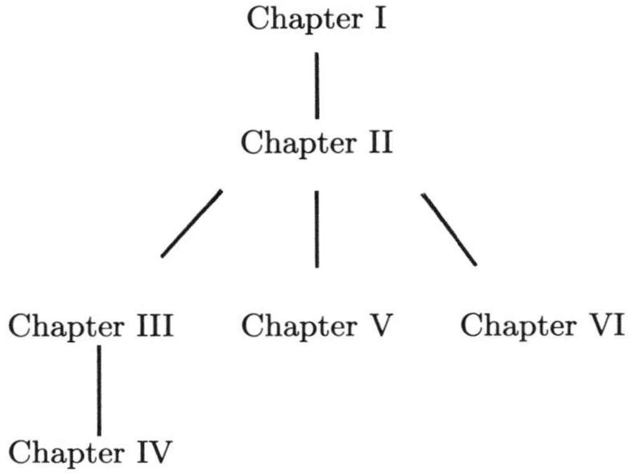


This book is based upon courses taught at the University of Washington (Seattle) in 1985–86 and at the Institute of Mathematical Sciences (Madras, India) in 1987. I would like to thank Gary Nelson and Douglas Lind for using the manuscript and making helpful corrections.

The frontispiece was drawn by Professor A. T. Fomenko of Moscow State University to illustrate the theme of the book. Notice that the coded decimal digits along the walls of the building are not random.

This book is dedicated to the memory of the students of Vietnam, Nicaragua and El Salvador who lost their lives in the struggle against U.S. aggression. The author's royalties from sales of the book will be used to buy mathematics and science books for the universities and institutes of those three countries.

Seattle, May 1987

# Preface to the Second Edition

As the field of cryptography expands to include new concepts and techniques, the cryptographic applications of number theory have also broadened. In addition to elementary and analytic number theory, increasing use has been made of algebraic number theory (primality testing with Gauss and Jacobi sums, cryptosystems based on quadratic fields, the number field sieve) and arithmetic algebraic geometry (elliptic curve factorization, cryptosystems based on elliptic and hyperelliptic curves, primality tests based on elliptic curves and abelian varieties). Some of the recent applications of number theory to cryptography — most notably, the number field sieve method for factoring large integers, which was developed since the appearance of the first edition — are beyond the scope of this book. However, by slightly increasing the size of the book, we were able to include some new topics that help convey more adequately the diversity of applications of number theory to this exciting multidisciplinary subject.

The following list summarizes the main changes in the second edition.

- Several corrections and clarifications have been made, and many references have been added.
- A new section on zero-knowledge proofs and oblivious transfer has been added to Chapter IV.
- A section on the quadratic sieve factoring method has been added to Chapter V.
- Chapter VI now includes a section on the use of elliptic curves for primality testing.
- Brief discussions of the following concepts have been added: $k$–threshold schemes, probabilistic encryption, hash functions, the Chor–Rivest knapsack cryptosystem, and the U.S. government's new Digital Signature Standard.

Seattle, May 1994

# Contents

# I

# Some Topics in Elementary Number Theory

Most of the topics reviewed in this chapter are probably well known to most readers. The purpose of the chapter is to recall the notation and facts from elementary number theory which we will need to have at our fingertips in our later work. Most proofs are omitted, since they can be found in almost any introductory textbook on number theory. One topic that will play a central role later — estimating the number of bit operations needed to perform various number theoretic tasks by computer — is not yet a standard part of elementary number theory textbooks. So we will go into most detail about the subject of time estimates, especially in §1.

## 1 Time estimates for doing arithmetic

**Numbers in different bases.** A nonnegative integer $n$ written to the *base* $b$ is a notation for $n$ of the form $(d_{k-1}d_{k-2}\cdots d_1d_0)_b$, where the $d$'s are *digits*, i.e., symbols for the integers between 0 and $b-1$; this notation means that $n = d_{k-1}b^{k-1} + d_{k-2}b^{k-2} + \cdots + d_1b + d_0$. If the first digit $d_{k-1}$ is not zero, we call $n$ a $k$-digit base-$b$ number. Any number between $b^{k-1}$ and $b^k$ is a $k$-digit number to the base $b$. We shall omit the parentheses and subscript $(\cdots)_b$ in the case of the usual decimal system ($b = 10$) and occasionally in other cases as well, if the choice of base is clear from the context, especially when we're using the binary system ($b = 2$). Since it is sometimes useful to work in bases other than 10, one should get used to doing arithmetic in an arbitrary base and to converting from one base to another. We now review this by doing some examples.

**Remarks.** (1) Fractions can also be expanded in any base, i.e., they can be represented in the form $(d_{k-1}d_{k-2}\cdots d_1d_0.d_{-1}d_{-2}\cdots)_b$. (2) When $b > 10$ it is customary to use letters for the digits beyond 9. One could also use letters for *all* of the digits.

**Example 1.** (a) $(11001001)_2 = 201$.

(b) When $b = 26$ let us use the letters A—Z for the digits 0—25, respectively. Then $(\mathrm{BAD})_{26}$=679, whereas $(\mathrm{B.AD})_{26} = 1\frac{3}{676}$.

**Example 2.** Multiply 160 and 199 in the base 7. **Solution:**

$$\begin{array}{r} 316 \\ \underline{403} \\ 1254 \\ \underline{16030\phantom{0}} \\ 161554 \end{array}$$

**Example 3.** Divide $(11001001)_2$ by $(100111)_2$, and divide $(\mathrm{HAPPY})_{26}$ by $(\mathrm{SAD})_{26}$.

**Solution:**

$$\begin{array}{rl} & 101\,\frac{110}{100111} \\ 100111 & \overline{11001001} \\ & \underline{100111}\phantom{00} \\ & 101101\phantom{0} \\ & \underline{100111}\phantom{0} \\ & 110 \end{array} \qquad\qquad \begin{array}{rl} & \mathrm{KD}\,\frac{\mathrm{MLP}}{\mathrm{SAD}} \\ \mathrm{SAD} & \overline{\mathrm{HAPPY}} \\ & \underline{\mathrm{GYBE}}\phantom{Y} \\ & \mathrm{COLY} \\ & \underline{\mathrm{CCAJ}} \\ & \mathrm{MLP} \end{array}$$

**Example 4.** Convert $10^6$ to the bases 2, 7 and 26 (using the letters A—Z as digits in the latter case).

**Solution.** To convert a number $n$ to the base $b$, one first gets the last digit (the ones' place) by dividing $n$ by $b$ and taking the remainder. Then replace $n$ by the quotient and repeat the process to get the second-to-last digit $d_1$, and so on. Here we find that

$$10^6 = (11110100001001000000)_2 = (11333311)_7 = (\mathrm{CEXHO})_{26}.$$

**Example 5.** Convert $\pi = 3.1415926\cdots$ to the base 2 (carrying out the computation 15 places to the right of the point) and to the base 26 (carrying out 3 places to the right of the point).

**Solution.** After taking care of the integer part, the fractional part is converted to the base $b$ by multiplying by $b$, taking the integer part of the result as $d_{-1}$, then starting over again with the fractional part of what you now have, successively finding $d_{-2}, d_{-3}, \ldots$. In this way one obtains:

$$3.1415926\cdots = (11.001001000011111\cdots)_2 = (\mathrm{D.DRS}\cdots)_{26}.$$

**Number of digits.** As mentioned before, an integer $n$ satisfying $b^{k-1} \leq n < b^k$ has $k$ digits to the base $b$. By the definition of logarithms, this gives the following formula for the number of base-$b$ digits (here "[ ]" denotes the greatest integer function):

$$\text{number of digits} = \left[log_b n\right] + 1 = \left[\frac{log\, n}{log\, b}\right] + 1,$$

where here (and from now on) "log" means the natural logarithm $log_e$.

**Bit operations.** Let us start with a very simple arithmetic problem, the addition of two binary integers, for example:

$$\begin{array}{r} {\scriptstyle 1111}\phantom{000} \\ 1111000 \\ +\underline{0011110} \\ 10010110 \end{array}$$

Suppose that the numbers are both $k$ bits long (the word "bit" is short for "binary digit"); if one of the two integers has fewer bits than the other, we fill in zeros to the left, as in this example, to make them have the same length. Although this example involves small integers (adding 120 to 30), we should think of $k$ as perhaps being very large, like 500 or 1000.

Let us analyze in complete detail what this addition entails. Basically, we must repeat the following steps $k$ times:

1. Look at the top and bottom bit, and also at whether there's a carry above the top bit.
2. If both bits are 0 and there is no carry, then put down 0 and move on.
3. If either (a) both bits are 0 and there is a carry, or (b) one of the bits is 0, the other is 1, and there is no carry, then put down 1 and move on.
4. If either (a) one of the bits is 0, the other is 1, and there is a carry, or else (b) both bits are 1 and there is no carry, then put down 0, put a carry in the next column, and move on.
5. If both bits are 1 and there is a carry, then put down 1, put a carry in the next column, and move on.

Doing this procedure once is called a *bit operation.* Adding two $k$-bit numbers requires $k$ bit operations. We shall see that more complicated tasks can also be broken down into bit operations. The amount of time a computer takes to perform a task is essentially proportional to the number of bit operations. Of course, the constant of proportionality — the number of nanoseconds per bit operation — depends on the particular computer system. (This is an over-simplification, since the time can be affected by "administrative matters," such as accessing memory.) When we speak of estimating the "time" it takes to accomplish something, we mean finding an estimate for the number of bit operations required. In these estimates we shall neglect the time required for "bookkeeping" or logical steps other than the bit operations; in general, it is the latter which takes by far the most time.

Next, let's examine the process of *multiplying* a $k$-bit integer by an $\ell$-bit integer in binary. For example,

$$\begin{array}{r} 11101\\ \underline{1101}\\ 11101\\ 111010\phantom{0}\\ \underline{11101\phantom{000}}\\ 101111001 \end{array}$$

Suppose we use this familiar procedure to multiply a $k$-bit integer $n$ by an $\ell$-bit integer $m$. We obtain at most $\ell$ rows (one row fewer for each 0-bit in $m$), where each row consists of a copy of $n$ shifted to the left a certain distance, i.e., with zeros put on at the end. Suppose there are $\ell' \le \ell$ rows. Because we want to break down all our computations into bit operations, we cannot simultaneously add together all of the rows. Rather, we move down from the 2nd row to the $\ell'$-th row, adding each new row to the partial sum of all of the earlier rows. At each stage, we note how many places to the left the number $n$ has been shifted to form the new row. We copy down the right-most bits of the partial sum, and then add to $n$ the integer formed from the rest of the partial sum — as explained above, this takes $k$ bit operations. In the above example $11101 \times 1101$, after adding the first two rows and obtaining 10010001, we copy down the last three bits 001 and add the rest (i.e., 10010) to $n = 11101$. We finally take this sum $10010 + 11101 = 101111$ and append 001 to obtain 101111001, the sum of the $\ell' = 3$ rows.

This description shows that the multiplication task can be broken down into $\ell' - 1$ additions, each taking $k$ bit operations. Since $\ell' - 1 < \ell' \le \ell$, this gives us the simple bound

$$\text{Time(multiply integer } k \text{ bits long by integer } \ell \text{ bits long)} < k\ell.$$

We should make several observations about this derivation of an estimate for the number of bit operations needed to perform a binary multiplication. In the first place, as mentioned before, we counted only the number of bit operations. We neglected to include the time it takes to shift the bits in $n$ a few places to the left, or the time it takes to copy down the right-most digits of the partial sum corresponding to the places through which $n$ has been shifted to the left in the new row. In practice, the shifting and copying operations are fast in comparison with the large number of bit operations, so we can safely ignore them. In other words, we shall *define* a "time estimate" for an arithmetic task to be an upper bound for the number of bit operations, without including any consideration of shift operations, changing registers ("copying"), memory access, etc. Note that this means that we would use the very same time estimate if we were multiplying a $k$-bit binary expansion of a fraction by an $\ell$-bit binary expansion; the only additional feature is that we must note the location of the point separating integer from fractional part and insert it correctly in the answer.

In the second place, if we want to get a time estimate that is simple and convenient to work with, we should assume at various points that we're in the "worst possible case." For example, if the binary expansion of $m$ has a lot of zeros, then $\ell'$ will be considerably less than $\ell$. That is, we could use the estimate Time(multiply $k$-bit integer by $\ell$-bit integer)$< k \cdot$(number of 1-bits in $m$). However, it is usually not worth the improvement (i.e., lowering) in our time estimate to take this into account, because it is more useful to have a simple uniform estimate that depends only on the size of $m$ and $n$ and not on the particular bits that happen to occur.

As a special case, we have: Time(multiply $k$-bit by $k$-bit)$< k^2$.

Finally, our estimate $k\ell$ can be written in terms of $n$ and $m$ if we remember the above formula for the number of digits, from which it follows that $k = [log_2 n] + 1 \le \frac{log\, n}{log\, 2} + 1$ and $\ell = [log_2 m] + 1 \le \frac{log\, m}{log\, 2} + 1$.

**Example 6.** Find an upper bound for the number of bit operations required to compute $n!$.

**Solution.** We use the following procedure. First multiply 2 by 3, then the result by 4, then the result of that by 5,..., until you get to $n$. At the $(j-1)$-th step $(j = 2, 3, \ldots, n-1)$, you are multiplying $j!$ by $j+1$. Hence you have $n-2$ steps, where each step involves multiplying a partial product (i.e., $j!$) by the next integer. The partial products will start to be very large. As a worst case estimate for the number of bits a partial product has, let's take the number of binary digits in the very last product, namely, in $n!$.

To find the number of bits in a product, we use the fact that the number of digits in the product of two numbers is either the sum of the number of digits in each factor or else 1 fewer than that sum (see the above discussion of multiplication). From this it follows that the product of $n$ $k$-bit integers will have at most $nk$ bits. Thus, if $n$ is a $k$-bit integer — which implies that every integer less than $n$ has at most $k$ bits — then $n!$ has at most $nk$ bits.

Hence, in each of the $n-2$ multiplications needed to compute $n!$, we are multiplying an integer with at most $k$ bits (namely $j+1$) by an integer with at most $nk$ bits (namely $j!$). This requires at most $nk^2$ bit operations. We must do this $n-2$ times. So the total number of bit operations is bounded by $(n-2)nk^2 = n(n-2)([log_2 n]+1)^2$. Roughly speaking, the bound is approximately $n^2(log_2 n)^2$.

**Example 7.** Find an upper bound for the number of bit operations required to multiply a polynomial $\sum a_i x^i$ of degree $\le n_1$ and a polynomial $\sum b_j x^j$ of degree $\le n_2$ whose coefficients are positive integers $\le m$. Suppose $n_2 \le n_1$.

**Solution.** To compute $\sum_{i+j=\nu} a_i b_j$, which is the coefficient of $x^\nu$ in the product polynomial (here $0 \le \nu \le n_1 + n_2$) requires at most $n_2 + 1$ multiplications and $n_2$ additions. The numbers being multiplied are bounded by $m$, and the numbers being added are each at most $m^2$; but since we have to add the partial sum of up to $n_2$ such numbers we should take $n_2m^2$ as our bound on the size of the numbers being added. Thus, in computing the coefficient of $x^\nu$ the number of bit operations required is at most

$$(n_2+1)(log_2 m+1)^2+n_2(log_2(n_2m^2)+1).$$

Since there are $n_1+n_2+1$ values of $\nu$, our time estimate for the polynomial multiplication is

$$(n_1+n_2+1)\big((n_2+1)(log_2 m+1)^2+n_2(log_2(n_2m^2)+1)\big).$$

A slightly less rigorous bound is obtained by dropping the 1's, thereby obtaining an expression having a more compact appearance:

$$\frac{n_2(n_1+n_2)}{log\,2}\left(\frac{(log\,m)^2}{log\,2}+(log\,n_2+2\log\,m)\right).$$

**Remark.** If we set $n=n_1\geq n_2$ and make the assumption that $m\geq 16$ and $m\geq\sqrt{n_2}$ (which usually holds in practice), then the latter expression can be replaced by the much simpler $4n^2(log_2 m)^2$. This example shows that there is generally no single "right answer" to the question of finding a bound on the time to execute a given task. One wants a function of the bounds on the imput data (in this problem, $n_1$, $n_2$ and $m$) which is fairly simple and at the same time gives an upper bound which for most input data is more-or-less the same order of magnitude as the number of bit operations that turns out to be required in practice. Thus, for example, in Example 7 we would not want to replace our bound by, say, $4n^2m$, because for large $m$ this would give a time estimate many orders of magnitude too large.

So far we have worked only with addition and multiplication of a $k$-bit and an $\ell$-bit integer. The other two arithmetic operations — subtraction and division — have the same time estimates as addition and multiplication, respectively: Time(subtract $k$-bit from $\ell$-bit)$\leq\max(k,\ell)$; Time(divide $k$-bit by $\ell$-bit)$\leq kl$. More precisely, to treat subtraction we must extend our definition of a bit operation to include the operation of subtracting a 0- or 1-bit from another 0- or 1-bit (with possibly a "borrow" of 1 from the previous column). See Exercise 8.

To analyze division in binary, let us orient ourselves by looking at an illustration, such as the one in Example 3. Suppose $k\geq\ell$ (if $k<\ell$, then the division is trivial, i.e., the quotient is zero and the entire dividend is the remainder). Finding the quotient and remainder requires at most $k-\ell+1$ subtractions. Each subtraction requires $\ell$ or $\ell+1$ bit operations; but in the latter case we know that the left-most column of the difference will always be a 0-bit, so we can omit that bit operation (thinking of it as "bookkeeping" rather than calculating). We similarly ignore other administrative details, such as the time required to compare binary integers (i.e., take just enough bits of the dividend so that the resulting integer is greater than the divisor), carry down digits, etc. So our estimate is simply $(k-\ell+1)\ell$, which is $\leq k\ell$.

**Example 8.** Find an upper bound for the number of bit operations it takes to compute the binomial coefficient $\binom{n}{m}$.

**Solution.** Since $\binom{n}{m} = \binom{n}{n-m}$, without loss of generality we may assume that $m \leq n/2$. Let us use the following procedure to compute $\binom{n}{m} = n(n-1)(n-2)\cdots(n-m+1)/(2\cdot3\cdots m)$. We have $m-1$ multiplications followed by $m-1$ divisions. In each case the maximum possible size of the first number in the multiplication or division is $n(n-1)(n-2)\cdots(n-m+1) < n^m$, and a bound for the second number is $n$. Thus, by the same argument used in the solution to Example 6, we see that a bound for the total number of bit operations is $2(m-1)m([log_2 n]+1)^2$, which for large m and n is essentially $2m^2(log_2 n)^2$.

We now discuss a very convenient notation for summarizing the situation with time estimates.

**The big-$O$ notation.** Suppose that $f(n)$ and $g(n)$ are functions of the positive integers $n$ which take *positive* (but not necessarily integer) values for all $n$. We say that $f(n) = O(g(n))$ (or simply that $f = O(g)$) if there exists a constant $C$ such that $f(n)$ is always less than $C \cdot g(n)$. For example, $2n^2 + 3n - 3 = O(n^2)$ (namely, it is not hard to prove that the left side is always less than $3n^2$).

Because we want to use the big-$O$ notation in more general situations, we shall give a more all-encompassing definition. Namely, we shall allow $f$ and $g$ to be functions of several variables, and we shall not be concerned about the relation between $f$ and $g$ for small values of $n$. Just as in the study of limits as $n \longrightarrow \infty$ in calculus, here also we shall only be concerned with large values of $n$.

**Definition.** Let $f(n_1, n_2, \ldots, n_r)$ and $g(n_1, n_2, \ldots, n_r)$ be two functions whose domains are subsets of the set of all $r$-tuples of positive integers. Suppose that there exist constants $B$ and $C$ such that whenever all of the $n_j$ are greater than $B$ the two functions are defined and positive, and $f(n_1, n_2, \ldots, n_r) < C\, g(n_1, n_2, \ldots, n_r)$. In that case we say that $f$ is *bounded* by $g$ and we write $f = O(g)$.

Note that the "=" in the notation $f = O(g)$ should be thought of as more like a "<" and the big-$O$ should be thought of as meaning "some constant multiple."

**Example 9.** (a) Let $f(n)$ be *any* polynomial of degree $d$ whose leading coefficient is positive. Then it is easy to prove that $f(n) = O(n^d)$. More generally, one can prove that $f = O(g)$ in any situation when $f(n)/g(n)$ has a finite limit as $n \longrightarrow \infty$.

(b) If $\epsilon$ is any positive number, no matter how small, then one can prove that $log\, n = O(n^\epsilon)$ (i.e., for large $n$, the log function is smaller than any power function, no matter how small the power). In fact, this follows because $lim_{n\to\infty} \frac{log\, n}{n^\epsilon} = 0$, as one can prove using l'Hôpital's rule.

(c) If $f(n)$ denotes the number $k$ of binary digits in $n$, then it follows from the above formulas for $k$ that $f(n) = O(log\, n)$. Also notice that the same relation holds if $f(n)$ denotes the number of base-$b$ digits, where $b$ is any fixed base. On the other hand, suppose that the base $b$ is not kept fixed but is allowed to increase, and we let $f(n, b)$ denote the number of base-$b$ digits. Then we would want to use the relation $f(n, b) = O(\frac{log\, n}{log\, b})$.

(d) We have: $\text{Time}(n \cdot m) = O(log\, n \cdot log\, m)$, where the left hand side means the number of bit operations required to multiply $n$ by $m$.

(e) In Exercise 6, we can write: $\text{Time}(n!) = O\big((n\, log\, n)^2\big)$.

(f) In Exercise 7, we have:

$$\text{Time}\Big(\sum a_i x^i \cdot \sum b_j x^j\Big) = O\Big(n_1 n_2 \big((log\, m)^2 + log(\min(n_1, n_2))\big)\Big).$$

In our use, the functions $f(n)$ or $f(n_1, n_2, \ldots, n_r)$ will often stand for the amount of time it takes to perform an arithmetic task with the integer $n$ or with the set of integers $n_1, n_2, \ldots, n_r$ as input. We will want to obtain fairly simple-looking functions $g(n)$ as our bounds. When we do this, however, we do not want to obtain functions $g(n)$ which are much larger than necessary, since that would give an exaggerated impression of how long the task will take (although, from a strictly mathematical point of view, it is not incorrect to replace $g(n)$ by any larger function in the relation $f = O(g)$).

Roughly speaking, the relation $f(n) = O(n^d)$ tells us that the function $f$ increases approximately like the $d$-th power of the variable. For example, if $d = 3$, then it tells us that doubling $n$ has the effect of increasing $f$ by about a factor of 8. The relation $f(n) = O(log^d n)$ (we write $log^d n$ to mean $(log\, n)^d$) tells us that the function increases approximately like the $d$-th power of the number of binary digits in $n$. That is because, up to a constant multiple, the number of bits is approximately $log\, n$ (namely, it is within 1 of being $log\, n/log\, 2 = 1.4427\, log\, n$). Thus, for example, if $f(n) = O(log^3 n)$, then doubling the number of bits in $n$ (which is, of course, a much more drastic increase in the size of $n$ than merely doubling $n$) has the effect of increasing $f$ by about a factor of 8.

Note that to write $f(n) = O(1)$ means that the function $f$ is bounded by some constant.

**Remark.** We have seen that, if we want to multiply two numbers of about the same size, we can use the estimate $\text{Time}(k\text{-bit}\cdot k\text{-bit}) = O(k^2)$. It should be noted that much work has been done on increasing the speed of multiplying two $k$-bit integers when $k$ is large. Using clever techniques of multiplication that are much more complicated than the grade-school method we have been using, mathematicians have been able to find a procedure for multiplying two $k$-bit integers that requires only $O(k\, log\, k\; log\, log\, k)$ bit operations. This is better than $O(k^2)$, and even better than $O(k^{1+\epsilon})$ for any $\epsilon > 0$, no matter how small. However, in what follows we shall always be content to use the rougher estimates above for the time needed for a multiplication.

In general, when estimating the number of bit operations required to do something, the first step is to decide upon and write down an outline of a detailed procedure for performing the task. An explicit step-by-step procedure for doing calculations is called an *algorithm.* Of course, there may be many different algorithms for doing the same thing. One may choose to use the one that is easiest to write down, or one may choose to use the fastest one known, or else one may choose to compromise and make a trade-off between simplicity and speed. The algorithm used above for multiplying $n$ by $m$ is far from the fastest one known. But it is certainly a lot faster than repeated addition (adding $n$ to itself $m$ times).

**Example 10.** Estimate the time required to convert a $k$-bit integer to its representation in the base 10.

**Solution.** Let $n$ be a $k$-bit integer written in binary. The conversion algorithm is as follows. Divide $10 = (1010)_2$ into $n$. The remainder — which will be one of the integers 0, 1, 10, 11, 100, 101, 110, 111, 1000, or 1001 — will be the ones digit $d_0$. Now replace $n$ by the quotient and repeat the process, dividing that quotient by $(1010)_2$, using the remainder as $d_1$ and the quotient as the next number into which to divide $(1010)_2$. This process must be repeated a number of times equal to the number of decimal digits in $n$, which is $\left[\frac{log\, n}{log\, 10}\right]+1 = O(k)$. Then we're done. (We might want to take our list of decimal digits, i.e., of remainders from all the divisions, and convert them to the more familiar notation by replacing $0, 1, 10, 11, \ldots, 1001$ by $0, 1, 2, 3, \ldots, 9$, respectively.) How many bit operations does this all take? Well, we have $O(k)$ divisions, each requiring $O(4k)$ operations (dividing a number with at most $k$ bits by the 4-bit number $(1010)_2$). But $O(4k)$ is the same as $O(k)$ (constant factors don't matter in the big-$O$ notation), so we conclude that the total number of bit operations is $O(k) \cdot O(k) = O(k^2)$. If we want to express this in terms of $n$ rather than $k$, then since $k = O(log\, n)$, we can write

$$\text{Time(convert } n \text{ to decimal)} = O(log^2 n).$$

**Example 11.** Estimate the time required to convert a $k$-bit integer $n$ to its representation in the base $b$, where $b$ might be very large.

**Solution.** Using the same algorithm as in Example 10, except dividing now by the $\ell$-bit integer $b$, we find that each division now takes longer (if $\ell$ is large), namely, $O(k\ell)$ bit operations. How many times do we have to divide? Here notice that the number of base-b digits in $n$ is $O(k/\ell)$ (see Example 9(c)). Thus, the total number of bit operations required to do all of the necessary divisions is $O(k/\ell) \cdot O(k\ell) = O(k^2)$. This turns out to be the same answer as in Example 10. That is, our estimate for the conversion time does not depend upon the base to which we're converting (no matter how large it may be). This is because the greater time required to find each digit is offset by the fact that there are fewer digits to be found.

**Example 12.** Express in terms of the $O$-notation the time required to compute (a) $n!$, (b) $\binom{n}{m}$ (see Examples 6 and 8).

**Solution.** (a) $O(n^2 log^2 n)$, (b) $O(m^2 log^2 n)$.

In concluding this section, we make a definition that is fundamental in computer science and the theory of algorithms.

**Definition.** An algorithm to perform a computation involving integers $n_1, n_2, \ldots, n_r$ of $k_1, k_2, \ldots, k_r$ bits, respectively, is said to be a *polynomial time* algorithm if there exist integers $d_1, d_2, \ldots, d_r$ such that the number of bit operations required to perform the algorithm is $O(k_1^{d_1} k_2^{d_2} \cdots k_r^{d_r})$.

Thus, the usual arithmetic operations $+$, $-$, $\times$, $\div$ are examples of polynomial time algorithms; so is conversion from one base to another. On the other hand, computation of $n!$ is not. (However, if one is satisfied with knowing $n!$ to only a certain number of significant figures, e.g., its first 1000 binary digits, then one can obtain that by a polynomial time algorithm using Stirling's approximation formula for $n!$.)

## *Exercises*

1. Multiply $(212)_3$ by $(122)_3$.
2. Divide $(40122)_7$ by $(126)_7$.
3. Multiply the binary numbers 101101 and 11001, and divide 10011001 by 1011.
4. In the base 26, with digits A—Z representing 0—25, (a) multiply YES by NO, and (b) divide JQVXHJ by WE.
5. Write $e = 2.7182818\cdots$ (a) in binary 15 places out to the right of the point, and (b) to the base 26 out 3 places beyond the point.
6. By a "pure repeating" fraction of "period" $f$ in the base $b$, we mean a number between 0 and 1 whose base-$b$ digits to the right of the point repeat in blocks of $f$. For example, 1/3 is pure repeating of period 1 and 1/7 is pure repeating of period 6 in the decimal system. Prove that a fraction $c/d$ (in lowest terms) between 0 and 1 is pure repeating of period $f$ in the base $b$ if and only if $b^f - 1$ is a multiple of $d$.
7. (a) The "hexadecimal" system means $b = 16$ with the letters A–F representing the tenth through fifteenth digits, respectively. Divide $(131B6C3)_{16}$ by $(1A2F)_{16}$.

   (b) Explain how to convert back and forth between binary and hexadecimal representations of an integer, and why the time required is far less than the general estimate given in Example 11 for converting from binary to base-$b$.
8. Describe a subtraction-type bit operation in the same way as was done for an addition-type bit operation in the text (the list of five alternatives).

9. (a) Using the big-$O$ notation, estimate in terms of a simple function of $n$ the number of bit operations required to compute $3^n$ in binary.
(b) Do the same for $n^n$.
10. Estimate in terms of a simple function of $n$ and $N$ the number of bit operations required to compute $N^n$.
11. The following formula holds for the sum of the first $n$ perfect squares:

$$\sum_{j=1}^{n} j^2 = n(n+1)(2n+1)/6.$$

(a) Using the big-$O$ notation, estimate (in terms of $n$) the number of bit operations required to perform the computations in the left side of this equality.
(b) Estimate the number of bit operations required to perform the computations on the right in this equality.
12. Using the big-0 notation, estimate the number of bit operations required to multiply an $r \times n$-matrix by an $n \times s$-matrix, where all matrix entries are $\leq m$.
13. The object of this exercise is to estimate as a function of $n$ the number of bit operations required to compute the product of all prime numbers less than $n$. Here we suppose that we have already compiled an extremely long list containing all primes up to $n$.
(a) According to the Prime Number Theorem, the number of primes less than or equal to $n$ (this is denoted $\pi(n)$) is asymptotic to $n/\log n$. This means that the following limit approaches 1 as $n \longrightarrow \infty$: $lim \frac{\pi(n)}{n/\log n}$. Using the Prime Number Theorem, estimate the number of binary digits in the product of all primes less than $n$.
(b) Find a bound for the number of bit operations in one of the multiplications that's required in the computation of this product.
(c) Estimate the number of bit operations required to compute the product of all prime numbers less than $n$.
14. (a) Suppose you want to test if a large odd number $n$ is a prime by trial division by all odd numbers $\leq \sqrt{n}$. Estimate the number of bit operations this will take.
(b) In part (a), suppose you have a list of prime numbers up to $\sqrt{n}$, and you test primality by trial division by those primes (i.e., no longer running through all odd numbers). Give a time estimate in this case. Use the Prime Number Theorem.
15. Estimate the time required to test if $n$ is divisible by a prime $\leq m$. Suppose that you have a list of all primes $\leq m$, and again use the Prime Number Theorem.
16. Let $n$ be a very large integer written in binary. Find a simple algorithm that computes $[\sqrt{n}]$ in $O(\log^3 n)$ bit operations (here $[\ ]$ denotes the greatest integer function).

## 2 Divisibility and the Euclidean algorithm

**Divisors and divisibility.** Given integers $a$ and $b$, we say that $a$ *divides* $b$ (or "$b$ is *divisible* by $a$") and we write $a|b$ if there exists an integer $d$ such that $b = ad$. In that case we call $a$ a *divisor* of $b$. Every integer $b > 1$ has at least two positive divisors: 1 and $b$. By a *proper divisor* of $b$ we mean a positive divisor not equal to $b$ itself, and by a *nontrivial divisor* of $b$ we mean a positive divisor not equal to 1 or $b$. A *prime* number, by definition, is an integer greater than one which has no positive divisors other than 1 and itself; a number is called *composite* if it has at least one nontrivial divisor. The following properties of divisibility are easy to verify directly from the definition:

1. If $a|b$ and $c$ is any integer, then $a|bc$.
2. If $a|b$ and $b|c$, then $a|c$.
3. If $a|b$ and $a|c$, then $a|b \pm c$.

If $p$ is a prime number and $\alpha$ is a nonnegative integer, then we use the notation $p^\alpha || b$ to mean that $p^\alpha$ is the highest power of $p$ dividing $b$, i.e., that $p^\alpha | b$ and $p^{\alpha+1} \nmid b$. In that case we say that $p^\alpha$ *exactly divides* $b$.

The *Fundamental Theorem of Arithmetic* states that any natural number $n$ can be written uniquely (except for the order of factors) as a product of prime numbers. It is customary to write this factorization as a product of distinct primes to the appropriate powers, listing the primes in increasing order. For example, $4200 = 2^3 \cdot 3 \cdot 5^2 \cdot 7$.

Two consequences of the Fundamental Theorem (actually, equivalent assertions) are the following properties of divisibility:

4. If a prime number $p$ divides $ab$, then either $p|a$ or $p|b$.
5. If $m|a$ and $n|a$, and if $m$ and $n$ have no divisors greater than 1 in common, then $mn|a$.

Another consequence of unique factorization is that it gives a systematic method for finding all divisors of $n$ once $n$ is written as a product of prime powers. Namely, any divisor $d$ of $n$ must be a product of the same primes raised to powers not exceeding the power that exactly divides $n$. That is, if $p^\alpha || n$, then $p^\beta || d$ for some $\beta$ satisfying $0 \le \beta \le \alpha$. To find the divisors of 4200, for example, one takes 2 to the 0-, 1-, 2- or 3-power, multiplied by 3 to the 0- or 1-power, times 5 to the 0-, 1- or 2-power, times 7 to the 0- or 1- power. The number of possible divisors is thus the product of the number of possibilities for each prime power, which, in turn, is $\alpha + 1$. That is, a number $n = p_1^{\alpha_1} p_2^{\alpha_2} \cdots p_r^{\alpha_r}$ has $(\alpha_1 + 1)(\alpha_2 + 1) \cdots (\alpha_r + 1)$ different divisors. For example, there are 48 divisors of 4200.

Given two integers $a$ and $b$, not both zero, the *greatest common divisor* of $a$ and $b$, denoted $g.c.d.(a, b)$ (or sometimes simply $(a, b)$) is the largest integer $d$ dividing both $a$ and $b$. It is not hard to show that another equivalent definition of $g.c.d.(a, b)$ is the following: it is the only positive integer $d$ which divides $a$ and $b$ and is divisible by any other number which divides both $a$ and $b$.

If you happen to have the prime factorization of $a$ and $b$ in front of you, then it's very easy to write down $g.c.d.(a, b)$. Simply take all primes which occur in both factorizations raised to the minimum of the two exponents. For example, comparing the factorization $10780 = 2^2 \cdot 5 \cdot 7^2 \cdot 11$ with the above factorization of 4200, we see that $g.c.d.(4200, 10780) = 2^2 \cdot 5 \cdot 7 = 140$.

One also occasionally uses the *least common multiple* of $a$ and $b$, denoted $l.c.m.(a, b)$. It is the smallest positive integer that both $a$ and $b$ divide. If you have the factorization of $a$ and $b$, then you can get $l.c.m.(a, b)$ by taking all of the primes which occur in *either* factorization raised to the *maximum* of the exponents. It is easy to prove that $l.c.m.(a, b) = |ab|/g.c.d.(a, b)$.

**The Euclidean algorithm.** If you're working with very large numbers, it's likely that you won't know their prime factorizations. In fact, an important area of research in number theory is the search for quicker methods of factoring large integers. Fortunately, there's a relatively quick way to find $g.c.d.(a, b)$ even when you have no idea of the prime factors of $a$ or $b$. It's called the *Euclidean algorithm.*

The Euclidean algorithm works as follows. To find $g.c.d.(a, b)$, where $a > b$, we first divide $b$ into $a$ and write down the quotient $q_1$ and the remainder $r_1$: $a = q_1 b + r_1$. Next, we perform a second division with $b$ playing the role of $a$ and $r_1$ playing the role of $b$: $b = q_2 r_1 + r_2$. Next, we divide $r_2$ into $r_1$: $r_1 = q_3 r_2 + r_3$. We continue in this way, each time dividing the last remainder into the second-to-last remainder, obtaining a new quotient and remainder. When we finally obtain a remainder that divides the previous remainder, we are done: that final nonzero remainder is the greatest common divisor of $a$ and $b$.

**Example 1.** Find $g.c.d.(1547, 560)$.

**Solution:**

$$\begin{aligned}
1547 &= 2 \cdot 560 + 427 \\
560 &= 1 \cdot 427 + 133 \\
427 &= 3 \cdot 133 + 28 \\
133 &= 4 \cdot 28 + 21 \\
28 &= 1 \cdot 21 + 7.
\end{aligned}$$

Since $7|21$, we are done: $g.c.d.(1547, 560) = 7$.

**Proposition I.2.1.** *The Euclidean algorithm always gives the greatest common divisor in a finite number of steps. In addition, for $a > b$*

$$\text{Time}(\text{finding } g.c.d.(a, b) \text{ by the Euclidean algorithm}) = O(log^3(a)).$$

**Proof.** The proof of the first assertion is given in detail in many elementary number theory textbooks, so we merely summarize the argument. First, it is easy to see that the remainders are strictly decreasing from one step to the next, and so must eventually reach zero. To see that the last remainder is the g.c.d., use the second definition of the g.c.d. That is, if any number divides both $a$ and $b$, it must divide $r_1$, and then, since it divides

$b$ and $r_1$, it must divide $r_2$, and so on, until you finally conclude that it must divide the last nonzero remainder. On the other hand, working from the last row up, one quickly sees that the last remainder must divide all of the previous remainders and also $a$ and $b$. Thus, it is the g.c.d., because the g.c.d. is the only number which divides both $a$ and $b$ and at the same time is divisible by any other number which divides $a$ and $b$.

We next prove the time estimate. The main question that must be resolved is how many divisions we're performing. We claim that the remainders are not only decreasing, but they're decreasing rather rapidly. More precisely:

**Claim.** $r_{j+2} < \frac{1}{2}r_j$.

**Proof of claim.** First, if $r_{j+1} \le \frac{1}{2}r_j$, then immediately we have $r_{j+2} < r_{j+1} \le \frac{1}{2}r_j$. So suppose that $r_{j+1} > \frac{1}{2}r_j$. In that case the next division gives: $r_j = 1 \cdot r_{j+1} + r_{j+2}$, and so $r_{j+2} = r_j - r_{j+1} < \frac{1}{2}r_j$, as claimed.

We now return to the proof of the time estimate. Since every two steps must result in cutting the size of the remainder at least in half, and since the remainder never gets below 1, it follows that there are at most $2 \cdot \left[log_2 a\right]$ divisions. This is $O(log\, a)$. Each division involves numbers no larger than $a$, and so takes $O(log^2 a)$ bit operations. Thus, the total time required is $O(log\, a) \cdot O(log^2 a) = O(log^3 a)$. This concludes the proof of the proposition.

**Remark.** If one makes a more careful analysis of the number of bit operations, taking into account the decreasing size of the numbers in the successive divisions, one can improve the time estimate for the Euclidean algorithm to $O(log^2 a)$.

**Proposition I.2.2.** *Let $d = g.c.d.(a, b)$, where $a > b$. Then there exist integers $u$ and $v$ such that $d = ua + bv$. In other words, the g.c.d. of two numbers can be expressed as a linear combination of the numbers with integer coefficients. In addition, finding the integers $u$ and $v$ can be done in $O(log^3 a)$ bit operations.*

**Outline of proof.** The procedure is to use the sequence of equalities in the Euclidean algorithm from the bottom up, at each stage writing $d$ in terms of earlier and earlier remainders, until finally you get to $a$ and $b$. At each stage you need a multiplication and an addition or subtraction. So it is easy to see that the number of bit operations is once again $O(log^3 a)$.

**Example 1 (continued).** To express 7 as a linear combination of 1547 and 560, we successively compute:

$$\begin{aligned} 7 &= 28 - 1 \cdot 21 = 28 - 1(133 - 4 \cdot 28) \\ &= 5 \cdot 28 - 1 \cdot 133 = 5(427 - 3 \cdot 133) - 1 \cdot 133 \\ &= 5 \cdot 427 - 16 \cdot 133 = 5 \cdot 427 - 16(560 - 1 \cdot 427) \\ &= 21 \cdot 427 - 16 \cdot 560 = 21(1547 - 2 \cdot 560) - 16 \cdot 560 \\ &= 21 \cdot 1547 - 58 \cdot 560. \end{aligned}$$

**Definition.** We say that two integers $a$ and $b$ are *relatively prime* (or that "$a$ is prime to $b$") if $g.c.d.(a, b) = 1$, i.e., if they have no common divisor greater than 1.

**Corollary.** *If $a > b$ are relatively prime integers, then 1 can be written as an integer linear combination of $a$ and $b$ in polynomial time, more precisely, in $O(\log^3 a)$ bit operations.*

**Definition.** Let $n$ be a positive integer. The *Euler phi-function* $\varphi(n)$ is defined to be the number of nonnegative integers $b$ less than $n$ which are prime to $n$:

$$\varphi(n) \underset{\text{def}}{=} \left|\{0 \le b < n \mid g.c.d.(b,n) = 1\}\right|.$$

It is easy to see that $\varphi(1) = 1$ and that $\varphi(p) = p - 1$ for any prime $p$. We can also see that for any prime power

$$\varphi(p^\alpha) = p^\alpha - p^{\alpha-1} = p^\alpha\left(1 - \frac{1}{p}\right).$$

To see this, it suffices to note that the numbers from 0 to $p^\alpha - 1$ which are *not* prime to $p^\alpha$ are precisely those that are divisible by $p$, and there are $p^{\alpha-1}$ of those.

In the next section we shall show that the Euler $\varphi$-function has a "multiplicative property" that enables us to evaluate $\varphi(n)$ quickly, provided that we have the prime factorization of $n$. Namely, if $n$ is written as a product of powers of distinct primes $p^\alpha$, then it turns out that $\varphi(n)$ is equal to the product of the $\varphi(p^\alpha)$.

## *Exercises*

1. (a) Prove the following properties of the relation $p^\alpha || b$: (i) if $p^\alpha || a$ and $p^\beta || b$, then $p^{\alpha+\beta} || ab$; (ii) if $p^\alpha || a$, $p^\beta || b$ and $\alpha < \beta$, then $p^\alpha || a \pm b$.
   (b) Find a counterexample to the assertion that, if $p^\alpha || a$ and $p^\alpha || b$, then $p^\alpha || a + b$.
2. How many divisors does 945 have? List them all.
3. Let $n$ be a positive odd integer.
   (a) Prove that there is a 1-to-1 correspondence between the divisors of $n$ which are $< \sqrt{n}$ and those that are $> \sqrt{n}$. (This part does not require $n$ to be odd.)
   (b) Prove that there is a 1-to-1 correspondence between all of the divisors of $n$ which are $\ge \sqrt{n}$ and all the ways of writing $n$ as a difference $s^2 - t^2$ of two squares of nonnegative integers. (For example, 15 has two divisors 6, 15 that are $\ge \sqrt{15}$, and $15 = 4^2 - 1^2 = 8^2 - 7^2$.)
   (c) List all of the ways of writing 945 as a difference of two squares of nonnegative integers.
4. (a) Show that the power of a prime $p$ which exactly divides $n!$ is equal to $[n/p] + [n/p^2] + [n/p^3] + \cdots$. (Notice that this is a finite sum.)
   (b) Find the power of each prime 2, 3, 5, 7 that exactly divides 100!, and then write out the entire prime factorization of 100!.

(c) Let $S_b(n)$ denote the sum of the base-$b$ digits in $n$. Prove that the exact power of 2 that divides $n!$ is equal to $n - S_2(n)$. Find and prove a similar formula for the exact power of an arbitrary prime $p$ that divides $n!$.

5. Find $d = g.c.d.(360, 294)$ in two ways: (a) by finding the prime factorization of each number, and from that finding the prime factorization of $d$; and (b) by means of the Euclidean algorithm.
6. For each of the following pairs of integers, find their greatest common divisor using the Euclidean algorithm, and express it as an integer linear combination of the two numbers:
(a) 26, 19; (b) 187, 34; (c) 841, 160; (d) 2613, 2171.
7. One can often speed up the Euclidean algorithm slightly by allowing divisions with negative remainders, i.e., $r_j = q_{j+2}r_{j+1} - r_{j+2}$ as well as $r_j = q_{j+2}r_{j+1} + r_{j+2}$, whichever gives the smallest $r_{j+2}$. In this way we always have $r_{j+2} \leq \frac{1}{2}r_{j+1}$. Do the four examples in Exercise 6 using this method.
8. (a) Prove that the following algorithm finds $d = g.c.d.(a, b)$ in finitely many steps. First note that $g.c.d.(a, b) = g.c.d.(|a|, |b|)$, so that without loss of generality we may suppose that $a$ and $b$ are positive. If $a$ and $b$ are both even, set $d = 2d'$ with $d' = g.c.d.(a/2, b/2)$. If one of the two is odd and the other (say $b$) is even, then set $d = d'$ with $d' = g.c.d.(a, b/2)$. If both are odd and they are unequal, say $a > b$, then set $d = d'$ with $d' = g.c.d.(a - b, b)$. Finally, if $a = b$, then set $d = a$. Repeat this process until you arrive at the last case (when the two integers are equal).
(b) Use the algorithm in part (a) to find $g.c.d.(2613, 2171)$ working in binary, i.e., find

$$g.c.d.((101000110101)_2, (100001111011)_2)$$

(c) Prove that the algorithm in part (a) takes only $O(log^2 a)$ bit operations (where $a > b$).
(d) Why is this algorithm in the form presented above not necessarily preferable to the Euclidean algorithm?
9. Suppose that $a$ is much greater than $b$. Find a big-$O$ time estimate for $g.c.d.(a, b)$ that is better than $O(log^3 a)$.
10. The purpose of this problem is to find a "best possible" estimate for the number of divisions required in the Euclidean algorithm. The *Fibonacci numbers* can be defined by the rule $f_1 = 1$, $f_2 = 1$, $f_{n+1} = f_n + f_{n-1}$ for $n \geq 2$, or, equivalently, by means of the matrix equation $\begin{pmatrix} f_{n+1} & f_n \\ f_n & f_{n-1} \end{pmatrix} = \begin{pmatrix} 1 & 1 \\ 1 & 0 \end{pmatrix}^n$.
(a) Suppose that $a > b > 0$, and it takes $k$ divisions to find $g.c.d.(a, b)$ by the Euclidean algorithm (the standard version given in the text, with nonnegative remainders). Show that $a \geq f_{k+2}$.

(b) Using the matrix definition of $f_n$, prove that

$$f_n = \frac{\alpha^n - \alpha'^n}{\sqrt{5}}, \qquad \text{where} \qquad \alpha = \frac{1+\sqrt{5}}{2}, \qquad \alpha' = \frac{1-\sqrt{5}}{2}.$$

(c) Using parts (a) and (b), find an upper bound for $k$ in terms of $a$. Compare with the estimate that follows from the proof of Proposition I.2.1.

11. The purpose of this problem is to find a general estimate for the time required to compute $g.c.d.(a,b)$ (where $a > b$) that is better than the estimate in Proposition I.2.1.
(a) Show that the number of bit operations required to perform a divison $a = qb + r$ is $O\big((log\, b)(1 + log\, q)\big)$.
(b) Applying part (a) to all of the $O(\log a)$ divisions of the form $r_{i-1} = q_{i+1}r_i + r_{i+1}$, derive the time estimate $O\big((log\, b)(log\, a)\big)$.
12. Consider polynomials with real coefficients. (This problem will apply as well to polynomials with coefficients in any *field*.) If $f$ and $g$ are two polynomials, we say that $f|g$ if there is a polynomial $h$ such that $g = fh$. We define $g.c.d.(f,g)$ in essentially the same way as for integers, namely, as a polynomial of greatest degree which divides both $f$ and $g$. The polynomial $g.c.d.(f,g)$ defined in this way is not unique, since we can get another polynomial of the same degree by multiplying by any nonzero constant. However, we can make it unique by requiring that the g.c.d. polynomial be *monic*, i.e., have leading coefficient 1. We say that $f$ and $g$ are relatively prime polynomials if their g.c.d. is the "constant polynomial" 1. Devise a procedure for finding g.c.d.'s of polynomials – namely, a Euclidean algorithm for polynomials — which is completely analogous to the Euclidean algorithm for integers, and use it to find (a) $g.c.d.(x^4 + x^2 + 1,\ x^2 + 1)$, and (b) $g.c.d.(x^4 - 4x^3 + 6x^2 - 4x + 1,\ x^3 - x^2 + x - 1)$. In each case find polynomials $u(x)$ and $v(x)$ such that the g.c.d. is expressed as $u(x)f(x) + v(x)g(x)$.
13. From algebra we know that a polynomial has a multiple root if and only if it has a common factor with its derivative; in that case the multiple roots of $f(x)$ are the roots of $g.c.d.(f,\ f')$. Find the multiple roots of the polynomial $x^4 - 2x^3 - x^2 + 2x + 1$.
14. (Before doing this exercise, recall how to do arithmetic with complex numbers. Remember that, since $(a+bi)(a-bi)$ is the real number $a^2+b^2$, one can divide by writing $(c+di)/(a+bi) = (c+di)(a-bi)/(a^2+b^2)$.) The *Gaussian integers* are the complex numbers whose real and imaginary parts are integers. In the complex plane they are the vertices of the squares that make up the grid. If $\alpha$ and $\beta$ are two Gaussian integers, we say that $\alpha|\beta$ if there is a Guassian integer $\gamma$ such that $\beta = \alpha\gamma$. We define $g.c.d.(\alpha, \beta)$ to be a Gaussian integer $\delta$ of maximum absolute value which divides both $\alpha$ and $\beta$ (recall that the absolute value $|\delta|$ is its distance from 0, i.e., the square root of the sum of the squares of its real and imaginary parts). The g.c.d. is not unique, because we

can multiply it by $\pm 1$ or $\pm i$ and obtain another $\delta$ of the same absolute value which also divides $\alpha$ and $\beta$. This gives four possibilities. In what follows we will consider any one of those four possibilities to be "the" g.c.d.

Notice that any complex number can be written as a Gaussian integer plus a complex number whose real and imaginary parts are each between $\frac{1}{2}$ and $-\frac{1}{2}$. Show that this means that we can divide one Gaussian integer $\alpha$ by another one $\beta$ and obtain a Gaussian integer quotient along with a remainder which is less than $\beta$ in absolute value. Use this fact to devise a Euclidean algorithm which finds the g.c.d. of two Gaussian integers. Use this Euclidean algorithm to find (a) *g.c.d.*$(5+6i,\ 3-2i)$, and (b) *g.c.d.*$(7-11i,\ 8-19i)$. In each case express the g.c.d. as a linear combination of the form $u\alpha + v\beta$, where $u$ and $v$ are Gaussian integers.

15. The last problem can be applied to obtain an efficient way to write certain large primes as a sum of two squares. For example, suppose that $p$ is a prime which divides a number of the form $b^6+1$. We want to write $p$ in the form $p = c^2 + d^2$ for some integers $c$ and $d$. This is equivalent to finding a nontrivial Gaussian integer factor of $p$, because $c^2 + d^2 = (c+di)(c-di)$. We can proceed as follows. Notice that

$$b^6+1 = (b^2+1)(b^4-b^2+1), \qquad \text{and} \qquad b^4-b^2+1 = (b^2-1)^2+b^2.$$

By property 4 of divisibility, the prime $p$ must divide one of the two factors on the right of the first equality. If $p|b^2+1 = (b+i)(b-i)$, then you will find that *g.c.d.*$(p,\ b+i)$ will give you the desired $c+di$. If $p|b^4-b^2+1 = \big((b^2-1)+bi\big)\big((b^2-1)-bi\big)$, then *g.c.d.*$(p,\ (b^2-1)+bi)$ will give you your $c+di$.

**Example.** The prime 12277 divides the second factor in the product $20^6+1 = (20^2+1)(20^4-20^2+1)$. So we find *g.c.d.*$(12277,\ 399+20i)$:

$$\begin{aligned} 12277 &= (31-2i)(399+20i) + (-132+178i), \\ 399+20i &= (-1-i)(-132+178i) + (89+66i), \\ -132+178i &= (2i)(89+66i), \end{aligned}$$

so that the g.c.d. is $89+66i$, i.e., $12277 = 89^2+66^2$.

(a) Using the fact that $19^6+1 = 2\cdot 13^2\cdot 181\cdot 769$ and the Euclidean algorithm for the Gaussian integers, express 769 as a sum of two squares.

(b) Similarly, express the prime 3877, which divides $15^6+1$, as a sum of two squares.

(c) Express the prime 38737, which divides $2^{36}+1$, as a sum of two squares.

# 3 Congruences

**Basic properties.** Given three integers $a$, $b$ and $m$, we say that "$a$ is *congruent* to $b$ *modulo* $m$" and write $a \equiv b \ mod \ m$, if the difference $a - b$ is divisible by $m$. $m$ is called the *modulus* of the congruence. The following properties are easily proved directly from the definition:

1. (i) $a \equiv a \ mod \ m$; (ii) $a \equiv b \ mod \ m$ if and only if $b \equiv a \ mod \ m$; (iii) if $a \equiv b \ mod \ m$ and $b \equiv c \ mod \ m$, then $a \equiv c \ mod \ m$. For fixed $m$, (i)–(iii) mean that congruence modulo $m$ is an *equivalence relation.*
2. For fixed $m$, each *equivalence class* with respect to congruence modulo $m$ has one and only one representative between 0 and $m - 1$. (This is just another way of saying that any integer is congruent modulo $m$ to one and only one integer between 0 and $m - 1$.) The set of equivalence classes (called *residue classes*) will be denoted $\mathbf{Z}/m\mathbf{Z}$. Any set of representatives for the residue classes is called a *complete set of residues modulo m.*
3. If $a \equiv b \ mod \ m$ and $c \equiv d \ mod \ m$, then $a \pm c \equiv b \pm d \ mod \ m$ and $ac \equiv bd \ mod \ m$. In other words, congruences (with the same modulus) can be added, subtracted, or multiplied. One says that the set of equivalence classes $\mathbf{Z}/m\mathbf{Z}$ is a *commutative ring*, i.e., residue classes can be added, subtracted or multiplied (with the result not depending on which representatives of the equivalence classes were used), and these operations satisfy the familiar axioms (associativity, commutativity, additive inverse, etc.).
4. If $a \equiv b \ mod \ m$, then $a \equiv b \ mod \ d$ for any divisor $d|m$.
5. If $a \equiv b \ mod \ m$, $a \equiv b \ mod \ n$, and $m$ and $n$ are relatively prime, then $a \equiv b \ mod \ mn$. (See Property 5 of divisibility in § I.2.)

**Proposition I.3.1.** *The elements of* $\mathbf{Z}/m\mathbf{Z}$ *which have multiplicative inverses are those which are relatively prime to m, i.e., the numbers* $a$ *for which there exists* $b$ *with* $ab \equiv 1 \ mod \ m$ *are precisely those* $a$ *for which* $g.c.d.(a, m) = 1$. *In addition, if* $g.c.d.(a, m) = 1$, *then such an inverse* $b$ *can be found in* $O(log^3 m)$ *bit operations.*

**Proof.** First, if $d = g.c.d.(a, m)$ were greater than 1, we could not have $ab \equiv 1 \ mod \ m$ for any $b$, because that would imply that $d$ divides $ab - 1$ and hence divides 1. Conversely, if $g.c.d.(a, m) = 1$, then by Property 2 above we may suppose that $a < m$. Then, by Proposition I.2.2, there exist integers $u$ and $v$ that can be found in $O(log^3 m)$ bit operations for which $ua + vm = 1$. Choosing $b = u$, we see that $m|1 - ua = 1 - ab$, as desired.

**Remark.** If $g.c.d.(a, m) = 1$, then by negative powers $a^{-n} \ mod \ m$ we mean the $n$-th power of the inverse residue class, i.e., it is represented by the $n$-th power of any integer $b$ for which $ab \equiv 1 \ mod \ m$.

**Example 1.** Find $160^{-1} \ mod \ 841$, i.e., the inverse of 160 modulo 841.

**Solution.** By Exercise 6(c) of the last section, the answer is 205.

**Corollary 1.** *If* $p$ *is a prime number, then every nonzero residue class has a multiplicative inverse which can be found in* $O(log^3 p)$ *bit operations.*

*We say that the ring* $\mathbf{Z}/p\mathbf{Z}$ *is a* field. *We often denote this field* $\mathbf{F}_p$, *the "field of* $p$ *elements."*

**Corollary 2.** *Suppose we want to solve a linear congruence* $ax \equiv b \bmod m$, *where without loss of generality we may assume that* $0 \le a, b < m$. *First, if* $g.c.d.(a, m) = 1$, *then there is a solution* $x_0$ *which can be found in* $O(\log^3 m)$ *bit operations, and all solutions are of the form* $x = x_0 + mn$ *for* $n$ *an integer. Next, suppose that* $d = g.c.d.(a, m)$. *There exists a solution if and only if* $d|b$, *and in that case our congruence is equivalent (in the sense of having the same solutions) to the congruence* $a'x \equiv b' \bmod m'$, *where* $a' = a/d$, $b' = b/d$, $m' = m/d$.

The first corollary is just a special case of Proposition I.3.1. The second corollary is easy to prove from Proposition I.3.1 and the definitions. As in the case of the familiar linear equations with real numbers, to solve linear equations in $\mathbf{Z}/m\mathbf{Z}$ one multiplies both sides of the equation by the multiplicative inverse of the coefficient of the unknown.

In general, when working modulo $m$, the analogy of "nonzero" is often "prime to $m$." We saw above that, like equations, congruences can be added, subtracted and multiplied (see Property 3 of congruences). They can also be divided, provided that the "denominator" is prime to $m$.

**Corollary 3.** *If* $a \equiv b \bmod m$ *and* $c \equiv d \bmod m$, *and if* $g.c.d.(c, m) = 1$ *(in which case also* $g.c.d.(d, m) = 1$*), then* $ac^{-1} \equiv bd^{-1} \bmod m$ *(where* $c^{-1}$ *and* $d^{-1}$ *denote any integers which are inverse to* $c$ *and* $d$ *modulo* $m$*).*

To prove Corollary 3, we have $c(ac^{-1} - bd^{-1}) \equiv (acc^{-1} - bdd^{-1}) \equiv a - b \equiv 0 \bmod m$, and since $m$ has no common factor with $c$, it follows that $m$ must divide $ac^{-1} - bd^{-1}$.

**Proposition I.3.2 (Fermat's Little Theorem).** *Let* $p$ *be a prime. Any integer* $a$ *satisfies* $a^p \equiv a \bmod p$, *and any integer* $a$ *not divisible by* $p$ *satisfies* $a^{p-1} \equiv 1 \bmod p$.

**Proof.** First suppose that $p \nmid a$. We first claim that the integers $0a, 1a, 2a, 3a, \ldots, (p-1)a$ are a complete set of residues modulo $p$. To see this, we observe that otherwise two of them, say $ia$ and $ja$, would have to be in the same residue class, i.e., $ia \equiv ja \bmod p$. But this would mean that $p|(i-j)a$, and since $a$ is not divisible by $p$, we would have $p|i-j$. Since $i$ and $j$ are both less than $p$, the only way this can happen is if $i = j$. We conclude that the integers $a, 2a, \ldots, (p-1)a$ are simply a rearrangement of $1, 2, \ldots, p-1$ when considered modulo $p$. Thus, it follows that the product of the numbers in the first sequence is congruent modulo $p$ to the product of the numbers in the second sequence, i.e., $a^{p-1}(p-1)! \equiv (p-1)! \bmod p$. Thus, $p|\big((p-1)!(a^{p-1}-1)\big)$. Since $(p-1)!$ is not divisible by $p$, we have $p|(a^{p-1}-1)$, as required. Finally, if we multiply both sides of the congruence $a^{p-1} \equiv 1 \bmod p$ by $a$, we get the first congruence in the statement of the proposition in the case when $a$ is not divisible by $p$. But if $a$ is divisible by $p$, then this congruence $a^p \equiv a \bmod p$ is trivial, since both sides are $\equiv 0 \bmod p$. This concludes the proof of the proposition.

**Corollary.** *If $a$ is not divisible by $p$ and if $n \equiv m$ mod $(p-1)$, then $a^n \equiv a^m$ mod $p$.*

**Proof of corollary.** Say $n > m$. Since $p-1|n-m$, we have $n = m + c(p-1)$ for some positive integer $c$. Then multiplying the congruence $a^{p-1} \equiv 1$ *mod* $m$ by itself $c$ times and then by $a^m \equiv a^m$ *mod* $p$ gives the desired result: $a^n \equiv a^m$ *mod* $p$.

**Example 2.** Find the last base–7 digit in $2^{1000000}$.

**Solution.** Let $p = 7$. Since 1000000 leaves a remainder of 4 when divided by $p-1 = 6$, we have $2^{1000000} \equiv 2^4 = 16 \equiv 2$ *mod* 7, so 2 is the answer.

**Proposition I.3.3 (Chinese Remainder Theorem).** *Suppose that we want to solve a system of congruences to different moduli:*

$$\begin{aligned} x &\equiv a_1 \ mod\ m_1, \\ x &\equiv a_2 \ mod\ m_2, \\ &\cdots \quad \cdots \\ x &\equiv a_r \ mod\ m_r. \end{aligned}$$

*Suppose that each pair of moduli is relatively prime: g.c.d.$(m_i, m_j) = 1$ for $i \neq j$. Then there exists a simultaneous solution $x$ to all of the congruences, and any two solutions are congruent to one another modulo $M = m_1 m_2 \cdots m_r$.*

**Proof.** First we prove uniqueness modulo $M$ (the last sentence). Suppose that $x'$ and $x''$ are two solutions. Let $x = x' - x''$. Then $x$ must be congruent to 0 modulo each $m_i$, and hence modulo $M$ (by Property 5 at the beginning of the section). We next show how to construct a solution $x$.

Define $M_i = M/m_i$ to be the product of all of the moduli *except for* the $i$-th. Clearly $g.c.d.(m_i, M_i) = 1$, and so there is an integer $N_i$ (which can be found by means of the Euclidean algorithm) such that $M_i N_i \equiv 1$ *mod* $m_i$. Now set $x = \sum_i a_i M_i N_i$. Then for each $i$ we see that the terms in the sum other than the $i$-th term are all divisible by $m_i$, because $m_i | M_j$ whenever $j \neq i$. Thus, for each $i$ we have: $x \equiv a_i M_i N_i \equiv a_i$ *mod* $m_i$, as desired.

**Corollary.** *The Euler phi-function is "multiplicative", meaning that $\varphi(mn) = \varphi(m)\varphi(n)$ whenever g.c.d.$(m, n) = 1$.*

**Proof of corollary.** We must count the number of integers between 0 and $mn-1$ which have no common factor with $mn$. For each $j$ in that range, let $j_1$ be its least nonnegative residue modulo $m$ (i.e., $0 \leq j_1 < m$ and $j \equiv j_1$ *mod* $m$) and let $j_2$ be its least nonnegative residue modulo $n$ (i.e., $0 \leq j_2 < n$ and $j \equiv j_2$ *mod* $n$). It follows from the Chinese Remainder Theorem that for each pair $j_1, j_2$ there is one and only one $j$ between 0 and $mn-1$ for which $j \equiv j_1$ *mod* $m$, $j \equiv j_2$ *mod* $n$. Notice that $j$ has no common factor with $mn$ if and only if it has no common factor with $m$ — which is equivalent to $j_1$ having no common factor with $m$ — and it has no common factor with $n$ — which is equivalent to $j_2$ having no common factor with $n$. Thus, the $j$'s which we must count are in 1-to-1 correspondence with the pairs $j_1, j_2$ for which $0 \leq j_1 < m$, $g.c.d.(j_1, m) = 1$; $0 \leq j_2 < n$,

$g.c.d.(j_2, n) = 1$. The number of possible $j_1$'s is $\varphi(m)$, and the number of possible $j_2$'s is $\varphi(n)$. So the number of pairs is $\varphi(m)\varphi(n)$. This proves the corollary.

Since every $n$ can be written as a product of prime powers, each of which has no common factors with the others, and since we know the formula $\varphi(p^\alpha) = p^\alpha(1 - \frac{1}{p})$, we can use the corollary to conclude that for $n = p_1^{\alpha_1} p_2^{\alpha_2} \cdots p_r^{\alpha_r}$:

$$\varphi(n) = p_1^{\alpha_1}\left(1 - \frac{1}{p_1}\right)p_2^{\alpha_2}\left(1 - \frac{1}{p_2}\right)\cdots p_r^{\alpha_r}\left(1 - \frac{1}{p_r}\right) = n\prod_{p|n}\left(1 - \frac{1}{p}\right).$$

As a consequence of the formula for $\varphi(n)$, we have the following fact, which we shall refer to later when discussing the RSA system of public key cryptography.

**Proposition I.3.4.** *Suppose that $n$ is known to be the product of two distinct primes. Then knowledge of the two primes $p$, $q$ is equivalent to knowledge of $\varphi(n)$. More precisely, one can compute $\varphi(n)$ from $p$, $q$ in $O(\log n)$ bit operations, and one can compute $p$ and $q$ from $n$ and $\varphi(n)$ in $O(\log^3 n)$ bit operations.*

**Proof.** The proposition is trivial if $n$ is even, because in that case we immediately know $p = 2$, $q = n/2$, and $\varphi(n) = n/2 - 1$; so we suppose that $n$ is odd. By the multiplicativity of $\varphi$, for $n = pq$ we have $\varphi(n) = (p-1)(q-1) = n+1-(p+q)$. Thus, $\varphi(n)$ can be found from $p$ and $q$ using one addition and one subtraction. Conversely, suppose that we know $n$ and $\varphi(n)$, but not $p$ or $q$. We regard $p$, $q$ as unknowns. We know their product $n$ and also their sum, since $p + q = n + 1 - \varphi(n)$. Call the latter expression $2b$ (notice that it is even). But two numbers whose sum is $2b$ and whose product is $n$ must be the roots of the quadratic equation $x^2 - 2bx + n = 0$. Thus, $p$ and $q$ equal $b \pm \sqrt{b^2 - n}$. The most time-consuming step is the evaluation of the square root, and by Exercise 16 of § I.1 this can be done in $O(\log^3 n)$ bit operations. This completes the proof.

We next discuss a generalization of Fermat's Little Theorem, due to Euler.

**Proposition I.3.5.** *If $g.c.d.(a, m) = 1$, then $a^{\varphi(m)} \equiv 1 \bmod m$.*

**Proof.** We first prove the proposition in the case when $m$ is a prime power: $m = p^\alpha$. We use induction on $\alpha$. The case $\alpha = 1$ is precisely Fermat's Little Theorem (Proposition I.3.2). Suppose that $\alpha \geq 2$, and the formula holds for the $(\alpha - 1)$-st power of $p$. Then $a^{p^{\alpha-1}-p^{\alpha-2}} = 1 + p^{\alpha-1}b$ for some integer $b$, by the induction assumption. Raising both sides of this equation to the $p$-th power and using the fact that the binomial coefficients in $(1+x)^p$ are each divisible by $p$ (except in the 1 and $x^p$ at the ends), we see that $a^{p^\alpha - p^{\alpha-1}}$ is equal to 1 plus a sum with each term divisible by $p^\alpha$. That is, $a^{\varphi(p^\alpha)} - 1$ is divisible by $p^\alpha$, as desired. This proves the proposition for prime powers.

Finally, by the multiplicativity of $\varphi$, it is clear that $a^{\varphi(m)} \equiv 1 \; mod \; p^\alpha$ (simply raise both sides of $a^{\varphi(p^\alpha)} \equiv 1 \; mod \; p^\alpha$ to the appropriate power). Since this is true for each $p^\alpha || m$, and since the different prime powers have no common factors with one another, it follows by Property 5 of congruences that $a^{\varphi(m)} \equiv 1 \; mod \; m$.

**Corollary.** *If g.c.d.$(a, m) = 1$ and if $n'$ is the least nonnegative residue of $n$ modulo $\varphi(m)$, then $a^n \equiv a^{n'} \; mod \; m$.*

This corollary is proved in the same way as the corollary of Proposition I.3.2.

**Remark.** As the proof of Proposition I.3.5 makes clear, there's a smaller power of $a$ which is guaranteed to give $1 \; mod \; m$: the least common multiple of the powers that give $1 \; mod \; p^\alpha$ for each $p^\alpha || m$. For example, $a^{12} \equiv 1 \; mod \; 105$ for $a$ prime to 105, because 12 is a multiple of $3 - 1$, $5 - 1$ and $7 - 1$. Note that $\varphi(105) = 48$. Here is another example:

**Example 3.** Compute $2^{1000000} \; mod \; 77$.

**Solution.** Because 30 is the least common multiple of $\varphi(7) = 6$ and $\varphi(11) = 10$, by the above remark we have $2^{30} \equiv 1 \; mod \; 77$. Since $1000000 = 30 \cdot 33333 + 10$, it follows that $2^{1000000} \equiv 2^{10} \equiv 23 \; mod \; 77$. A second method of solution would be first to compute $2^{1000000} \; mod \; 7$ (since $1000000 = 6 \cdot 166666 + 4$, this is $2^4 \equiv 2$) and also $2^{1000000} \; mod \; 11$ (since 1000000 is divisible by $11 - 1$, this is 1), and then use the Chinese Remainder Theorem to find an $x$ between 0 and 76 which is $\equiv 2 \; mod \; 7$ and $\equiv 1 \; mod \; 11$.

**Modular exponentiation by the repeated squaring method.** A basic computation one often encounters in modular arithmetic is finding $b^n \; mod \; m$ (i.e., finding the least nonnegative residue) when both $m$ and $n$ are very large. There is a clever way of doing this that is much quicker than repeated multiplication of $b$ by itself. In what follows we shall assume that $b < m$, and that whenever we perform a multiplication we then immediately reduce $mod \; m$ (i.e., replace the product by its least nonnegative residue). In that way we never encounter any integers greater than $m^2$. We now describe the algorithm.

Use $a$ to denote the partial product. When we're done, we'll have $a$ equal to the least nonnegative residue of $b^n \; mod \; m$. We start out with $a = 1$. Let $n_0, n_1, \ldots, n_{k-1}$ denote the binary digits of $n$, i.e., $n = n_0 + 2n_1 + 4n_2 + \cdots + 2^{k-1}n_{k-1}$. Each $n_j$ is 0 or 1. If $n_0 = 1$, change $a$ to $b$ (otherwise keep $a = 1$). Then square $b$, and set $b_1 = b^2 \; mod \; m$ (i.e., $b_1$ is the least nonnegative residue of $b^2 \; mod \; m$). If $n_1 = 1$, multiply $a$ by $b_1$ (and reduce $mod \; m$); otherwise keep $a$ unchanged. Next square $b_1$, and set $b_2 = b_1^2 \; mod \; m$. If $n_2 = 1$, multiply $a$ by $b_2$; otherwise keep $a$ unchanged. Continue in this way. You see that in the $j$-th step you have computed $b_j \equiv b^{2^j} \; mod \; m$. If $n_j = 1$, i.e., if $2^j$ occurs in the binary expansion of $n$, then you include $b_j$ in the product for $a$ (if $2^j$ is absent from $n$, then you do not). It is easy to see that after the $(k-1)$–st step you'll have the desired $a \equiv b^n \; mod \; m$.

How many bit operations does this take? In each step you have either 1 or 2 multiplications of numbers which are less than $m^2$. And there are $k-1$ steps. Since each step takes $O(log^2(m^2))= O(log^2 m)$ bit operations, we end up with the following estimate:

**Proposition I.3.6.** Time($b^n$ $mod$ $m$) $= O((log\, n)(log^2 m))$.

**Remark.** If $n$ is very large in Proposition I.3.6, you might want to use the corollary of Proposition I.3.5, replacing $n$ by its least nonnegative residue modulo $\varphi(m)$. But this requires that you know $\varphi(m)$. If you do know $\varphi(m)$, and if $g.c.d.(b, m) = 1$, so that you can replace $n$ by its least nonnegative residue modulo $\varphi(m)$, then the estimate on the right in Proposition I.3.6 can be replaced by $O(log^3 m)$.

As a final application of the multiplicativity of the Euler $\varphi$-function, we prove a formula that will be used at the beginning of Chapter II.

**Proposition I.3.7.** $\sum_{d|n} \varphi(d) = n$.

**Proof.** Let $f(n)$ denote the left side of the equality in the proposition, i.e., $f(n)$ is the sum of $\varphi(d)$ taken over all divisors $d$ of $n$ (including 1 and $n$). We must show that $f(n) = n$. We first claim that $f(n)$ is multiplicative, i.e., that $f(mn) = f(m)f(n)$ whenever $g.c.d.(m, n) = 1$. To see this, we note that any divisor $d|mn$ can be written (in one and only one way) in the form $d_1 \cdot d_2$, where $d_1|m$, $d_2|n$. Since $g.c.d.(d_1, d_2) = 1$, we have $\varphi(d) = \varphi(d_1)\varphi(d_2)$, because of the multiplicativity of $\varphi$. We get all possible divisors $d$ of $mn$ by taking all possible pairs $d_1$, $d_2$ where $d_1$ is a divisor of $m$ and $d_2$ is a divisor of $n$. Thus, $f(mn) = \sum_{d_1|m} \sum_{d_2|n} \varphi(d_1)\varphi(d_2) = \left(\sum_{d_1|m} \varphi(d_1)\right)\left(\sum_{d_2|n} \varphi(d_2)\right) = f(m)f(n)$, as claimed. Now to prove the proposition suppose that $n = p_1^{\alpha_1} \cdots p_r^{\alpha_r}$ is the prime factorization of $n$. By the multiplicativity of $f$, we find that $f(n)$ is a product of terms of the form $f(p^\alpha)$. So it suffices to prove the proposition for $p^\alpha$, i.e., to prove that $f(p^\alpha) = p^\alpha$. But the divisors of $p^\alpha$ are $p^j$ for $0 \le j \le \alpha$, and so $f(p^\alpha) = \sum_{j=0}^{\alpha} \varphi(p^j) = 1 + \sum_{j=1}^{\alpha}(p^j - p^{j-1}) = p^\alpha$. This proves the proposition for $p^\alpha$, and hence for all $n$.

## *Exercises*

1. Describe all of the solutions of the following congruences:

| | |
|---|---|
| (a) $3x \equiv 4$ $mod$ 7; | (d) $27x \equiv 25$ $mod$ 256; |
| (b) $3x \equiv 4$ $mod$ 12; | (e) $27x \equiv 72$ $mod$ 900; |
| (c) $9x \equiv 12$ $mod$ 21; | (f) $103x \equiv 612$ $mod$ 676. |

2. What are the possibilities for the last hexadecimal digit of a perfect square? (See Exercise 7 of § I.1.)
3. What are the possibilities for the last base-12 digit of a product of two consecutive positive odd numbers?

4. Prove that a decimal integer is divisible by 3 if and only if the sum of its digits is divisible by 3, and that it is divisible by 9 if and only if the sum of its digits is divisible by 9.
5. Prove that $n^5 - n$ is always divisible by 30.
6. Suppose that in tiling a floor that is 8 ft × 9 ft, you bought 72 tiles at a price you cannot remember. Your receipt gives the total cost before taxes as some amount under \$100, but the first and last digits are illegible. It reads \$?0.6?. How much did the tiles cost?
7. (a) Suppose that $m$ is either a power $p^\alpha$ of a prime $p > 2$ or else twice an odd prime power. Prove that, if $x^2 \equiv 1 \ mod\ m$, then either $x \equiv 1 \ mod\ m$ or $x \equiv -1 \ mod\ m$.
   (b) Prove that part (a) is always false if $m$ is not of the form $p^\alpha$ or $2p^\alpha$, and $m \neq 4$.
   (c) Prove that if $m$ is an odd number which is divisible by $r$ different primes, then the congruence $x^2 \equiv 1 \ mod\ m$ has $2^r$ different solutions between 0 and $m$.
8. Prove "Wilson's Theorem," which states that for any prime $p$: $(p-1)! \equiv -1 \ mod\ p$. Prove that $(n-1)!$ is *not* congruent to $-1 \ mod\ n$ if $n$ is *not* prime.
9. Find a 3-digit (decimal) number which leaves a remainder of 4 when divided by 7, 9, or 11.
10. Find the smallest positive integer which leaves a remainder of 1 when divided by 11, a remainder of 2 when divided by 12, and a remainder of 3 when divided by 13.
11. Find the smallest nonnegative solution of each of the following systems of congruences:

| (a) | (b) | (c) |
|---|---|---|
| $x \equiv 2 \ mod\ 3$ | $x \equiv 12 \ mod\ 31$ | $19x \equiv 103 \ mod\ 900$ |
| $x \equiv 3 \ mod\ 5$ | $x \equiv 87 \ mod\ 127$ | $10x \equiv 511 \ mod\ 841$ |
| $x \equiv 4 \ mod\ 11$ | $x \equiv 91 \ mod\ 255$ | |
| $x \equiv 5 \ mod\ 16$ | | |

12. Suppose that a 3-digit (decimal) positive integer which leaves a remainder of 7 when divided by 9 or 10 and 3 when divided by 11 goes evenly into a six-digit natural number which leaves a remainder of 8 when divided by 9, 7 when divided by 10, and 1 when divided by 11. Find the quotient.
13. In the situation of Proposition I.3.3, suppose that $0 \le a_j < m_j < B$ for all $j$, where $B$ is some large bound on the size of the moduli. Suppose that $r$ is also large. Find an estimate for the number of bit operations required to solve the system. Your time estimate should be a function of $B$ and $r$, and should allow for the possibility that $r$ is either very large or very small compared to the number of bits in $B$.
14. Use the repeated squaring method to find $38^{75} \ mod\ 103$.

15. In exact integer arithmetic (rather than modular arithmetic) does the repeated squaring method save time? Explain, using big-$O$ estimates.
16. Notice that for $a$ prime to $p$, $a^{p-2}$ is an inverse of $a$ modulo $p$. Suppose that $p$ is very large. Compare using the repeated squaring method to find $a^{p-2}$ with the Euclidean algorithm as an efficient means to find $a^{-1}$ *mod* $p$ when (a) $a$ has almost as many digits as $p$, and (b) when $a$ is much smaller than $p$.
17. Find $\varphi(n)$ for all $m$ from 90 to 100.
18. Make a list showing all $n$ for which $\varphi(n) \leq 12$, and prove that your list is complete.
19. Suppose that $n$ is not a perfect square, and that $n-1 > \varphi(n) > n-n^{2/3}$. Prove that $n$ is a product of two distinct primes.
20. If $m \geq 8$ is a power of 2, show that the exponent in Proposition I.3.5 can be replaced by $\varphi(m)/2$.
21. Let $m = 7785562197230017200 = 2^4 \cdot 3^3 \cdot 5^2 \cdot 7 \cdot 11 \cdot 13 \cdot 19 \cdot 31 \cdot 37 \cdot 41 \cdot 61 \cdot 73 \cdot 181$.
(a) Find the least nonnegative residue of $6647^{362}$ *mod* $m$.
(b) Let $a$ be a positive integer less than $m$ which is prime to $m$. First, find a positive power of $a$ less than 500 which is certain to give $a^{-1}$ *mod* $m$. Next, describe an algorithm for finding this power of $a$ working modulo $m$. How many multiplications and divisions are needed to carry out this algorithm? (Reducing a number modulo $m$ counts as one division.) What is the maximum number of bits you could encounter in the integers that you work with? Finally, give a good estimate of the number of bit operations needed to find $a^{-1}$ *mod* $m$ by this method. (Your answer should be a specific number — do not use the big-$O$ notation here.)
22. Give another proof of Proposition I.3.7 as follows. For each divisor $d$ of $n$, let $S_d$ denote the subset (actually a so-called "subgroup") of $\mathbf{Z}/n\mathbf{Z}$ consisting of all multiples of $n/d$. Thus, $S_d$ has $d$ elements.
(a) Prove that $S_d$ has $\varphi(d)$ different elements $x$ which *generate* $S_d$, meaning that the multiples of $x$ (considered modulo $n$) give all elements of $S_d$.
(b) Prove that every element of $x$ generates one of the $S_d$, and hence that the number of elements in $\mathbf{Z}/n\mathbf{Z}$ is equal to the sum (taken over divisors $d$) of the number of elements that generate $S_d$. In light of part (a), this gives Proposition I.3.7.
23. (a) Using the Fundamental Theorem of Arithmetic, prove that

$$\prod_{\text{all primes } p} \frac{1}{1-\frac{1}{p}}$$

diverges to infinity.
(b) Using part (a), prove that the sum of the reciprocals of the primes diverges.

(c) Find a sequence $n_j$ approaching $\infty$ for which $lim_{j \longrightarrow \infty} \frac{\varphi(n_j)}{n_j} = 1$ and a sequence $n_j$ for which $lim_{j \longrightarrow \infty} \frac{\varphi(n_j)}{n_j} = 0$.

24. Let $N$ be an extremely large secret integer used to unlock a missile system, i.e., knowing $N$ would enable one to launch the missiles. Suppose you have a commanding general and $n$ different lieutenant generals. In the event that the commanding general (who knows $N$) is incapacitated, you want the lieutenant generals each to have enough partial information about $N$ so that any three of them (but never two of them) can agree to launch the missiles.
(a) Let $p_1, \ldots, p_n$ be $n$ different primes, all of which are greater than $\sqrt[3]{N}$ but much smaller than $\sqrt{N}$. Using the $p_i$, describe the partial information about $N$ that should be given to the lieutenant generals.
(b) Generalize this system to the situation where you want any set of $k$ ($k \geq 2$) of the lieutenant generals, working together, to be able to launch the missiles (but a set of $k - 1$ of them can never unlock the system). Such a set-up is called a *$k$-threshold system for sharing a secret.*

# 4 Some applications to factoring

**Proposition I.4.1.** *For any integer $b$ and any positive integer $n$, $b^n - 1$ is divisible by $b - 1$ with quotient $b^{n-1} + b^{n-2} + \cdots + b^2 + b + 1$.*

**Proof.** We have a polynomial identity coming from the following fact: 1 is a root of $x^n - 1$, and so the linear term $x - 1$ must divide $x^n - 1$. Namely, polynomial division gives $x^n - 1 = (x-1)(x^{n-1} + x^{n-2} + \cdots + x^2 + x + 1)$. (Alternately, we can derive this by multiplying $x$ by $x^{n-1} + x^{n-2} + \cdots + x^2 + x + 1$, then subtracting $x^{n-1} + x^{n-2} + \cdots + x^2 + x + 1$, and finally obtaining $x^n - 1$ after all the canceling.) Now we get the proposition by replacing $x$ by $b$.

A second proof is to use arithmetic in the base $b$. Written to the base $b$, the number $b^n - 1$ consists of $n$ digits $b - 1$ (for example, $10^6 - 1 = 999999$). On the other hand, $b^{n-1} + b^{n-2} + \cdots + b^2 + b + 1$ consists of $n$ digits all 1. Multiplying $111 \cdots 111$ by the 1-digit number $b - 1$ gives $(b-1)(b-1)(b-1)\cdots(b-1)(b-1)(b-1)_b = b^n - 1$.

**Corollary.** *For any integer $b$ and any positive integers $m$ and $n$, we have $b^{mn} - 1 = (b^m - 1)(b^{m(n-1)} + b^{m(n-2)} + \cdots + b^{2m} + b^m + 1)$.*

**Proof.** Simply replace $b$ by $b^m$ in the last proposition.

As an example of the use of this corollary, we see that $2^{35} - 1$ is divisible by $2^5 - 1 = 31$ and by $2^7 - 1 = 127$. Namely, we set $b = 2$ and either $m = 5$, $n = 7$ or else $m = 7$, $n = 5$.

**Proposition I.4.2.** *Suppose that $b$ is prime to $m$, and $a$ and $c$ are positive integers. If $b^a \equiv 1 \bmod m$ and $b^c \equiv 1 \bmod m$, and if $d = g.c.d.(a, c)$, then $b^d \equiv 1 \bmod m$.*

**Proof.** Using the Euclidean algorithm, we can write $d$ in the form $ua + vc$, where $u$ and $v$ are integers. It is easy to see that one of the two numbers $u$, $v$ is positive and the other is negative or zero. Without loss of generality, we may suppose that $u > 0$, $v \leq 0$. Now raise both sides of the congruence $b^a \equiv 1 \; mod \; m$ to the $u$-th power, and raise both sides of the congruence $b^c \equiv 1 \; mod \; m$ to the $(-v)$-th power. Now divide the resulting two congruences, obtaining: $b^{au-c(-v)} \equiv 1 \; mod \; m$. But $au + cv = d$, so the proposition is proved.

**Proposition I.4.3.** *If $p$ is a prime dividing $b^n - 1$, then either* (i) $p|b^d - 1$ *for some* **proper** *divisor $d$ of $n$, or else* (ii) $p \equiv 1 \; mod \; n$. *If $p > 2$ and $n$ is odd, then in case (ii) one has $p \equiv 1 \; mod \; 2n$.*

**Proof.** We have $b^n \equiv 1 \; mod \; p$ and also, by Fermat's Little Theorem, we have $b^{p-1} \equiv 1 \; mod \; p$. By the above proposition, this means that $b^d \equiv 1 \; mod \; p$, where $d = g.c.d.(n, \; p-1)$. First, if $d < n$, then this says that $p \,|\, b^d - 1$ for a proper divisor $d$ of $n$, i.e., case (i) holds. On the other hand, if $d = n$, then, since $d|p-1$, we have $p \equiv 1 \; mod \; n$. Finally, if $p$ and $n$ are both odd and $n \,|\, p-1$ (i.e., we're in case (ii)), then obviously $2n|\, p-1$.

We now show how this proposition can be used to factor certain types of large integers.

**Examples**

1. Factor $2^{11} - 1 = 2047$. If $p \mid 2^{11} - 1$, by the theorem we must have $p \equiv 1 \; mod \; 22$. Thus, we test $p = 23, 67, 89, \ldots$ (actually, we need go no farther than $\sqrt{2047} = 45.\cdots$). We immediately obtain the prime factorization of 2047: $2047 = 23 \cdot 89$. In a very similar way, one can quickly show that $2^{13} - 1 = 8191$ is prime. A prime of the form $2^n - 1$ is called a "Mersenne prime."
2. Factor $3^{12} - 1 = 531440$. By the proposition above, we first try the factors of the much smaller numbers $3^1 - 1$, $3^2 - 1$, $3^3 - 1$, $3^4 - 1$, and the factors of $3^6 - 1 = (3^3 - 1)(3^3 + 1)$ which do not already occur in $3^3 - 1$. This gives us $2^4 \cdot 5 \cdot 7 \cdot 13$. Since $531440/(2^4 \cdot 5 \cdot 7 \cdot 13) = 73$, which is prime, we are done. Note that, as expected, any prime that did not occur in $3^d - 1$ for $d$ a proper divisor of 12 — namely, 73 – must be $\equiv 1 \; mod \; 12$.
3. Factor $2^{35} - 1 = 34359738367$. First we consider the factors of $2^d - 1$ for $d = 1, 5, 7$. This gives the prime factors 31 and 127. Now $(2^{35} - 1)/(31 \cdot 127) = 8727391$. According to the proposition, any remaining prime factor must be $\equiv 1 \; mod \; 70$. So we check 71, 211, 281,..., looking for divisors of 8727391. At first, we might be afraid that we'll have to check all such primes less than $\sqrt{8727391} = 2954.\cdots$. However, we immediately find that $8727391 = 71 \cdot 122921$, and then it remains to check only up to $\sqrt{122921} = 350.\cdots$. We find that 122921 is prime. Thus, $2^{35} - 1 = 31 \cdot 71 \cdot 127 \cdot 122921$ is the prime factorization.

**Remark.** In Example 3, how can one do the arithmetic on a calculator that only shows, say, 8 decimal places? Simply break up the numbers into sections. For example, when we compute $2^{35}$ we reach the limit of our calculator display with $2^{26} = 67108864$. To multiply this by $2^9 = 512$, we write $2^{35} = 512 \cdot (67108 \cdot 1000 + 864) = 34359296 \cdot 1000 + 442368 = 34359738368$. Later, when we divide $2^{35}-1$ by $31 \cdot 127 = 3937$, we first divide 3937 into 34359738, taking the integer part of the quotient: $\left[\frac{34359738}{3937}\right] = 8727$. Next, we write $34359738 = 3937 \cdot 8727 + 1539$. Then

$$\begin{aligned}\frac{34359738367}{3937} &= \frac{(3937 \cdot 8727 + 1539) \cdot 1000 + 367}{3937} \\ &= 8727000 + \frac{1539367}{3937} \\ &= 8727391.\end{aligned}$$

## *Exercises*

1. Give two different proofs that if $n$ is odd, then $b^n + 1 = (b+1)(b^{n-1} - b^{n-2} + \cdots + b^2 - b + 1)$. In one proof use a polynomial identity. In the other proof use arithmetic to the base $b$.
2. Prove that if $2^n - 1$ is a prime, then $n$ is a prime, and that if $2^n + 1$ is a prime, then $n$ is a power of 2. The first type of prime is called a "Mersenne prime," as mentioned above, and the second type is called a "Fermat prime." The first few Mersenne primes are 3, 7, 31, 127; the first few Fermat primes are 3, 5, 17, 257.
3. Suppose that $b$ is prime to $m$, where $m > 2$, and $a$ and $c$ are positive integers. Prove that, if $b^a \equiv -1 \; mod \; m$ and $b^c \equiv \pm 1 \; mod \; m$, and if $d = g.c.d.(a, c)$, then $b^d \equiv -1 \; mod \; m$, and $a/d$ is odd.
4. Prove that, if $p \mid b^n + 1$, then either (i) $p \mid b^d + 1$ for some proper divisor $d$ of $n$ for which $n/d$ is odd, or else (ii) $p \equiv 1 \; mod \; 2n$.
5. Let $m = 2^{24} + 1 = 16777217$.
   (a) Find a Fermat prime which divides $m$.
   (b) Prove that any other prime is $\equiv 1 \; mod \; 48$.
   (c) Find the complete prime factorization of $m$.
6. Factor $3^{15} - 1$ and $3^{24} - 1$.
7. Factor $5^{12} - 1$.
8. Factor $10^5 - 1$, $10^6 - 1$ and $10^8 - 1$.
9. Factor $2^{33} - 1$ and $2^{21} - 1$.
10. Factor $2^{15} - 1$, $2^{30} - 1$, and $2^{60} - 1$.
11. (a) Prove that if $d = g.c.d.(m, n)$ and $a > 1$ is an integer, then $g.c.d.(a^m - 1, \; a^n - 1) = a^d - 1$.
    (b) Suppose you want to multiply two $k$-bit integers $a$ and $b$, where $k$ is very large. Let $\ell$ be a fixed integer much smaller than $k$. Choose a set of $m_i$, $1 \le i \le r$, such that $\frac{\ell}{2} < m_i < \ell$ for all $i$ and $g.c.d.(m_i, m_j) = 1$ for $i \ne j$. Choose $r = \lceil 4k/\ell \rceil + 1$. Suppose that a large integer such as

$a$ is stored as an $r$-tuple $(a_1, \ldots, a_r)$, where $a_i$ is the least nonnegative residue of $a$ mod $2^{m_i} - 1$. Prove that $a$, $b$ and $ab$ are each uniquely determined by the corresponding $r$-tuple, and estimate the number of bit operations required to find the $r$-tuple corresponding to $ab$ from the $r$-tuples corresponding to $a$ and $b$.

## References for Chapter I


1. J. Brillhart, D. H. Lehmer, J. L. Selfridge, B. Tuckerman, and S. S. Wagstaff, Jr., *Factorizations of* $b^n \pm 1$, $b = 2, 3, 5, 6, 7, 10, 11, 12$, *up to High Powers*, Amer. Math. Society, 1983.
2. L. E. Dickson, *History of the Theory of Numbers*, three volumes, Chelsea, 1952.
3. R. K. Guy, *Unsolved Problems in Number Theory*, Springer–Verlag, 1982.
4. G. H. Hardy and E. M. Wright, *An Introduction to the Theory of Numbers*, 5th ed., Oxford University Press, 1979.
5. W. J. LeVeque, *Fundamentals of Number Theory*, Addison–Wesley, 1977.
6. H. Rademacher, *Lectures on Elementary Number Theory*, Krieger, 1977.
7. K. H. Rosen, *Elementary Number Theory and Its Applications*, 3rd ed., Addison–Wesley, 1993.
8. M. R. Schroeder, *Number Theory in Science and Communication*, 2nd ed., Springer–Verlag, 1986.
9. D. Shanks, *Solved and Unsolved Problems in Number Theory*, 3rd ed., Chelsea Publ. Co., 1985.
10. W. Sierpínski, *A Selection of Problems in the Theory of Numbers*, Pergamon Press, 1964.
11. D. D. Spencer, *Computers in Number Theory*, Computer Science Press, 1982.

# II

# Finite Fields and Quadratic Residues

In this chapter we shall assume familiarity with the basic definitions and properties of a field. We now briefly recall what we need.

1. A *field* is a set $\mathbf{F}$ with a *multiplication* and *addition* operation which satisfy the familiar rules — associativity and commutativity of both addition and multiplication, the distributive law, existence of an additive identity 0 and a multiplicative identity 1, additive inverses, and multiplicative inverses for everything except 0. The following examples of fields are basic in many areas of mathematics: (1) the field $\mathbf{Q}$ consisting of all rational numbers; (2) the field $\mathbf{R}$ of real numbers; (3) the field $\mathbf{C}$ of complex numbers; (4) the field $\mathbf{Z}/p\mathbf{Z}$ of integers modulo a prime number $p$.
2. A *vector space* can be defined over any field $\mathbf{F}$ by the same properties that are used to define a vector space over the real numbers. Any vector space has a *basis*, and the number of elements in a basis is called its *dimension*. An *extension field*, i.e., a bigger field containing $\mathbf{F}$, is automatically a vector space over $\mathbf{F}$. We call it a *finite extension* if it is a finite dimensional vector space. By the *degree* of a finite extension we mean its dimension as a vector space. One common way of obtaining extension fields is to *adjoin* an element to $\mathbf{F}$: we say that $\mathbf{K} = \mathbf{F}(\alpha)$ if $\mathbf{K}$ is the field consisting of all rational expressions formed using $\alpha$ and elements of $\mathbf{F}$.
3. Similarly, the *polynomial ring* can be defined over any field $\mathbf{F}$. It is denoted $\mathbf{F}[X]$; it consists of all finite sums of powers of $X$ with coefficients in $\mathbf{F}$. One adds and multiplies polynomials in $\mathbf{F}[X]$ in the same way as one does with polynomials over the reals. The *degree* $d$ of a polynomial

is the largest power of $X$ which occurs with nonzero coefficient; in a *monic* polynomial the coefficient of $X^d$ is 1. We say that $g$ *divides* $f$, where $f$, $g \in \mathbf{F}[X]$, if there exists a polynomial $h \in \mathbf{F}[X]$ such that $f = gh$. The *irreducible* polynomials $f \in \mathbf{F}[X]$ are those that are not divisible by any polynomials of lower degree except for constants; they play the role among the polynomials that the primes play among the integers. The polynomial ring has *unique factorization*, meaning that every monic polynomial can be written in one and only one way (except for the order of factors) as a product of monic irreducible polynomials. (A non-monic polynomial can be uniquely written as a constant times such a product.)

4. An element $\alpha$ in some extension field $\mathbf{K}$ containing $\mathbf{F}$ is said to be *algebraic* over $\mathbf{F}$ if it satisfies a polynomial with coefficients in $\mathbf{F}$. In that case there is a *unique* monic irreducible polynomial in $\mathbf{F}[X]$ of which $\alpha$ is a root (and any other polynomial which $\alpha$ satisfies must be divisible by this monic irreducible polynomial). If this monic irreducible polynomial has degree $d$, then any element of $\mathbf{F}(\alpha)$ (i.e., any rational expression involving powers of $\alpha$ and elements in $\mathbf{F}$) can actually be expressed as a linear combination of the powers $1, \alpha, \alpha^2, \ldots, \alpha^{d-1}$. Thus, those powers of $\alpha$ form a basis of $F(\alpha)$ over $F$, and so the degree of the extension obtained by adjoining $\alpha$ is the same as the degree of the monic irreducible polynomial of $\alpha$. Any other root $\alpha'$ of the same irreducible polynomial is called a *conjugate* of $\alpha$ over $\mathbf{F}$. The fields $\mathbf{F}(\alpha)$ and $\mathbf{F}(\alpha')$ are *isomorphic* by means of the map that takes any expression in terms of $\alpha$ to the same expression with $\alpha$ replaced by $\alpha'$. The word "isomorphic" means that we have a 1-to-1 correspondence that preserves addition and multiplication. In some cases the fields $\mathbf{F}(\alpha)$ and $\mathbf{F}(\alpha')$ are the same, in which case we obtain an *automorphism* of the field. For example, $\sqrt{2}$ has one conjugate, namely $-\sqrt{2}$, over $\mathbf{Q}$, and the map $a+b\sqrt{2} \mapsto a-b\sqrt{2}$ is an automorphism of the field $\mathbf{Q}(\sqrt{2})$ (which consists of all real numbers of the form $a + b\sqrt{2}$ with $a$ and $b$ rational). If all of the conjugates of $\alpha$ are in the field $\mathbf{F}(\alpha)$, then $\mathbf{F}(\alpha)$ is called a *Galois* extension of $\mathbf{F}$.

5. The *derivative* of a polynomial is defined using the $nX^{n-1}$ rule (not as a limit, since limits don't make sense in $\mathbf{F}$ unless there is a concept of distance or a topology in $\mathbf{F}$). A polynomial $f$ of degree $d$ may or may not have a root $r \in \mathbf{F}$, i.e., a value which gives 0 when substituted in place of $X$ in the polynomial. If it does, then the degree–1 polynomial $X-r$ divides $f$; if $(X-r)^m$ is the highest power of $X-r$ which divides $f$, then we say that $r$ is a root of *multiplicity* $m$. Because of unique factorization, the total number of roots of $f$ in $\mathbf{F}$, counting multiplicity, cannot exceed $d$. If a polynomial $f \in \mathbf{F}[X]$ has a multiple root $r$, then $r$ will be a root of the *greatest common divisor* of $f$ and its derivative $f'$(see Exercise 13 of § I.2).

6. Given any polynomial $f(X) \in \mathbf{F}[X]$, there is an extension field $\mathbf{K}$ of

**F** such that $f(X)$ splits into a product of linear factors (equivalently, has $d$ roots in **K**, counting multiplicity, where $d$ is its degree) and such that **K** is the smallest extension field containing those roots. **K** is called the *splitting field* of $f$. The splitting field is unique *up to isomorphism*, meaning that if we have any other field $\mathbf{K}'$ with the same properties, then there must be a 1-to-1 correspondence $\mathbf{K}\xrightarrow{\sim}\mathbf{K}'$ which preserves addition and multiplication. For example, $\mathbf{Q}(\sqrt{2})$ is the splitting field of $f(X) = X^2 - 2$, and to obtain the splitting field of $f(X) = X^3 - 2$ one must adjoin to **Q** both $\sqrt[3]{2}$ and $\sqrt{-3}$.

7. If adding the multiplicative identity 1 to itself in **F** never gives 0, then we say that **F** has *characteristic zero*; in that case **F** contains a copy of the field of rational numbers. Otherwise, there is a prime number $p$ such that $1 + 1 + \cdots + 1$ ($p$ times) equals 0, and $p$ is called the *characteristic* of the field $F$. In that case $F$ contains a copy of the field $\mathbf{Z}/p\mathbf{Z}$ (see Corollary 1 of Proposition I.3.1), which is called its *prime field.*

# 1 Finite fields

Let $\mathbf{F}_q$ denote a field which has a finite number $q$ of elements in it. Clearly a finite field cannot have characteristic zero; so let $p$ be the characteristic of $\mathbf{F}_q$. Then $\mathbf{F}_q$ contains the prime field $\mathbf{F}_p = \mathbf{Z}/p\mathbf{Z}$, and so is a vector space — necessarily finite dimensional — over $\mathbf{F}_p$. Let $f$ denote its dimension as an $\mathbf{F}_p$–vector space. Since choosing a basis enables us to set up a 1-to-1 correspondence between the elements of this $f$-dimensional vector space and the set of all $f$-tuples of elements in $\mathbf{F}_p$, it follows that there must be $p^f$ elements in $\mathbf{F}_q$. That is, *$q$ is a power of the characteristic $p$.*

We shall soon see that for every prime power $q = p^f$ there is a field of $q$ elements, and it is unique (up to isomorphism).

But first we investigate the multiplicative *order* of elements in $\mathbf{F}_q^*$, the set of nonzero elements of our finite field. By the "order" of a nonzero element we mean the least positive power which is 1.

**Existence of multiplicative generators of finite fields.** There are $q - 1$ nonzero elements, and, by the definition of a field, they form an *abelian group* with respect to multiplication. This means that the product of two nonzero elements is nonzero, the associative law and commutative law hold, there is an identity element 1, and any nonzero element has an inverse. It is a general fact about finite groups that the order of any element must divide the number of elements in the group. For the sake of completeness, we give a proof of this in the case of our group $\mathbf{F}_q^*$.

**Proposition II.1.1.** *The order of any $a \in \mathbf{F}_q^*$ divides $q - 1$.*

**First proof.** Let $d$ be the smallest power of $a$ which equals 1. (Note that there is a finite power of $a$ that is 1, since the powers of $a$ in the finite set $\mathbf{F}_q^*$ cannot all be distinct, and as soon as $a^i = a^j$ for $j > i$ we have

$a^{j-i} = 1$.) Let $S = \{1, a, a^2, \ldots, a^{d-1}\}$ denote the set of all powers of $a$, and for any $b \in \mathbf{F}_q^*$ let $bS$ denote the "coset" consisting of all elements of the form $ba^j$ (for example, $1S = S$). It is easy to see that any two cosets are either identical or distinct (namely: if some $b_1a^i$ in $b_1S$ is also in $b_2S$, i.e., if it is of the form $b_2a^j$, then *any* element $b_1a^{i'}$ in $b_1S$ is of the form to be in $b_2S$, because $b_1a^{i'} = b_1a^ia^{i'-i} = b_2a^{j+i'-i}$). And each coset contains exactly $d$ elements. Since the union of all the cosets exhausts $\mathbf{F}_q^*$, this means that $\mathbf{F}_q^*$ is a disjoint union of $d$-element sets; hence $d|(q-1)$.

**Second proof.** First we show that $a^{q-1} = 1$. To see this, write the product of all nonzero elements in $\mathbf{F}_q$. There are $q - 1$ of them. If we multiply each of them by $a$, we get a rearrangement of the same elements (since any two distinct elements remain distinct after multiplication by $a$). Thus, the product is not affected. But we have multiplied this product by $a^{q-1}$. Hence $a^{q-1} = 1$. (Compare with the proof of Proposition I.3.2.) Now let $d$ be the order of $a$, i.e., the smallest positive power which gives 1. If $d$ did not divide $q - 1$, we could find a smaller positive number $r$ — namely, the remainder when $q - 1 = bd + r$ is divided by $d$ — such that $a^r = a^{q-1-bd} = 1$. But this contradicts the minimality of $d$. This concludes the proof.

**Definition.** A *generator* $g$ of a finite field $\mathbf{F}_q$ is an element of order $q-1$; equivalently, the powers of $g$ run through all of the elements of $\mathbf{F}_q^*$.

The next proposition is one of the very basic facts about finite fields. It says that the nonzero elements of any finite field form a *cyclic group*, i.e., they are all powers of a single element.

**Proposition II.1.2.** *Every finite field has a generator. If $g$ is a generator of $\mathbf{F}_q^*$, then $g^j$ is also a generator if and only if g.c.d.$(j, q-1) = 1$. In particular, there are a total of $\varphi(q-1)$ different generators of $\mathbf{F}_q^*$.*

**Proof.** Suppose that $a \in \mathbf{F}_q^*$ has order $d$, i.e., $a^d = 1$ and no lower power of $a$ gives 1. By Proposition II.1.1, $d$ divides $q - 1$. Since $a^d$ is the smallest power which equals 1, it follows that the elements $a, a^2, \ldots, a^d = 1$ are distinct. We claim that the elements of order $d$ are precisely the $\varphi(d)$ values $a^j$ for which $g.c.d.(j, d) = 1$. First, since the $d$ distinct powers of $a$ all satisfy the equation $x^d = 1$, these are all of the roots of the equation (see paragraph 5 in the list of facts about fields). Any element of order $d$ must thus be among the powers of $a$. However, not all powers of $a$ have order $d$, since if $g.c.d.(j, d) = d' > 1$, then $a^j$ has lower order: because $d/d'$ and $j/d'$ are integers, we can write $(a^j)^{(d/d')} = (a^d)^{j/d'} = 1$. Conversely, we now show that $a^j$ does have order $d$ whenever $g.c.d.(j, d) = 1$. If $j$ is prime to $d$, and if $a^j$ had a smaller order $d''$, then $a^{d''}$ raised to either the $j$–th or the $d$–th power would give 1, and hence $a^{d''}$ raised to the power $g.c.d.(j, d) = 1$ would give 1 (this is proved in exactly the same way as Proposition I.4.2). But this contradicts the fact that $a$ is of order $d$ and so $a^{d''} \neq 1$. Thus, $a^j$ has order $d$ if and only if $g.c.d.(j, d) = 1$.

This means that, if there is any element $a$ of order $d$, then there are exactly $\varphi(d)$ elements of order $d$. So for every $d|(q-1)$ there are only two possibilities: *no* element has order $d$, or exactly $\varphi(d)$ elements have order $d$.

Now every element has some order $d|(q-1)$. And there are either 0 or $\varphi(d)$ elements of order $d$. But, by Proposition I.3.7, $\sum_{d|(q-1)} \varphi(d) = q-1$, which is the number of elements in $\mathbf{F}_q^*$. Thus, the only way that every element can have some order $d|(q-1)$ is if there are always $\varphi(d)$ (and never 0) elements of order $d$. In particular, there are $\varphi(q-1)$ elements of order $q-1$; and, as we saw in the previous paragraph, if $g$ is any element of order $q-1$, then the other elements of order $q-1$ are precisely the powers $g^j$ for which $g.c.d.(j,\ q-1) = 1$. This completes the proof.

**Corollary.** *For every prime $p$, there exists an integer $g$ such that the powers of $g$ exhaust all nonzero residue classes modulo $p$.*

**Example 1.** We can get all residues *mod* 19 from 1 to 18 by taking powers of 2. Namely, the successive powers of 2 reduced *mod* 19 are: 2, 4, 8, 16, 13, 7, 14, 9, 18, 17, 15, 11, 3, 6, 12, 5, 10, 1.

In many situations when working with finite fields, such as $\mathbf{F}_p$ for some prime $p$, it is useful to find a generator. What if a number $g \in \mathbf{F}_p^*$ is chosen at random? What is the probability that it will be a generator? In other words, what proportion of all of the nonzero elements consists of generators? According to Proposition II.1.2, the proportion is $\varphi(p-1)/(p-1)$. But by our formula for $\varphi(n)$ following the corollary of Proposition I.3.3, this fraction is equal to the $\prod(1-\frac{1}{\ell})$, where the product is over all primes $\ell$ dividing $p-1$. Thus, the odds of getting a generator by a random guess depend heavily on the factorization of $p-1$. For example, we can prove:

**Proposition II.1.3.** *There exists a sequence of primes $p$ such that the probability that a random $g \in \mathbf{F}_p^*$ is a generator approaches zero.*

**Proof.** Let $\{n_j\}$ be any sequence of positive integers which is divisible by more and more of the successive primes 2, 3, 5, 7,... as $j \longrightarrow \infty$. For example, we could take $n_j = j!$. Choose $p_j$ to be any prime such that $p_j \equiv 1 \; mod \; n_j$. How do we know that such a prime exists? That follows from *Dirichlet's theorem on primes in an arithmetic progression*, which states: *If $n$ and $k$ are relatively prime, then there are infinitely many primes which are $\equiv k \; mod \; n$.* (In fact, more is true: the primes are "evenly distributed" among the different possible $k \; mod \; n$, i.e., the proportion of primes $\equiv k \; mod \; n$ is $1/\varphi(n)$; but we don't need that fact here.) Then the primes dividing $p_j - 1$ include all of the primes dividing $n_j$, and so $\frac{\varphi(p_j-1)}{p_j-1} \le \prod_{\text{primes } \ell | n_j} (1-\frac{1}{\ell})$. But as $j \longrightarrow \infty$ this product approaches $\prod_{\text{all primes } \ell} (1-\frac{1}{\ell})$, which is zero (see Exercise 23 of § I.3). This proves the proposition.

**Existence and uniqueness of finite fields with prime power number of elements.** We prove both existence and uniqueness by showing that a finite field of $q = p^f$ elements is the splitting field of the polynomial $X^q - X$. The following proposition shows that for every prime power $q$ there is one and (up to isomorphism) only one finite field with $q$ elements.

**Proposition II.1.4.** *If $\mathbf{F}_q$ is a field of $q = p^f$ elements, then every element satisfies the equation $X^q - X = 0$, and $\mathbf{F}_q$ is precisely the set* *of roots of that equation. Conversely, for every prime power* $q = p^f$ *the splitting field over* $\mathbf{F}_p$ *of the polynomial* $X^q - X$ *is a field of* $q$ *elements.*

**Proof.** First suppose that $\mathbf{F}_q$ is a finite field. Since the order of any nonzero element divides $q-1$, it follows that any nonzero element satisfies the equation $X^{q-1} = 1$, and hence, if we multiply both sides by $X$, the equation $X^q = X$. Of course, the element 0 also satisfies the latter equation. Thus, all $q$ elements of $\mathbf{F}_q$ are roots of the degree-$q$ polynomial $X^q - X$. Since this polynomial cannot have more than $q$ roots, its roots are precisely the elements of $\mathbf{F}_q$. Notice that this means that $\mathbf{F}_q$ is the splitting field of the polynomial $X^q - X$, that is, the smallest field extension of $\mathbf{F}_p$ which contains all of its roots.

Conversely, let $q = p^f$ be a prime power, and let $\mathbf{F}$ be the splitting field over $\mathbf{F}_p$ of the polynomial $X^q - X$. Note that $X^q - X$ has derivative $qX^{q-1} - 1 = -1$ (because the integer $q$ is a multiple of $p$ and so is zero in the field $\mathbf{F}_p$); hence, the polynomial $X^q - X$ has no common roots with its derivative (which has no roots at all), and therefore has no multiple roots. Thus, $\mathbf{F}$ must contain at least the $q$ distinct roots of $X^q - X$. But we claim that the set of $q$ roots is already a field. The key point is that a sum or product of two roots is again a root. Namely, if $a$ and $b$ satisfy the polynomial, we have $a^q = a$, $b^q = b$, and hence $(ab)^q = ab$, i.e., the product is also a root. To see that the sum $a+b$ also satisfies the polynomial $X^q - X = 0$, we note a fundamental fact about any field of characteristic $p$:

**Lemma.** $(a+b)^p = a^p + b^p$ *in any field of characteristic* $p$.

The lemma is proved by observing that all of the intermediate terms vanish in the binomial expansion $\sum_{j=0}^{p} \binom{p}{j} a^{p-j} b^j$, because $p!/(p-j)!j!$ is divisible by $p$ for $0 < j < p$.

Repeated application of the lemma gives us: $a^p + b^p = (a+b)^p$, $a^{p^2} + b^{p^2} = (a^p + b^p)^p = (a+b)^{p^2}, \ldots, a^q + b^q = (a+b)^q$. Thus, if $a^q = a$ and $b^q = b$ it follows that $(a+b)^q = a+b$, and so $a+b$ is also a root of $X^q - X$. We conclude that the set of $q$ roots is the smallest field containing the roots of $X^q - X$, i.e., the splitting field of this polynomial is a field of $q$ elements. This completes the proof.

In the proof we showed that raising to the $p$-th power preserves addition and multiplication. We derive another important consequence of this in the next proposition.

**Proposition II.1.5.** *Let* $\mathbf{F}_q$ *be the finite field of* $q = p^f$ *elements, and let* $\sigma$ *be the map that sends every element to its* $p$*-th power:* $\sigma(a) = a^p$*. Then* $\sigma$ *is an* **automorphism** *of the field* $\mathbf{F}_q$ *(a 1-to-1 map of the field to itself which preserves addition and multiplication). The elements of* $\mathbf{F}_q$ *which are kept fixed by* $\sigma$ *are precisely the elements of the prime field* $\mathbf{F}_p$*. The* $f$*-th power (and no lower power) of the map* $\sigma$ *is the identity map.*

**Proof.** A map that raises to a power always preserves multiplication. The fact that $\sigma$ preserves addition comes from the lemma in the proof of Proposition II.1.4. Notice that for any $j$ the $j$-th power of $\sigma$ (the result of repeating $\sigma$ $j$ times) is the map $a \mapsto a^{p^j}$. Thus, the elements left fixed by $\sigma^j$ are the roots of $X^{p^j} - X$. If $j = 1$, these are precisely the $p$ elements of the prime field (this is the special case $q = p$ of Proposition II.1.4, namely, Fermat's Little Theorem). The elements left fixed by $\sigma^f$ are the roots of $X^q - X$, i.e., all of $\mathbf{F}_q$. Since the $f$-th power of $\sigma$ is the identity map, $\sigma$ must be 1-to-1 (its inverse map is $\sigma^{f-1}: \ a \mapsto a^{p^{f-1}}$). No lower power of $\sigma$ gives the identity map, since for $j < f$ not all of the elements of $\mathbf{F}_q$ could be roots of the polynomial $X^{p^j} - X$. This completes the proof.

**Proposition II.1.6.** *In the notation of Proposition II.1.5, if $\alpha$ is any element of $\mathbf{F}_q$, then the conjugates of $\alpha$ over $\mathbf{F}_p$ (the elements of $\mathbf{F}_q$ which satisfy the same monic irreducible polynomial with coefficients in $\mathbf{F}_p$) are the elements $\sigma^j(\alpha) = \alpha^{p^j}$.*

**Proof.** Let $d$ be the degree of $\mathbf{F}_p(\alpha)$ as an extension of $\mathbf{F}_p$. That is, $\mathbf{F}_p(\alpha)$ is a copy of $\mathbf{F}_{p^d}$. Then $\alpha$ satisfies $X^{p^d} - X$ but does not satisfy $X^{p^j} - X$ for any $j < d$. Thus, one obtains $d$ distinct elements by repeatedly applying $\sigma$ to $\alpha$. It now suffices to show that each of these elements satisfies the same monic irreducible polynomial $f(X)$ that $\alpha$ does, in which case they must be the $d$ roots. To do this, it is enough to prove that, if $\alpha$ satisfies a polynomial $f(X) \in \mathbf{F}_p[X]$, then so does $\alpha^p$. Let $f(X) = \sum a_j X^j$, where $a_j \in \mathbf{F}_p$. Then $0 = f(\alpha) = \sum a_j \alpha^j$. Raising both sides to the $p$-th power gives $0 = \sum (a_j \alpha^j)^p$ (where we use the fact that raising a sum $a + b$ to the $p$-th power gives $a^p + b^p$). But $a_j^p = a_j$, by Fermat's Little Theorem, and so we have: $0 = \sum a_j (\alpha^p)^j = f(\alpha^p)$, as desired. This completes the proof.

**Explicit construction.** So far our discussion of finite fields has been rather theoretical. Our only practical experience has been with the finite fields of the form $\mathbf{F}_p = \mathbf{Z}/p\mathbf{Z}$. We now discuss how to work with finite extensions of $\mathbf{F}_p$. At this point we should recall how in the case of the rational numbers $\mathbf{Q}$ we work with an extension such as $\mathbf{Q}(\sqrt{2})$. Namely, we get this field by taking a root $\alpha$ of the equation $X^2 - 2$ and looking at expressions of the form $a + b\alpha$, which are added and multiplied in the usual way, except that $\alpha^2$ should always be replaced by 2. (In the case of $\mathbf{Q}(\sqrt[3]{2})$ we work with expressions of the form $a + b\alpha + c\alpha^2$, and when we multiply we always replace $\alpha^3$ by 2.) We can take the same general approach with finite fields.

**Example 2.** To construct $\mathbf{F}_9$ we take any monic quadratic polynomial in $\mathbf{F}_3[X]$ which has no roots in $\mathbf{F}_3$. By trying all possible choices of coefficients and testing whether the elements $0, \pm 1 \in \mathbf{F}_3$ are roots, we find that there are three monic irreducible quadratics: $X^2 + 1$, $X^2 \pm X - 1$. If, for example, we take $\alpha$ to be a root of $X^2 + 1$ (let's call it $i$ rather than $\alpha$ — after all, we are simply adjoining a square root of $-1$), then the elements of $\mathbf{F}_9$ are all combinations $a + bi$, where $a$ and $b$ are 0, 1, or $-1$. Doing arithmetic in $\mathbf{F}_9$ is thus a lot like doing arithmetic in the Gaussian integers (see Exercise 14 of § I.2), except that our arithmetic with the coefficients $a$ and $b$ occurs in the tiny field $\mathbf{F}_3$.

Notice that the element $i$ that we adjoined is *not* a generator of $\mathbf{F}_9^*$, since it has order 4 rather than $q-1=8$. If, however, we adjoin a root $\alpha$ of $X^2-X-1$, we can get all nonzero elements of $\mathbf{F}_9$ by taking the successive powers of $\alpha$ (remember that $\alpha^2$ must always be replaced by $\alpha+1$, since $\alpha$ satisfies $X^2=X+1$): $\alpha^1=\alpha$, $\alpha^2=\alpha+1$, $\alpha^3=-\alpha+1$, $\alpha^4=-1$, $\alpha^5=-\alpha$, $\alpha^6=-\alpha-1$, $\alpha^7=\alpha-1$, $\alpha^8=1$. We sometimes say that the polynomial $X^2-X-1$ is *primitive*, meaning that any root of the irreducible polynomial is a generator of the group of nonzero elements of the field. There are $4=\varphi(8)$ generators of $\mathbf{F}_9^*$, by Proposition II.1.2: two are the roots of $X^2-X-1$ and two are the roots of $X^2+X-1$. (The second root of $X^2-X-1$ is the conjugate of $\alpha$, namely, $\sigma(\alpha)=\alpha^3=-\alpha+1$.) Of the remaining four nonzero elements, two are the roots of $X^2+1$ (namely $\pm i=\pm(\alpha+1)$) and the other two are the two nonzero elements $\pm 1$ of $\mathbf{F}_3$ (which are roots of the degree-1 monic irreducible polynomials $X-1$ and $X+1$).

In general, in any finite field $\mathbf{F}_q$, $q=p^f$, each element $\alpha$ satisfies a unique monic irreducible polynomial over $\mathbf{F}_p$ of some degree $d$. Then the field $\mathbf{F}_p(\alpha)$ obtained by adjoining this element to the prime field is an extension of degree $d$ that is contained in $\mathbf{F}_q$. That is, it is a copy of the field $\mathbf{F}_{p^d}$. Since the big field $\mathbf{F}_{p^f}$ contains $\mathbf{F}_{p^d}$, and so is an $\mathbf{F}_{p^d}$−vector space of some dimension $f'$, it follows that the number of elements in $\mathbf{F}_{p^f}$ must be $(p^d)^{f'}$, i.e., $f=df'$. Thus, $d|f$. Conversely, for any $d|f$ the finite field $\mathbf{F}_{p^d}$ is contained in $\mathbf{F}_q$, because any solution of $X^{p^d}=X$ is also a solution of $X^{p^f}=X$. (To see this, note that for any $d'$, if you repeatedly replace $X$ by $X^{p^d}$ on the left in the equation $X^{p^d}=X$, you can obtain $X^{p^{dd'}}=1$.) Thus, we have proved:

**Proposition II.1.7.** *The subfields of $\mathbf{F}_{p^f}$ are the $\mathbf{F}_{p^d}$ for $d$ dividing $f$. If an element of $\mathbf{F}_{p^f}$ is adjoined to $\mathbf{F}_p$, one obtains one of these fields.*

It is now easy to prove a formula that is useful in determining the number of irreducible polynomials of a given degree.

**Proposition II.1.8.** *For any $q=p^f$ the polynomial $X^q-X$ factors in $\mathbf{F}_p[X]$ into the product of all monic irreducible polynomials of degrees $d$ dividing $f$.*

**Proof.** If we adjoin to $\mathbf{F}_p$ a root $\alpha$ of any monic irreducible polynomial of degree $d|f$, we obtain a copy of $\mathbf{F}_{p^d}$, which is contained in $\mathbf{F}_{p^f}$. Since $\alpha$ then satisfies $X^q-X=0$, the monic irreducible must divide that polynomial. Conversely, let $f(X)$ be a monic irreducible polynomial which divides $X^q-X$. Then $f(X)$ must have its roots in $\mathbf{F}_q$ (since that's where all of the roots of $X^q-X$ are). Thus $f(X)$ must have degree dividing $f$, by Proposition II.1.7, since adjoining a root gives a subfield of $\mathbf{F}_q$. Thus, the monic irreducible polynomials which divide $X^q-X$ are precisely all of the ones of degree dividing $f$. Since we saw that $X^q-X$ has no multiple factors, this means that $X^q-X$ is equal to the product of all such irreducible polynomials, as was to be proved.

**Corollary.** *If $f$ is a prime number, then there are $(p^f - p)/f$ distinct monic irreducible polynomials of degree $f$ in $\mathbf{F}_p[X]$.*

Notice that $(p^f - p)/f$ is an integer because of Fermat's Little Theorem for the prime $f$, which guarantees that $p^f \equiv p \bmod f$. To prove the corollary, let $n$ be the number of monic irreducible polynomials of degree $f$. According to the proposition, the degree-$p^f$ polynomial $X^{p^f} - X$ is the product of $n$ polynomials of degree $f$ and the $p$ degree-1 irreducible polynomials $X - a$ for $a \in \mathbf{F}_p$. Thus, equating degrees gives: $p^f = nf + p$, from which the desired equality follows.

More generally, suppose that $f$ is not necessarily prime. Then, letting $n_d$ denote the number of monic irreducible polynomials of degree $d$ over $\mathbf{F}_p$, we have $n_f = (p^f - \sum d\, n_d)/f$, where the summation is over all $d < f$ which divide $f$.

We now extend the time estimates in Chapter I for arithmetic modulo $p$ to general finite fields.

**Proposition II.1.9.** *Let $\mathbf{F}_q$, where $q = p^f$, be a finite field, and let $F(X)$ be an irreducible polynomial of degree $f$ over $\mathbf{F}_p$. Then two elements of $\mathbf{F}_q$ can be multiplied or divided in $O(log^3 q)$ bit operations. If $k$ is a positive integer, then an element of $\mathbf{F}_q$ can be raised to the $k$-th power in $O(log\, k\, log^3 q)$ bit operations.*

**Proof.** An element of $\mathbf{F}_q$ is a polynomial with coefficients in $\mathbf{F}_p = \mathbf{Z}/p\mathbf{Z}$ regarded modulo $F(X)$. To multiply two such elements, we multiply the polynomials — this requires $O(f^2)$ multiplications of integers modulo $p$ (and some additions of integers modulo $p$, which take much less time) — and then divide the polynomial $F(X)$ into the product, taking the remainder polynomial as our answer. The polynomial division involves $O(f)$ divisions of integers modulo $p$ and $O(f^2)$ multiplications of integers modulo $p$. Since a multiplication modulo $p$ takes $O(log^2 p)$ bit operations, and a division (using the Euclidean algorithm, for example) takes $O(log^3 p)$ bit operations (see the corollary to Proposition I.2.2), the total number of bit operations is: $O(f^2 log^2 p + f\, log^3 p) = O((f\, log\, p)^3) = O(log^3 q)$. To prove the same result for division, it suffices to show that the reciprocal of an element can be found in time $O(log^3 q)$. Using the Euclidean algorithm for polynomials over the field $\mathbf{F}_p$ (see Exercise 12 of § I.2), we must write 1 as a linear combination of our given element in $\mathbf{F}_q$ (i.e., a given polynomial of degree $< f$) and the fixed degree-$f$ polynomial $F(X)$. This involves $O(f)$ divisions of polynomials of degree $< f$, and each polynomial division requires $O(f^2 log^2 p + f\, log^3 p) = O(f^2 log^3 p)$ bit operations. Thus, the total time required is $O(f^3 log^3 p) = O(log^3 q)$. Finally, a $k$-th power can be computed by the repeated squaring method in the same way as modular exponentiation (see the end of § I.3). This takes $O(log\, k)$ multiplications (or squarings) of elements of $\mathbf{F}_q$, and hence $O(log\, k\, log^3 q)$ bit operations. This completes the proof.

We conclude this section with an example of computation with polynomials over finite fields. We illustrate by an example over the very smallest (and perhaps the most important) finite field, the 2-element field

$\mathbf{F}_2 = \{0, 1\}$. A polynomial in $\mathbf{F}_2[X]$ is simply a sum of powers of $X$. In some ways, polynomials over $\mathbf{F}_p$ are like integers expanded to the base $p$, where the digits are analogous to the coefficients of the polynomial. For example, in its binary expansion an integer is written as a sum of powers of 2 (with coefficients 0 or 1), just as a polynomial over $\mathbf{F}_2$ is a sum of powers of $X$. But the comparison is often misleading. For example, the sum of any number of polynomials of degree $d$ is a polynomial of degree (at most) $d$; whereas a sum of several $d$-bit integers will be an integer having more than $d$ binary digits.

**Example 3.** Let $f(X) = X^4 + X^3 + X^2 + 1$, $g = X^3 + 1 \in \mathbf{F}_2[X]$. Find $g.c.d.(f, g)$ using the Euclidean algorithm for polynomials, and express the g.c.d. in the form $u(X)f(X) + v(X)g(X)$.

**Solution.** Polynomial division gives us the sequence of equalities below, which lead to the conclusion that $g.c.d.(f, g) = X+1$, and the next sequence of equalities enables us, working backwards, to express $X + 1$ as a linear combination of $f$ and $g$. (Note, by the way, that in a field of characteristic 2 adding is the same as subtracting, i.e., $a - b = a + b - 2b = a + b$.) We have:

$$\begin{aligned} f &= (X+1)g + (X^2 + X) \\ g &= (X+1)(X^2+X) + (X+1) \\ X^2 + X &= X(X+1) \end{aligned}$$

and then

$$\begin{aligned} X + 1 &= g + (X+1)(X^2+X) \\ &= g + (X+1)(f + (X+1)g) \\ &= (X+1)f + (X^2)g. \end{aligned}$$

## *Exercises*

1. For $p = 2, 3, 5, 7, 11, 13$ and 17, find the smallest positive integer which generates $\mathbf{F}_p^*$, and determine how many of the integers $1, 2, 3, \ldots, p-1$ are generators.
2. Let $(\mathbf{Z}/p^\alpha\mathbf{Z})^*$ denote all residues modulo $p^\alpha$ which are *invertible*, i.e., are not divisible by $p$. **Warning:** Be sure not to confuse $\mathbf{Z}/p^\alpha\mathbf{Z}$ (which has $p^\alpha - p^{\alpha-1}$ invertible elements) with $\mathbf{F}_{p^\alpha}$ (in which all elements except 0 are invertible). The two are the same only when $\alpha = 1$.
(a) Let $g$ be an integer which generates $\mathbf{F}_p^*$, where $p > 2$. Let $\alpha$ be any integer greater than 1. Prove that either $g$ or $(p+1)g$ generates $(\mathbf{Z}/p^\alpha\mathbf{Z})^*$. Thus, the latter is also a *cyclic group*.
(b) Prove that if $\alpha > 2$, then $(\mathbf{Z}/2^\alpha\mathbf{Z})^*$ is *not* cyclic, but that the number 5 generates a *subgroup* consisting of half of its elements, namely those which are $\equiv 1$ *mod* 4.
3. How many elements are in the smallest field extension of $\mathbf{F}_5$ which contains all of the roots of the polynomials $X^2+X+1$ and $X^3+X+1$?

4. For each degree $d \le 6$, find the number of irreducible polynomials over $\mathbf{F}_2$ of degree $d$, and make a list of them.
5. For each degree $d \le 6$, find the number of monic irreducible polynomials over $\mathbf{F}_3$ of degree $d$, and for $d \le 3$ make a list of them.
6. Suppose that $f$ is a power of a prime $\ell$. Find a simple formula for the number of monic irreducible polynomials of degree $f$ over $\mathbf{F}_p$.
7. Use the polynomial version of the Euclidean algorithm (see Exercise 12 of § I.2) to find $g.c.d.(f,\, g)$ for $f,\, g \in \mathbf{F}_p[X]$ in each of the following examples. In each case express the g.c.d. polynomial as a combination of $f$ and $g$, i.e., in the form $d(X) = u(X)f(X) + v(X)g(X)$.
   (a) $f = X^3 + X + 1$, $g = X^2 + X + 1$, $p = 2$;
   (b) $f = X^6 + X^5 + X^4 + X^3 + X^2 + X + 1$, $g = X^4 + X^2 + X + 1$, $p = 2$;
   (c) $f = X^3 - X + 1$, $g = X^2 + 1$, $p = 3$;
   (d) $f = X^5 + X^4 + X^3 - X^2 - X + 1$, $g = X^3 + X^2 + X + 1$, $p = 3$;
   (e) $f = X^5 + 88x^4 + 73X^3 + 83X^2 + 51X + 67$, $g = X^3 + 97X^2 + 40X + 38$, $p = 101$.
8. By computing $g.c.d.(f, f')$ (see Exercise 13 of § I.2), find all multiple roots of $f(X) = X^7 + X^5 + X^4 - X^3 - X^2 - X + 1 \in \mathbf{F}_3[X]$ in its splitting field.
9. Suppose that $\alpha \in \mathbf{F}_{p^2}$ satisfies the polynomial $X^2 + aX + b$, where $a, b \in \mathbf{F}_p$.
   (a) Prove that $\alpha^p$ also satisfies this polynomial.
   (b) Prove that if $\alpha \notin \mathbf{F}_p$, then $a = -\alpha - \alpha^p$ and $b = \alpha^{p+1}$.
   (c) Prove that if $\alpha \notin \mathbf{F}_p$ and $c, d \in \mathbf{F}_p$, then $(c\alpha + d)^{p+1} = d^2 - acd + bc^2$ (which is $\in \mathbf{F}_p$).
   (d) Let $i$ be a square root of $-1$ in $\mathbf{F}_{19^2}$. Use part (c) to find $(2+3i)^{101}$ (i.e., write it in the form $a + bi$, $a, b \in \mathbf{F}_{19}$).
10. Let $d$ be the maximum degree of two polynomials $f,\, g \in \mathbf{F}_p[X]$. Give an estimate in terms of $d$ and $p$ for the number of bit operations needed to compute $g.c.d.(f, g)$ using the Euclidean algorithm.
11. For each of the following fields $\mathbf{F}_q$, where $q = p^f$, find an irreducible polynomial with coefficients in the prime field whose root $\alpha$ is primitive (i.e., generates $\mathbf{F}_q^*$), and write all of the powers of $\alpha$ as polynomials in $\alpha$ of degree $< f$: (a) $\mathbf{F}_4$; (b) $\mathbf{F}_8$; (c) $\mathbf{F}_{27}$; (d) $\mathbf{F}_{25}$.
12. Let $F(X) \in \mathbf{F}_2[X]$ be a primitive irreducible polynomial of degree $f$. If $\alpha$ denotes a root of $F(X)$, this means that the powers of $\alpha$ exhaust all of $\mathbf{F}_{2^f}^*$. Using the big-$O$ notation, estimate (in terms of $f$) the number of bit operations required to write every power of $\alpha$ as a polynomial in $\alpha$ of degree less than $f$.
13. (a) Under what conditions on $p$ and $f$ is *every* element of $\mathbf{F}_{p^f}$ besides 0, 1 a generator of $\mathbf{F}_{p^f}^*$?
    (b) Under what conditions is every element $\neq 0, 1$ either a generator or the square of a generator?

14. For any fixed $p$, show that there is a sequence $q_j = p^{f_j}$ of powers of $p$ such that the probability that a random element of $\mathbf{F}_{q_j}$ is a generator of $\mathbf{F}^*_{q_j}$ approaches 0 as $j \longrightarrow \infty$.
15. Which polynomials in $\mathbf{F}_p[X]$ have derivative identically zero?
16. Let $\sigma$ be the automorphism of $\mathbf{F}_q$ in Proposition II.1.5. Prove that the set of elements left fixed by $\sigma^j$ is the field $\mathbf{F}_{p^d}$, where $d = g.c.d.(j, f)$.
17. Prove that if $b$ is a generator of $\mathbf{F}^*_{p^n}$ and if $d|n$, then $b^{(p^n-1)/(p^d-1)}$ is a generator of $\mathbf{F}^*_{p^d}$.

# 2 Quadratic residues and reciprocity

**Roots of unity.** In many situations it is useful to have solutions of the equation $x^n = 1$. Suppose we are working in a finite field $\mathbf{F}_q$. We now answer the question: How many $n$-th roots of unity are there in $\mathbf{F}_q$?

**Proposition II.2.1.** *Let $g$ be a generator of $\mathbf{F}^*_q$. Then $g^j$ is an $n$-th root of unity if and only if $nj \equiv 0 \ mod\ q-1$. The number of $n$-th roots of unity is $g.c.d.(n, q-1)$. In particular, $\mathbf{F}_q$ has a* **primitive** *$n$-th root of unity (i.e., an element $\xi$ such that the powers of $\xi$ run through $n$ $n$-th roots of unity) if and only if $n|\, q-1$. If $\xi$ is a primitive $n$-th root of unity in $\mathbf{F}_q$, then $\xi^j$ is also a primitive $n$-th root if and only if $g.c.d.(j, n) = 1$.*

**Proof.** Any element of $\mathbf{F}^*_q$ can be written as a power $g^j$ of the generator $g$. A power of $g$ is 1 if and only if the power is divisible by $q-1$. Thus, an element $g^j$ is an $n$-th root of unity if and only if $nj \equiv 0 \ mod\ q-1$. Next, let $d = g.c.d.(n, q-1)$. According to Corollary 2 of Proposition I.3.1, the equation $nj \equiv 0 \ mod\ q-1$ (with $j$ the unknown) is equivalent to the equation $\frac{n}{d} j \equiv 0 \ mod\ \frac{q-1}{d}$. Since $n/d$ is prime to $(q-1)/d$, the latter congruence is equivalent to requiring $j$ to be a multiple of $(q-1)/d$. In other words, the $d$ distinct powers of $g^{(q-1)/d}$ are precisely the $n$-th roots of unity. There are $n$ such roots if and only if $d = n$, i.e., $n|\, q-1$. Finally, if $n$ does divide $q-1$, let $\xi = g^{(q-1)/n}$. Then $\xi^j$ equals 1 if and only if $n|j$. The $k$-th power of $\xi^j$ equals 1 if and only if $kj \equiv 0 \ mod\ n$. It is easy to see that $\xi^j$ has order $n$ (i.e., this equation does not hold for any positive $k < n$) if and only if $j$ is prime to $n$. Thus, there are $\varphi(n)$ different primitive $n$-th roots of unity if $n|\, q-1$. This completes the proof.

**Corollary 1.** *If $g.c.d.(n, q-1) = 1$, then 1 is the only $n$-th root of unity.*

**Corollary 2.** *The element $-1 \in \mathbf{F}_q$ has a square root in $\mathbf{F}_q$ if and only if $q \equiv 1 \ mod\ 4$.*

The first corollary is a special case of the proposition. To prove Corollary 2, note that a square root of $-1$ is the same thing as a primitive 4-th root of 1, and our field has a primitive 4-th root if and only if $4|\, q-1$.

Corollary 2 says that if $q \equiv 3 \ mod\ 4$, we can always get the quadratic extension $\mathbf{F}_{q^2}$ by adjoining a root of $X^2+1$, i.e., by considering "Gaussian integer" type expressions $a + bi$. We did this for $q = 3$ in the last section.

Let us suppose, for example, that $p$ is a prime which is $\equiv 3 \ mod\ 4$. There is a nice way to think of the field $\mathbf{F}_{p^2}$ which generalizes to other situations. Let $R$ denote the Gaussian integer ring (see Exercise 14 of § I.2). Sometimes we write $R = \mathbf{Z}+\mathbf{Z}i$, meaning the set of all integer combinations of 1 and $i$. If $m$ is any Gaussian integer, and $\alpha = a + bi$ and $\beta = c + di$ are two Gaussian integers, we write $\alpha \equiv \beta \ mod\ m$ if $\alpha - \beta$ is divisible by $m$, i.e., if the quotient is a Gaussian integer. We can then look at the set $R/mR$ of residue classes modulo $m$; just as in the case of ordinary integers, residue classes can be added or multiplied, and the residue class of the result does not depend on which representatives were chosen for the residue class factors. Now if $m = p + 0i$ is a prime number which is $\equiv 3 \ mod\ 4$, it is not hard to show that $R/pR$ is the field $\mathbf{F}_{p^2}$.

**Quadratic residues.** Suppose that $p$ is an odd prime, i.e., $p > 2$. We are interested in knowing which of the nonzero elements $\{1, 2, \ldots, p-1\}$ of $\mathbf{F}_p$ are squares. If some $a \in \mathbf{F}_p^*$ is a square, say $b^2 = a$, then $a$ has precisely two square roots $\pm b$ (since the equation $X^2 - a = 0$ has at most two solutions in a field). Thus, the squares in $\mathbf{F}_p^*$ can all be found by computing $b^2 \ mod\ p$ for $b = 1, 2, 3, \ldots, (p-1)/2$ (since the remaining integers up to $p-1$ are all $\equiv -b$ for one of these $b$), and precisely half of the elements in $\mathbf{F}_p^*$ are squares. For example, the squares in $\mathbf{F}_{11}$ are $1^2 = 1$, $2^2 = 4$, $3^2 = 9$, $4^2 = 5$, and $5^2 = 3$. The squares in $\mathbf{F}_p$ are called *quadratic residues* modulo $p$. The remaining nonzero elements are called *nonresidues.* For $p = 11$ the nonresidues are 2, 6, 7, 8, 10. There are $(p-1)/2$ residues and $(p-1)/2$ nonresidues.

If $g$ is a generator of $\mathbf{F}_p$, then any element can be written in the form $g^j$. Thus, the square of any element is of the form $g^j$ with $j$ even. Conversely, any element of the form $g^j$ with $j$ even is the square of some element, namely $\pm g^{j/2}$.

**The Legendre symbol.** Let $a$ be an integer and $p > 2$ a prime. We define the *Legendre symbol* $(\frac{a}{p})$ to equal 0, 1 or $-1$, as follows:

$$\left(\frac{a}{p}\right) = \begin{cases} 0, & \text{if } p|a; \\ 1, & \text{if } a \text{ is a quadratic residue } mod\ p; \\ -1, & \text{if } a \text{ is a nonresidue } mod\ p. \end{cases}$$

Thus, the Legendre symbol is simply a way of identifying whether or not an integer is a quadratic residue modulo $p$.

**Proposition II.2.2.**

$$\left(\frac{a}{p}\right) \equiv a^{(p-1)/2} \ mod\ p.$$

**Proof.** If $a$ is divisible by $p$, then both sides are $\equiv 0 \ mod\ p$. Suppose $p \nmid a$. By Fermat's Little Theorem, in $\mathbf{F}_p$ the square of $a^{(p-1)/2}$ is 1, so $a^{(p-1)/2}$ itself is $\pm 1$. Let $g$ be a generator of $\mathbf{F}_p^*$, and let $a = g^j$. As we saw, $a$ is a residue if and only if $j$ is even. And $a^{(p-1)/2} = g^{j(p-1)/2}$ is 1 if and only if $j(p-1)/2$ is divisible by $p-1$, i.e., if and only if $j$ is even. Thus, both sides of the congruence in the proposition are $\pm 1$ in $\mathbf{F}_p$, and each side is $+1$ if and only if $j$ is even. This completes the proof.

**Proposition II.2.3.** *The Legendre symbol satisfies the following properties:*

(a) $(\frac{a}{p})$ *depends only on the residue of* a *modulo* p;

(b) $(\frac{ab}{p}) = (\frac{a}{p})(\frac{b}{p})$;

(c) *for* b *prime to* p, $(\frac{ab^2}{p}) = (\frac{a}{p})$;

(d) $(\frac{1}{p}) = 1$ *and* $(\frac{-1}{p}) = (-1)^{(p-1)/2}$.

**Proof.** Part (a) is obvious from the definition. Part (b) follows from Proposition II.2.2, because the right side is congruent modulo $p$ to $a^{(p-1)/2} \cdot b^{(p-1)/2} = (ab)^{(p-1)/2}$, as is the left side. Part (c) follows immediately from part (b). The first equality in part (d) is obvious, because $1^2 = 1$, and the second equality comes from Corollary 2 of Proposition II.2.1 (or by taking $a = -1$ in Proposition II.2.2). This completes the proof.

Part (b) of Proposition II.2.3 shows that one can determine if a number $a$ is a quadratic residue modulo $p$, i.e., one can evaluate $(\frac{a}{p})$, if one factors $a$ and knows the Legendre symbol for the factors. The first step in doing this is to write $a$ as a power of 2 times an odd number. We then want to know how to evaluate $(\frac{2}{p})$.

**Proposition II.2.4.**

$$\left(\frac{2}{p}\right) = (-1)^{(p^2-1)/8} = \begin{cases} 1 & \text{if } p \equiv \pm 1 \ mod\ 8; \\ -1 & \text{if } p \equiv \pm 3 \ mod\ 8. \end{cases}$$

**Proof.** Let $f(n) = (-1)^{(n^2-1)/8}$ for $n$ odd, $f(n) = 0$ for $n$ even. We want to show that $(\frac{2}{p}) = f(p)$. Of the various ways of proving this, we shall use an efficient method based on what we already know about finite fields. Since $p^2 \equiv 1 \ mod\ 8$ for any odd prime $p$, we know that the field $\mathbf{F}_{p^2}$ contains a primitive 8-th root of unity. Let $\xi \in \mathbf{F}_{p^2}$ denote a primitive 8-th root of 1. Note that $\xi^4 = -1$. Define $G = \sum_{j=0}^{7} f(j)\xi^j$. ($G$ is an example of what is called a *Gauss sum.*) Then $G = \xi - \xi^3 - \xi^5 + \xi^7 = 2(\xi - \xi^3)$ (because $\xi^5 = \xi^4\xi = -\xi$ and $\xi^7 = -\xi^3$), and $G^2 = 4(\xi^2 - 2\xi^4 + \xi^6) = 8$. Thus, in $\mathbf{F}_{p^2}$ we have

$$G^p = (G^2)^{(p-1)/2}G = 8^{(p-1)/2}G = \left(\frac{8}{p}\right)G = \left(\frac{2}{p}\right)G,$$

by Proposition II.2.2 and Proposition II.2.3(c). On the other hand, using the definition of $G$, the fact that $(a+b)^p = a^p + b^p$ in $\mathbf{F}_{p^2}$, and the obvious observation that $f(j)^p = f(j)$, we compute: $G^p = \sum_{j=0}^{7} f(j)\xi^{pj}$. Notice that $f(j) = f(p)f(pj)$, as we easily check. Then, making the change of variables $j' = pj$ (i.e., modulo 8 we have $j'$ running through $0, \ldots, 7$ when $j$ does), we obtain:

$$G^p = \sum_{j=0}^{7} f(p)f(pj)\xi^{pj} = f(p)\sum_{j'=0}^{7} f(j')\xi^{j'} = f(p)G.$$

Comparing the two equalities for $G^p$ gives the desired result. (Notice that we can divide by $G$, since it is not 0 in $\mathbf{F}_{p^2}$, as is clear from the fact that its square is 8.)

Next, we must deal with the odd prime factors of $a$. Let $q$ stand for such an odd prime factor. **Warning:** for the remainder of this section, $q$ will stand for an odd prime distinct from $p$, *not* for a power of $p$ as in the last section.

Since $a$ can be assumed to be smaller than $p$ (by part (a) of Proposition II.2.3), the prime factors $q$ will be smaller than $p$. The next proposition — the fundamental Law of Quadratic Reciprocity — tells us how to relate $(\frac{q}{p})$ to $(\frac{p}{q})$. The latter Legendre symbol will be easier to evaluate, since we can immediately replace $p$ by its least positive residue modulo $q$, thereby reducing ourselves to a Legendre symbol involving smaller numbers. The quadratic reciprocity law states that $(\frac{q}{p})$ and $(\frac{p}{q})$ are the same unless $p$ and $q$ are both $\equiv 3$ *mod* 4, in which case they are the negatives of one another. This can be expressed as a formula using the fact that $(p-1)(q-1)/4$ is even unless both primes are $\equiv 3$ *mod* 4, in which case it is odd.

**Proposition II.2.5 (Law of Quadratic Reciprocity).** *Let $p$ and $q$ be two odd primes. Then*

$$\left(\frac{q}{p}\right) = (-1)^{(p-1)(q-1)/4}\left(\frac{p}{q}\right) = \begin{cases} -\left(\frac{p}{q}\right) & \text{if } p \equiv q \equiv 3 \textit{ mod } 4; \\ \left(\frac{p}{q}\right) & \text{otherwise.} \end{cases}$$

**Proof.** There are several dozen proofs of quadratic reciprocity in print. We shall give a particularly short proof along the lines of the proof of the last proposition, using finite fields. Let $f$ be any power of $p$ such that $p^f \equiv 1$ *mod* $q$. For example, we can always take $f = q-1$. Then, as we saw at the beginning of the section (Proposition II.2.1), the field $\mathbf{F}_{p^f}$ contains a primitive $q$-th root of unity, which we denote $\xi$. (Remember that $q$ here denotes another prime besides $p$; it does *not* denote $p^f$.) We define the "Gauss sum" $G$ by the formula $G = \sum_{j=0}^{q-1}(\frac{j}{q})\xi^j$. In the next paragraph we shall prove that $G^2 = (-1)^{(q-1)/2}q$. Before proving that lemma, we show how to use it to prove our proposition. The proof is very similar to the proof of Proposition II.2.4. We first obtain (using the lemma to be proved below):

$$\begin{aligned} G^p &= (G^2)^{(p-1)/2}G = \left((-1)^{(q-1)/2}q\right)^{(p-1)/2}G \\ &= (-1)^{(p-1)(q-1)/4}q^{(p-1)/2}G = (-1)^{(p-1)(q-1)/4}\left(\frac{q}{p}\right)G, \end{aligned}$$

by Proposition II.2.2 with $a$ replaced by $q$ (recall that we're working in a field of characteristic $p$, namely $\mathbf{F}_{p^f}$, and so congruence modulo $p$ becomes equality). On the other hand, using the definition of $G$, the fact that $(a+b)^p = a^p + b^p$ in $\mathbf{F}_{p^f}$, and the obvious observation that $(\frac{j}{q})^p = (\frac{j}{q})$, we compute:

$$G^p = \sum_{j=0}^{q-1} \left(\frac{j}{q}\right) \xi^{pj} = \sum_{j=0}^{q-1} \left(\frac{p}{q}\right)\left(\frac{pj}{q}\right) \xi^{pj},$$

by parts (b) and (c) of Proposition II.2.3. Pulling $(\frac{p}{q})$ outside the summation and making the change of variables $j' = pj$ in the summation, we finally obtain: $G^p = (\frac{p}{q})G$. Equating our two expressions for $G^p$ and dividing by $G$ (which is possible, since $G^2 = \pm q$ and so is not zero in $\mathbf{F}_{p^f}$), we obtain the quadratic reciprocity law. Thus, it remains to prove the following lemma.

**Lemma.** $G^2 = (-1)^{(q-1)/2} q$.

**Proof.** Using the definition of $G$, where in one copy of $G$ we replace the variable of summation $j$ by $-k$ (and note that the summation can start at 1 rather than 0, since $(\frac{0}{q}) = 0$), we have:

$$G^2 = \sum_{j,k=1}^{q-1} \left(\frac{j}{q}\right) \xi^j \left(\frac{-k}{q}\right) \xi^{-k} = \left(\frac{-1}{q}\right) \sum_{j=1}^{q-1} \sum_{k=1}^{q-1} \left(\frac{jk}{q}\right) \xi^{j-k}$$

$$= (-1)^{(q-1)/2} \sum_{j=1}^{q-1} \sum_{k=1}^{q-1} \left(\frac{j^2 k}{q}\right) \xi^{j(1-k)},$$

where we have used Part (d) of Proposition II.2.3 to replace $(\frac{-1}{q})$ by $(-1)^{(q-1)/2}$, and for each value of $j$ we have made a change of variable in the inner summation $k \longleftrightarrow kj$ (i.e., for each fixed $j$, $kj$ runs through the residues modulo $q$ as $k$ does, and the summands depend only on the residue modulo $q$). We next use part (c) of Proposition II.2.3, interchange the order of summation, and pull the $(\frac{k}{q})$ outside the inner sum over $j$. The double sum then becomes $\sum_k (\frac{k}{q}) \sum_j \xi^{j(1-k)}$. Here both sums go from 1 to $q-1$, but if we want we can insert the terms with $j = 0$, since that simply adds to the double sum $\sum_k (\frac{k}{q})$, which is zero (because there are equally many residues and nonresidues modulo $q$). Thus, the double sum can be written $\sum_{k=1}^{q-1} (\frac{k}{q}) \sum_{j=0}^{q-1} \xi^{j(1-k)}$. But for each $k$ other than 1, the inner sum vanishes. This is because the sum of the distinct powers of a nontrivial ($\neq 1$) root of unity $\xi'$ is zero (the simplest way to see this is to note that multiplying the sum by $\xi'$ just rearranges it, and so the sum multiplied by $\xi' - 1$ is zero). So we are left with the contribution when $k = 1$, and we finally obtain:

$$G^2 = (-1)^{(q-1)/2} \left(\frac{1}{q}\right) \sum_{j=0}^{q-1} \xi^0 = (-1)^{(q-1)/2} q.$$

This completes the proof of the lemma, and hence also the proof of the Law of Quadratic Reciprocity.

**Example 1.** Determine whether 7411 is a residue modulo the prime 9283.

**Solution.** Since 7411 and 9283 are both primes which are $\equiv 3$ *mod* 4, we have $\left(\frac{7411}{9283}\right) = -\left(\frac{9283}{7411}\right) = -\left(\frac{1872}{7411}\right)$ by part (a) of Proposition II.2.3. Since $1872 = 2^4 \cdot 3^2 \cdot 13$, by part (c) of Proposition II.2.3 we find that the desired Legendre symbol is $-\left(\frac{13}{7411}\right)$. But we can now apply quadratic reciprocity again: since $13 \equiv 1$ *mod* 4 we find that $-\left(\frac{13}{7411}\right) = -\left(\frac{7411}{13}\right) = -\left(\frac{1}{13}\right) = -1$. In other words, 7411 is a quadratic nonresidue.

One difficulty with this method of evaluating Legendre symbols is that at each stage we must factor the number on top in order to apply Proposition II.2.5. If our numbers are astronomically large, this will be very time-consuming. Fortunately, it is possible to avoid any need for factoring (except taking out powers of 2, which is very easy), once we prove a generalization of the quadratic reciprocity law that applies to all positive odd integers, not necessarily prime. But we first need a definition which generalizes the definition of the Legendre symbol.

**The Jacobi symbol.** Let $a$ be an integer, and let $n$ be any positive odd number. Let $n = p_1^{\alpha_1} \cdots p_r^{\alpha_r}$ be the prime factorization of $n$. Then we define the *Jacobi symbol* $\left(\frac{a}{n}\right)$ as the product of the Legendre symbols for the prime factors of $n$:

$$\left(\frac{a}{n}\right) = \left(\frac{a}{p_1}\right)^{\alpha_1} \cdots \left(\frac{a}{p_r}\right)^{\alpha_r}.$$

A word of warning is in order here. If $\left(\frac{a}{n}\right) = 1$ for $n$ composite, it is *not* necessarily true that $a$ is a square modulo $n$. For example, $\left(\frac{2}{15}\right) = \left(\frac{2}{3}\right)\left(\frac{2}{5}\right) = (-1)(-1) = 1$, but there is no integer $x$ such that $x^2 \equiv 2$ *mod* 15.

We now generalize Propositions II.2.4–5 to the Jacobi symbol.

**Proposition II.2.6.** *For any positive odd $n$ we have $\left(\frac{2}{n}\right) = (-1)^{(n^2-1)/8}$.*

**Proof.** Let $f(n)$ denote the function on the right side of the equality, as in the proof of Proposition II.2.4. It is easy to see that $f(n_1 n_2) = f(n_1)f(n_2)$ for any two odd numbers $n_1$ and $n_2$. (Just consider the different possibilities for $n_1$ and $n_2$ modulo 8.) This means that the right side of the equality in the proposition equals $f(p_1)^{\alpha_1} \cdots f(p_r)^{\alpha_r} = \left(\frac{2}{p_1}\right)^{\alpha_1} \cdots \left(\frac{2}{p_r}\right)^{\alpha_r}$ by Proposition II.2.4. But this is $\left(\frac{2}{n}\right)$, by definition.

**Proposition II.2.7.** *For any two positive odd integers $m$ and $n$ we have $\left(\frac{m}{n}\right) = (-1)^{(m-1)(n-1)/4}\left(\frac{n}{m}\right)$.*

**Proof.** First note that if $m$ and $n$ have a common factor, then it follows from the definition of the Legendre and Jacobi symbols that both sides are zero. So we can suppose that $g.c.d.(m, n) = 1$. Next, we write $m$ and $n$ as products of primes: $m = p_1 p_2 \cdots p_r$ and $n = q_1 q_2 \cdots q_s$. (The $p$'s and $q$'s include repetitions if $m$ or $n$ has a square factor.) In converting from $\left(\frac{m}{n}\right) = \prod_{i,j}\left(\frac{p_i}{q_j}\right)$ to $\left(\frac{n}{m}\right) = \prod_{i,j}\left(\frac{q_j}{p_i}\right)$ we must apply the quadratic reciprocity law for the Legendre symbol $rs$ times. The number of $(-1)$'s we get is the number of times both $p_i$ and $q_j$ are $\equiv 3$ *mod* 4, i.e., it is the product of the number of primes $\equiv 3$ *mod* 4 in the factorization of $m$ and in the factorization of $n$. Thus, $\left(\frac{m}{n}\right) = \left(\frac{n}{m}\right)$ unless there are an odd number of primes $\equiv 3 \; mod \; 4$ in both factorizations, in which case $(\frac{m}{n}) = -(\frac{n}{m})$. But a product of odd primes, such as $m$ or $n$, is $\equiv 3 \; mod \; 4$ if and only if it contains an odd number of primes which are $\equiv 3 \; mod \; 4$. We conclude that $(\frac{m}{n}) = (\frac{n}{m})$ unless both $m$ and $n$ are $\equiv 3 \; mod \; 4$, as was to be proved. This gives us the reciprocity law for the Jacobi symbol.

**Example 2.** We return to Example 1, and show how to evaluate the Legendre symbol without factoring 1872, except to take out the power of 2. By the reciprocity law for the Jacobi symbol we have

$$-\left(\frac{1872}{7411}\right) = -\left(\frac{16}{7411}\right)\left(\frac{117}{7411}\right) = -\left(\frac{7411}{117}\right) = -\left(\frac{40}{117}\right),$$

and this is equal to $-(\frac{2}{117})(\frac{5}{117}) = (\frac{5}{117}) = (\frac{117}{5}) = (\frac{2}{5}) = -1$.

**Square roots modulo $p$.** Using quadratic reciprocity, one can quickly determine whether or not an integer $a$ is a quadratic residue modulo $p$. However, if it is a residue, that does not tell us how to find a solution to the congruence $x^2 \equiv a \; mod \; p$ — it tells us only that a solution exists. We conclude this section by giving an algorithm for finding a square root of a residue $a$ once we know any *non*residue $n$.

Let $p$ be an odd prime, and suppose that we somehow know a quadratic nonresidue $n$. Let $a$ be an integer such that $(\frac{a}{p}) = 1$. We want to find an integer $x$ such that $x^2 \equiv a \; mod \; p$. Here is how we proceed. First write $p-1$ in the form $2^\alpha \cdot s$, where $s$ is odd. Then compute $n^s$ modulo $p$, and call that $b$. Next compute $a^{(s+1)/2}$ modulo $p$, and call that $r$. Our first claim is that $r$ comes reasonably close to being a square root of $a$. More precisely, if we take the ratio of $r^2$ to $a$, we claim that we get a $2^{\alpha-1}$-th root of unity modulo $p$. Namely, we compute (for brevity, we shall use equality to mean congruence modulo $p$, and we use $a^{-1}$ to mean the inverse of $a$ modulo $p$):

$$(a^{-1}r^2)^{2^{\alpha-1}} = a^{s2^{\alpha-1}} = a^{(p-1)/2} = \left(\frac{a}{p}\right) = 1.$$

We must then modify $r$ by a suitable $2^\alpha$-th root of unity to get an $x$ such that $x^2/a$ is 1. To do this, we claim that $b$ is a *primitive* $2^\alpha$-th root of unity, which means that all $2^\alpha$-th roots of unity are powers of $b$. To see this, first we note that $b$ is a $2^\alpha$-th root of 1, because $b^{2^\alpha} = n^{2^\alpha s} = n^{p-1} = 1$. If $b$ weren't primitive, there would be a lower power (a divisor of $2^\alpha$) of $b$ that gives 1. But then $b$ would be an even power of a primitive $2^\alpha$-th root of unity, and so would be a square in $\mathbf{F}_p^*$. This is impossible, because $(\frac{b}{p}) = (\frac{n}{p})^s = -1$ (since $s$ is odd and $n$ is a nonresidue). Thus, $b$ is a primitive $2^\alpha$-th root of unity. So it remains to find a suitable power $b^j$, $0 \le j < 2^\alpha$, such that $x = b^j r$ gives the desired square root of $a$. To do that, we write $j$ in binary as $j = j_0 + 2j_1 + 4j_2 + \cdots + 2^{\alpha-2}j_{\alpha-2}$, and show how one successively determines whether $j_0, j_1, \ldots$ is 0 or 1. (Note that we may suppose that $j < 2^{\alpha-1}$, since $b^{2^{\alpha-1}} = -1$, and so $j$ can be modified by $2^{\alpha-1}$ to give another $j$ for which $b^j r$ is the other square root of $a$.) Here is the inductive procedure for determining the binary digits of $j$:

1. Raise $(r^2/a)$ to the $2^{\alpha-2}$-th power. We proved that the square of this is 1. Hence, you get either $\pm 1$. If you get 1, take $j_0 = 0$; if you get $-1$, take $j_0 = 1$. Notice that $j_0$ has been chosen so that $((b^{j_0}r)^2/a)$ is a $2^{\alpha-2}$-th root of unity.
2. Suppose you've found $j_0, \ldots, j_{k-1}$ such that $(b^{j_0+2j_1+\cdots+2^{k-1}j_{k-1}}r)^2/a$ is a $2^{\alpha-k-1}$-th root of unity, and you want to find $j_k$. Raise this number to half the power that gives 1, and choose $j_k$ according to whether you get $+1$ or $-1$:

$$\text{if} \qquad \left(\frac{\left(b^{j_0+2j_1+\cdots+2^{k-1}j_{k-1}}r\right)^2}{a}\right)^{2^{\alpha-k-2}} = \begin{cases} 1 \\ -1 \end{cases},$$

$$\text{then take} \qquad j_k = \begin{cases} 0 \\ 1 \end{cases}, \quad \text{respectively.}$$

We easily check that with this choice of $j_k$ the "corrected" value comes closer to being a square root of $a$, i.e., we find that $(b^{j_0+2j_1+\cdots+2^k j_k}r)^2/a$ is a $2^{\alpha-k-2}$-th root of unity.

When we get to $k = \alpha - 2$ and find $j_{\alpha-2}$, we then have

$$(b^{j_0+2j_1+\cdots+2^{\alpha-2}j_{\alpha-2}}r)^2/a = 1,$$

i.e., $b^j r$ is a square root of $a$, as desired.

**Example 3.** Use the above algorithm to find a square root of $a = 186$ modulo $p = 401$.

**Solution.** The first nonresidue is $n = 3$. We have $p - 1 = 2^4 \cdot 25$, and so $b = 3^{25} = 268$ and $r = a^{13} = 103$ (where we use equality to denote congruence modulo $p$). After first computing $a^{-1} = 235$, we note that $r^2/a = 98$, which must be an 8-th root of 1. We compute that $98^4 = -1$, and so $j_0 = 1$. Next, we compute $(br)^2/a = -1$. Since the 2-nd power of this is 1, we have $j_1 = 0$, and then $j_2 = 1$. Thus, $j = 5$ and the desired square root is $b^5 r = 304$.

**Remarks.** **1.** The easiest case of this algorithm occurs when $p$ is a prime which is $\equiv 3 \; mod \; 4$. Then $\alpha = 1$, $s = (p-1)/2$, so $(s+1)/2 = (p+1)/4$, and we see that $x = r = a^{(p+1)/4}$ is already the desired square root.

**2.** We now discuss the time estimate for this algorithm. We suppose that we start already knowing the information that $n$ is a nonresidue. The steps in finding $s$, $b$, and $r = a^{(s+1)/2}$ (working modulo $p$, of course) take at most $O(log^3 p)$ bit operations (see Proposition I.3.6). Then in finding $j$ the most time-consuming part of the $k$-th induction step is raising a number to the $2^{\alpha-k-2}$-th power, and this means $\alpha - k - 2$ squarings mod $p$ of integers less than $p$. Since $\alpha - k - 2 < \alpha$, we have the estimate $O(\alpha \, log^2 p)$ for each step. Thus, since there are $\alpha - 1$ steps, the final estimate is $O(log^3 p + \alpha^2 log^2 p) = O(log^2 p(log\, p + \alpha^2))$. At worst (if almost all of $p-1$ is a power of 2), this is $O(log^4 p)$, since $\alpha < log_2 p = O(log\, p)$. Thus, given a nonresidue modulo $p$, we can extract square roots mod $p$ in polynomial time (bounded by the fourth power of the number of bits in $p$).

**3.** Strictly speaking, it is not known (unless one assumes the validity of the so-called "Riemann Hypothesis") whether there is an algorithm for finding a nonresidue modulo $p$ in polynomial time. However, given any $\epsilon > 0$ there is a polynomial time algorithm that finds a nonresidue with probability greater than $1-\epsilon$. Namely, a randomly chosen number $n$, $0 < n < p$, has a 50% chance of being a nonresidue, and this can be checked in polynomial time (see Exercise 17 below). If we do this for more than $log_2(1/\epsilon)$ different randomly chosen $n$, then with probability $> 1-\epsilon$ at least one of them will be a nonresidue.

## *Exercises*

1. Make a table showing all quadratic residues and nonresidues modulo $p$ for $p = 3, 5, 7, 13, 17, 19$.
2. Suppose that $p|2^{2^k}+1$, where $k > 1$.
   (a) Use Exercise 4 of § I.4 to prove that $p \equiv 1 \ mod\ 2^{k+1}$.
   (b) Use Proposition II.2.4 to prove that $p \equiv 1 \ mod\ 2^{k+2}$.
   (c) Use part (b) to prove that $2^{16}+1$ is prime.
3. How many 84-th roots of 1 are there in the field of $11^3$ elements?
4. Prove that $(\frac{-2}{p}) = 1$ if $p \equiv 1$ or $3 \ mod\ 8$, and $(\frac{-2}{p}) = -1$ if $p \equiv 5$ or $7 \ mod\ 8$.
5. Find $(\frac{91}{167})$ using quadratic reciprocity.
6. Find the Gauss sum $G = \sum_{j=1}^{q-1}(\frac{j}{q})\xi^j$ (here $\xi$ is a $q$-th root of 1 in $\mathbf{F}_{p^f}$, where $p^f \equiv 1 \ mod\ q$) when:
   (a) $q = 7$, $p = 29$, $f = 1$, $\xi = 7$;
   (b) $q = 5$, $p = 19$, $f = 2$, $\xi = 2-4i$, where $i$ is a root of $X^2+1$;
   (c) $q = 7$, $p = 13$, $f = 2$, $\xi = 4+\alpha$, where $\alpha$ is a root of $X^2-2$.
7. Let $m = a^4+1$, $a \geq 2$. Find a positive integer $x$ between 0 and $m/2$ such that $x^2 \equiv 2 \ mod\ m$. Use this to find $\sqrt{2}$ in $\mathbf{F}_p$ when $p$ is each of the following: the Fermat primes 17, 257, 65537; $p = 41 = (3^4+1)/2$, $p = 1297$, and $p = 1201$. (Hint: see the proof of Proposition II.2.4.)
8. Let $p$ and $q$ be two primes with $q \equiv 1 \ mod\ p$. Let $\xi$ be a primitive $p$-th root of unity in $\mathbf{F}_q$. Find a formula in terms of $\xi$ for a square root of $(\frac{-1}{p})p$ in $\mathbf{F}_q$.
9. (a) Let $m = a^p - 1$, where $p$ is an odd prime and $a \geq 2$. Find a positive integer $x$ between 0 and $m/2$ such that $x^2 \equiv (\frac{-1}{p})p \ mod\ m$. Use this to find $\sqrt{5}$ in $\mathbf{F}_{31}$, $\sqrt{-7}$ in $\mathbf{F}_{127}$, $\sqrt{13}$ in $\mathbf{F}_{8191}$, and $\sqrt{-7}$ in $\mathbf{F}_{1093}$.
   (b) If $q = 2^p - 1$ is a Mersenne prime, find an expression for the least positive integer whose square is $\equiv (\frac{-1}{p})p \ mod\ q$.
10. Evaluate the Legendre symbol $(\frac{1801}{8191})$ (a) using the reciprocity law only for the Legendre symbol (i.e., factoring all numbers that arise), and (b)

without factoring any odd integers, instead using the reciprocity law for the Jacobi symbol.

11. Evaluate the following Legendre symbols:
(a) $(\frac{11}{37})$; (b) $(\frac{19}{31})$; (c) $(\frac{97}{101})$; (d) $(\frac{31}{167})$; (e) $(\frac{5}{160465489})$; (f) $(\frac{3083}{3911})$; (g) $(\frac{43691}{65537})$.
12. (a) Let $p$ be an odd prime. Prove that $-3$ is a residue in $\mathbf{F}_p$ if and only if $p \equiv 1 \ mod\ 3$.
(b) Prove that 3 is a quadratic nonresidue modulo any Mersenne prime greater than 3.
13. Find a condition on the last decimal digit of $p$ which is equivalent to 5 being a square in $\mathbf{F}_p$.
14. Prove that a quadratic residue can never be a generator of $\mathbf{F}_p^*$.
15. Let $p$ be a Fermat prime.
(a) Show that any quadratic nonresidue is a generator of $\mathbf{F}_p^*$.
(b) Show that 5 is a generator of $\mathbf{F}_p^*$, except in the case $p = 5$.
(c) Show that 7 is a generator of $\mathbf{F}_p^*$, except in the case $p = 3$.
16. Let $p$ be a Mersenne prime, let $q = p^2$, and let $i$ be a root of $X^2+1 = 0$, so that $\mathbf{F}_q = \mathbf{F}_p(i)$.
(a) Suppose that the integer $a^2 + b^2$ is a generator of $\mathbf{F}_p^*$. Prove that $a + bi$ is a generator of $\mathbf{F}_q^*$.
(b) Show that either $4 + i$ or $3 + 2i$ will serve as a generator of $\mathbf{F}_{31^2}^*$.
17. Let $p$ be an odd prime and $a$ be an integer between 1 and $p - 1$. Estimate in terms of $p$ the number of bit operations needed to compute $(\frac{a}{p})$ (a) using the reciprocity law for the Jacobi symbol, and (b) using Proposition II.2.2 and Proposition I.3.6.
18. (a) Let $p$ be an odd prime, and let $a$, $b$, $c$ be integers with $p \nmid a$. Prove that the number of solutions $x \in \{0, 1, 2, \ldots, p-1\}$ to the congruence $ax^2 + bx + c \equiv 0 \ mod\ p$ is given by the formula $1 + (\frac{D}{p})$, where $D = b^2 - 4ac$ is the discriminant.
(b) How many solutions in $\mathbf{F}_{83}$ are there to each of the following equations: (i) $x^2 + 1 = 0$; (ii) $x^2 + x + 1 = 0$; (iii) $x^2 + 21x - 11 = 0$; (iv) $x^2 + x + 21 = 0$; (v) $x^2 - 4x - 13 = 0$?
(c) How many solutions in $\mathbf{F}_{97}$ are there to each of the equations in part (b)?
19. Let $p = 2081$, and let $n$ be the smallest positive nonresidue modulo $p$. Find $n$, and use the method in the text to find a square root of 302 modulo $p$.
20. Let $m = p_1^{\alpha_1} \cdots p_r^{\alpha_r}$ be an odd integer, and suppose that $a$ is prime to $m$ and is the square of some integer modulo $m$. Your object is to find $x$ such that $x^2 \equiv a \ mod\ m$. Suppose that for each $j$ you know a nonresidue modulo $p_j$, i.e., an integer $n_j$ such that $(\frac{n_j}{p_j}) = -1$.
(a) For each fixed $p = p_j$ and $\alpha = \alpha_j$, suppose you use the algorithm in the text to find some $x_0$ such that $x_0^2 \equiv a \ mod\ p$. Show how you can then find some $x = x_0 + x_1 p + \cdots + x_{\alpha-1} p^{\alpha-1}$ such that $x^2 \equiv a \ mod\ p^{\alpha}$.

(b) Describe how to find an $x$ such that $x^2 \equiv a \bmod m$.
The technique in parts (a)–(b) of this exercise is known as "lifting" a square root from $\mathbf{F}_{p_j}$ $(1 \le j \le r)$ to $\mathbf{Z}/m\mathbf{Z}$.

21. In the text we saw that if $n$ is an odd prime and $g.c.d.(b, n) = 1$, then

$$b^{(n-1)/2} \equiv \left(\frac{b}{n}\right) \bmod n. \qquad (*)$$

The purpose of this exercise is to show that, if $n$ is an odd composite integer, then the relation $(*)$ is false for at least 50% of all $b$ for which $g.c.d.(b, n) = 1$.
(a) Prove that if $(*)$ is true for $b_1$ and is false for $b_2$, then it is false for the product $b_1b_2$. Use this to prove that if $(*)$ is false for even a single $b$, then the number of $b$'s for which it is false is at least as great as the number of $b$'s for which it is true.
(b) If $n$ is divisible by the square of a prime $p$, show how to find an integer $b$ prime to $n$ such that $b^{(n-1)/2}$ is not $\equiv \pm 1 \bmod n$.
(c) If $n$ is a product of distinct primes, if $p$ is one of those primes, and if $b$ has the property that $(\frac{b}{p}) = -1$ and $b \equiv 1 \bmod n/p$, prove that $(*)$ fails for $b$. Then show that such a $b$ always exists.

22. Explain why the following probabilistic algorithm gives a square root of $a$ modulo $p$: Choose $t$ in $\mathbf{F}_p$ at random until you find $t$ such that $t^2 - a$ is a *non*square modulo $p$. Let $\alpha$ denote the element $\sqrt{t^2 - a}$ in the quadratic extension $\mathbf{F}_{p^2}$. Then compute $b = (t + \alpha)^{(p+1)/2}$. Show that $b$ is in $\mathbf{F}_p$ and has the property that $b^2 = a$.

23. Suppose that $p$ is a prime $\equiv 1 \bmod 4$, and suppose you have found a quadratic nonresidue $n$. Describe an algorithm for expressing $p$ as a sum of two squares $p = c^2 + d^2$ that takes time $O(log^3 p)$.

## References for Chapter II


1. L. Adleman, K. Manders, and G. Miller, "On taking roots in finite fields," *Proc. 20th Annual Symposium on the Foundations of Computer Science* (1979), 175–178.
2. E. R. Berlekamp, "Factoring polynomials over large finite fields," *Math. Comp.*, **24** (1970), 713–735.
3. I. Blake, X. Gao, A. Menezes, R. Mullen, S. Vanstone, and T. Yaghoobian, *Applications of Finite Fields*, Kluwer Acad. Publ., 1992.
4. C. F. Gauss, *Disquisitiones Arithmeticae*, Yale Univ. Press, 1966.
5. E. Grosswald, *Topics from the Theory of Numbers*, 2nd ed., Birkhäuser, 1984.
6. I. N. Herstein, *Topics in Algebra*, 2nd ed., Wiley, 1975.
7. K. Ireland and M. I. Rosen, *A Classical Introduction to Modern Number Theory*, 2nd ed., Springer–Verlag, 1990.

8. S. Lang, *Algebra*, 2nd ed., Addison–Wesley, 1984.
9. R. Lidl and H. Niederreiter, *Introduction to Finite Fields and Their Applications*, Cambridge Univ. Press, 1986.
10. V. Pless, *Introduction to the Theory of Error–Correcting Codes*, Wiley, 1982.
11. D. Shanks, *Solved and Unsolved Problems in Number Theory*, 3rd ed., Chelsea Publ. Co., 1985.

# III
# Cryptography

## 1 Some simple cryptosystems

**Basic notions.** Cryptography is the study of methods of sending messages in disguised form so that only the intended recipients can remove the disguise and read the message. The message we want to send is called the *plaintext* and the disguised message is called the *ciphertext.* The plaintext and ciphertext are written in some *alphabet* (usually, but not always, they are written in the same alphabet) consisting of a certain number $N$ of *letters.* The term "letter" (or "character") can refer not only to the familiar A—Z, but also to numerals, blanks, punctuation marks, or any other symbols that we allow ourselves to use when writing the messages. (If we don't include a blank, for example, then all of the words are run together, and the messages are harder to read.) The process of converting a plaintext to a ciphertext is called *enciphering* or *encryption,* and the reverse process is called *deciphering* or *decryption.*

The plaintext and ciphertext are broken up into *message units.* A message unit might be a single letter, a pair of letters (*digraph*), a triple of letters (*trigraph*), or a block of 50 letters. An *enciphering transformation* is a function that takes any plaintext message unit and gives us a ciphertext message unit. In other words, it is a map $f$ from the set $\mathcal{P}$ of all possible plaintext message units to the set $\mathcal{C}$ of all possible ciphertext message units. We shall always assume that $f$ is a 1-to-1 correspondence. That is, given a ciphertext message unit, there is one and only one plaintext message unit for which it is the encryption. The *deciphering transformation* is the map $f^{-1}$ which goes back and recovers the plaintext from the ciphertext. We can represent the situation schematically by the diagram

$$\mathcal{P} \xrightarrow{f} \mathcal{C} \xrightarrow{f^{-1}} \mathcal{P}.$$

Any such set-up is called a *cryptosystem.*

The first step in inventing a cryptosystem is to "label" all possible plaintext message units and all possible ciphertext message units by means of mathematical objects from which functions can be easily constructed. These objects are often simply the integers in some range. For example, if our plaintext and ciphertext message units are single letters from the 26-letter alphabet A—Z, then we can label the letters using the integers $0, 1, 2, \ldots, 25$, which we call their "numerical equivalents." Thus, in place of A we write 0, in place of S we write 18, in place of X we write 23, and so on. As another example, if our message units are digraphs in the 27-letter alphabet consisting of A—Z and a blank, we might first let the blank have numerical equivalent 26 (one beyond Z), and then label the digraph whose two letters correspond to $x, y \in \{0, 1, 2, \ldots, 26\}$ by the integer

$$27x + y \in \{0, 1, \ldots, 728\}.$$

Thus, we view the individual letters as digits to the base 27 and we view the digraph as a 2-digit integer to that base. For example, the digraph "NO" corresponds to the integer $27 \cdot 13 + 14 = 365$. Analogously, if we were using trigraphs as our message units, we could label them by integers $729x+27y+z \in \{0, 1, \ldots, 19682\}$. In general, we can label blocks of $k$ letters in an $N$-letter alphabet by integers between 0 and $N^k - 1$ by regarding each such block as a $k$-digit integer to the base $N$.

In some situations, one might want to label message units using other mathematical objects besides integers — for example, vectors or points on some curve. But for the duration of this section we shall use integers.

**Examples.** Let us start with the case when we take a message unit (of plaintext or of ciphertext) to be a single letter in an $N$-letter alphabet labeled by the integers $0, 1, 2, \ldots, N-1$. Then, by definition, an enciphering transformation is a rearrangement of these $N$ integers.

To facilitate rapid enciphering and deciphering, it is convenient to have a relatively simple rule for performing such a rearrangement. One way is to think of the set of integers $\{0, 1, 2, \ldots, N-1\}$ as $\mathbf{Z}/N\mathbf{Z}$, and make use of the operations of addition and multiplication modulo $N$.

**Example 1.** Suppose we are using the 26-letter alphabet A—Z with numerical equivalents 0—25. Let the letter $P \in \{0, 1, \ldots, 25\}$ stand for a plaintext message unit. Define a function $f$ from the set $\{0, 1, \ldots, 25\}$ to itself by the rule

$$f(P) = \begin{cases} P+3, & \text{if } x < 23, \\ P-23, & \text{if } x \geq 23. \end{cases}$$

In other words, $f$ simply adds 3 modulo 26: $f(P) \equiv P + 3 \; mod \; 26$. The definition using modular arithmetic is easier to write down and work with.

Thus, with this system, to encipher the word "YES" we first convert to numbers: 24 4 18, then add 3 modulo 26: 1 7 21, then translate back to letters: "BHV." To decipher a message, one subtracts 3 modulo 26. For example, the ciphertext "ZKB" yields the plaintext "WHY." This cryptosystem was apparently used in ancient Rome by Julius Caesar, who supposedly invented it himself.

Example 1 can be generalized as follows. Suppose we are using an $N$-letter alphabet with numerical equivalents $0, 1, \ldots, N-1$. Let $b$ be a fixed integer. By a *shift* transformation we mean the enciphering function $f$ defined by the rule $C = f(P) \equiv P + b \; mod \; N$. Julius Caesar's cryptosystem was the case $N = 26$, $b = 3$. To decipher a ciphertext message unit $C \in \{0, 1, \ldots, N-1\}$, we simply compute $P = f^{-1}(C) \equiv C - b \; mod \; N$.

Now suppose that you are not privy to the enciphering and deciphering information, but you would nevertheless like to be able to read the coded messages. This is called *breaking* the code, and the science of breaking codes is called *cryptanalysis*.

In order to break a cryptosystem, one needs two types of information. The first is the general nature (the *structure*) of the system. For example, suppose we know that the cryptosystem uses a shift transformation on single letters of the 26-letter alphabet A—Z with numerical equivalents 0—25, respectively. The second type of information is knowledge of a specific choice of certain parameters connected with the given type of cryptosystem. In our example, the second type of information one needs to know is the choice of the shift parameter $b$. Once one has that information, one can encipher and decipher by the formulas $C \equiv P + b \; mod \; N$ and $P \equiv C - b \; mod \; N$.

We shall always assume that the general structural information is already known. In practice, users of cryptography often have equipment for enciphering and deciphering which is constructed to implement only one type of cryptosystem. Over a period of time the information about what type of system they're using might leak out. To increase their security, therefore, they frequently change the choice of parameters used with the system. For example, suppose that two users of the shift cryptosystem are able to meet once a year. At that time they agree on a list of 52 choices of the parameter $b$, one for each week of the coming year.

The parameter $b$ (more complicated cryptosystems usually have several parameters) is called a *key*, or, more precisely, the *enciphering key*.

**Example 2.** So suppose that we intercept the message "FQOCUDEM", which we know was enciphered using a shift transformation on single letters of the 26-letter alphabet, as in the example above. It remains for us to find the $b$. One way to do this is by *frequency analysis*. This works as follows. Suppose that we have already intercepted a long string of ciphertext, say several hundred letters. We know that "E" is the most frequently occurring letter in the English language. So it is reasonable to assume that the most frequently occurring letter in the ciphertext is the encryption of E. Suppose that we find that "U" is the most frequently occurring character in the ciphertext. That means that the shift takes "E"=4 to "U"=20, i.e., $20 \equiv 4 + b \bmod 26$, so that $b = 16$. To decipher the message, then, it remains for us to subtract 16 (working modulo 26) from the numerical equivalents of "FQOCUDEM":

$$\text{“FQOCUDEM”} = 5\ 16\ 14\ 2\ 20\ 3\ 4\ 12 \quad \mapsto \quad 15\ 0\ 24\ 12\ 4\ 13\ 14\ 22 = \text{“PAYMENOW”}.$$

In the case of a shift encryption of single letters of a 26-letter alphabet, it is not even necessary to have a long string of ciphertext to find the most frequently occurring letter. After all, there are only 26 possibilities for $b$, and one can simply run through all of them. Most likely, only one will give a message that makes any sense, and that $b$ is the enciphering key.

Thus, this type of cryptosystem is too simple to be much good. It is too easy to break. An improvement is to use a more general type of transformation of $\mathbf{Z}/N\mathbf{Z}$, called an **affine** map: $C \equiv aP + b \bmod N$, where $a$ and $b$ are fixed integers (together they form the enciphering key). For example, working again in the 26-letter alphabet, if we want to encipher our message "PAYMENOW" using the affine transformation with enciphering key $a = 7$, $b = 12$, we obtain: $15\ 0\ 24\ 12\ 4\ 13\ 14\ 22 \mapsto 13\ 12\ 24\ 18\ 14\ 25\ 6\ 10 =$ "NMYSOZGK".

To decipher a message that was enciphered by means of the affine map $C \equiv aP + b \bmod N$, one simply solves for $P$ in terms of $C$, obtaining $P \equiv a'C + b' \bmod N$, where $a'$ is the inverse of $a$ modulo $N$ and $b'$ is equal to $-a^{-1}b$. Note that this works only if $g.c.d.(a, N) = 1$; otherwise, we cannot solve for $P$ in terms of $C$. If $g.c.d.(a, N) > 1$, then it is easy to see that more than one plaintext letter will give the same ciphertext letter, so we cannot uniquely recover the plaintext from the ciphertext. By definition, that is not an enciphering transformation: we always require that the map be 1-to-1, i.e., that the plaintext be uniquely determined from the ciphertext. To summarize, an affine cryptosystem in an $N$-letter alphabet with parameters $a \in (\mathbf{Z}/N\mathbf{Z})^*$ and $b \in \mathbf{Z}/N\mathbf{Z}$ consists of the rules:

$$C \equiv aP + b \bmod N, \qquad P \equiv a'C + b' \bmod N,$$

where

$$a' = a^{-1} \text{ in } (\mathbf{Z}/N\mathbf{Z})^*, \quad b' = -a^{-1}b.$$

As a special case of the affine cryptosystems we can set $a = 1$, thereby obtaining the shift transformations. Another special case is when $b = 0$: $P \equiv aC \bmod N$, $C \equiv a^{-1}P \bmod N$. The case $b = 0$ is called a *linear* transformation, meaning that the map takes a sum to a sum, i.e., if $C_1$ is the encryption of $P_1$ and $C_2$ is the encryption of $P_2$, then $C_1 + C_2$ is the encryption of $P_1 + P_2$ (where, of course, we are adding modulo $N$).

Now suppose that we know that an intercepted message was enciphered using an affine map of single letters in an $N$-letter alphabet. We would like to determine the enciphering key $a$, $b$ so that we can read the message. We need two bits of information to do this.

**Example 3.** Still working in our 26-letter alphabet, suppose that we know the most frequently occurring letter of ciphertext is "K", and the second most frequently occurring letter is "D". It is reasonable to assume that these are the encryptions of "E" and "T", respectively, which are the two most frequently occurring letters in the English language. Thus, replacing the letters by their numerical equivalents and substituting for $P$ and $C$ in the deciphering formula, we obtain:

$$10a' + b' \equiv 4 \ mod \ 26,$$
$$3a' + b' \equiv 19 \ mod \ 26.$$

We have two congruences with two unknowns, $a'$ and $b'$. The quickest way to solve is to subtract the two congruences to eliminate $b'$. We obtain $7a' \equiv 11 \ mod \ 26$, and $a' \equiv 7^{-1}11 \equiv 9 \ mod \ 26$. Finally, we obtain $b'$ by substituting this value for $a'$ in one of the congruences: $b' \equiv 4 - 10a' \equiv 18 \ mod \ 26$. So messages can be deciphered by means of the formula $P \equiv 9C + 18 \ mod \ 26$.

Recall from linear algebra that $n$ equations suffice to find $n$ unknowns only if the equations are independent (i.e., if the determinant is nonzero). For example, in the case of 2 equations in 2 unknowns this means that the straight line graphs of the equations intersect in a single point (are not parallel). In our situation, when we try to cryptanalyze an affine system from the knowledge of the two most frequently occurring letters of ciphertext, we might find that we cannot solve the two congruences uniquely for $a'$ and $b'$.

**Example 4.** Suppose that we have a string of ciphertext which we know was enciphered using an affine transformation of single letters in a 28-letter alphabet consisting of A—Z, a blank, and ?, where A—Z have numerical equivalents 0—25, blank=26, ?=27. A frequency analysis reveals that the two most common letters of ciphertext are "B" and "?", in that order. Since the most common letters in an English language text written in this 28-letter alphabet are " " (blank) and "E", in that order, we suppose that "B" is the encryption of " " and "?" is the encryption of "E". This leads to the two congruences: $a' + b' \equiv 26 \ mod \ 28$, $27a' + b' \equiv 4 \ mod \ 28$. Subtracting the two congruences, we obtain: $2a' \equiv 22 \ mod \ 28$, which is equivalent to the congruence $a' \equiv 11 \ mod \ 14$. This means that $a' \equiv 11$ or $25 \ mod \ 28$, and then $b' \equiv 15$ or $1 \ mod \ 28$, respectively. The fact of the matter is that both of the possible affine deciphering transformations $11C + 15$ and $25C + 1$ give " " and "E" as the plaintext letters corresponding to "B" and "?", respectively. At this point we could try both possibilities, and see which gives an intelligible message. Or we could continue our frequency analysis. Suppose we find that "I" is the third most frequently occurring letter of ciphertext. Using the fact that "T" is the third most common letter in the English language (of our 28 letters), we obtain a third congruence: $8a' + b' \equiv 19 \ mod \ 28$. This extra bit of information is enough to determine which of the affine maps is the right one. We find that it is $11C + 15$.

**Digraph transformations.** We now suppose that our plaintext and ciphertext message units are *two*-letter blocks, called *digraphs.* This means that the plaintext is split up into two-letter segments. If the entire plaintext has an odd number of letters, then in order to obtain a whole number of digraphs we add on an extra letter at the end; we choose a letter which is not likely to cause confusion, such as a blank if our alphabet contains a blank, or else "X" or "Q" if we are using just the 26-letter alphabet.

Each digraph is then assigned a numerical equivalent. The simplest way to do this is to take $xN+y$, where $x$ is the numerical equivalent of the first letter in the digraph, $y$ is the numerical equivalent of the second letter in the digraph, and $N$ is the number of letters in the alphabet. Equivalently, we think of a digraph as a 2-digit base-N integer. This gives a 1-to-1 correspondence between the set of all digraphs in the $N$-letter alphabet and the set of all nonnegative integers less than $N^2$. We described this "labeling" of digraphs before in the special case when $N = 27$.

Next, we decide upon an enciphering transformation, i.e., a rearrangement of the integers $\{0, 1, 2, \ldots, N^2-1\}$. Among the simplest enciphering transformations are the *affine* ones, where we view this set of integers as $\mathbf{Z}/N^2\mathbf{Z}$, and define the encryption of $P$ to be the nonnegative integer less than $N^2$ satisfying the congruence $C \equiv aP + b \ mod\ N^2$. Here, as before, $a$ must have no common factor with $N$ (which means it has no common factor with $N^2$), in order that we have an inverse transformation telling us how to decipher: $P \equiv a'C + b' \ mod\ N^2$, where $a' \equiv a^{-1} \ mod\ N^2$, $b' \equiv -a^{-1}b \ mod\ N^2$. We translate $C$ into a two-letter block of ciphertext by writing it in the form $C = x'N + y'$, and then looking up the letters with numerical equivalents $x'$ and $y'$.

**Example 5.** Suppose we are working in the 26-letter alphabet and using the digraph enciphering transformation $C \equiv 159P+580 \ mod\ 676$. Then the digraph "NO" has numerical equivalent $13 \cdot 26 + 14 = 352$ and is taken to the ciphertext digraph $159 \cdot 352 + 580 \equiv 440 \ mod\ 676$, which is "QY". The digraph "ON" has numerical equivalent 377, and is taken to 359="NV". Notice that the digraphs change as a unit, and there is no relation between the encryption of one digraph and that of another one that has a letter in common with it or even consists of the same letters in the reverse order.

To break a digraphic encryption system which uses an affine transformation $C \equiv aP+b \ mod\ N^2$, we need to know the ciphertext corresponding to two different plaintext message units. Since the message units are digraphs, a frequency analysis means counting which two-letter blocks occur most often in a long string of ciphertext (of course, counting only those occurrences where the first letter begins a message unit, ignoring the occurrences of the two letters which straddle two message units), and comparing with the known frequency of digraphs in English language texts (written in the same alphabet). For example, if we use the 26-letter alphabet, statistical analyses seem to show that "TH" and "HE" are the two most frequently occurring digraphs, in that order. Knowing two plaintext–ciphertext pairs of digraphs is often (but not always) enough to determine $a$ and $b$.

**Example 6.** You know that your adversary is using a cryptosystem with a 27-letter alphabet, in which the letters A—Z have numerical equivalents 0—25, and blank=26. Each digraph then corresponds to an integer between 0 and $728 = 27^2 - 1$ according to the rule that, if the two letters in the digraph have numerical equivalents $x$ and $y$, then the digraph has numerical equivalent $27x + y$, as explained earlier. Suppose that a study of a large sample of ciphertext reveals that the most frequently occurring digraphs are (in order) "ZA", "IA", and "IW". Suppose that the most common digraphs in the English language (for text written in our 27-letter alphabet) are "E " (i.e., "E blank"), "S ", " T". You know that the cryptosystem uses an affine enciphering transformation modulo 729. Find the deciphering key, and read the message "NDXBHO". Also find the enciphering key.

**Solution.** We know that plaintexts are enciphered by means of the rule $C \equiv aP + b \; mod \; 729$, and that ciphertexts can be deciphered by means of the rule $P \equiv a'C + b' \; mod \; 729$; here $a$, $b$ form the enciphering key, and $a'$, $b'$ form the deciphering key. We first want to find $a'$ and $b'$. We know how three digraphs are deciphered, and, after we replace the digraphs by their numerical equivalents, this gives us the three congruences:

$$\begin{aligned} 675a' + b' &\equiv 134 \; mod \; 729, \\ 216a' + b' &\equiv 512 \; mod \; 729, \\ 238a' + b' &\equiv 721 \; mod \; 729. \end{aligned}$$

If we try to eliminate $b'$ by subtracting the first two congruences, we arrive at $459a' \equiv 351 \; mod \; 729$, which does not have a unique solution $a' \; mod \; 729$ (there are 27 solutions). We do better if we subtract the third congruence from the first, obtaining $437a' \equiv 142 \; mod \; 729$. To solve this, we must find the inverse of 437 modulo 729. By way of review of the Euclidean algorithm, let's go through that in detail:

$$\begin{aligned} 729 &= 437 + 292 \\ 437 &= 292 + 145 \\ 292 &= 2 \cdot 145 + 2 \\ 145 &= 72 \cdot 2 + 1 \end{aligned}$$

and then

$$\begin{aligned} 1 &= 145 - 72 \cdot 2 \\ &= 145 - 72(292 - 2 \cdot 145) \\ &= 145 \cdot 145 - 72 \cdot 292 \\ &= 145(437 - 292) - 72 \cdot 292 \\ &= 145 \cdot 437 - 217 \cdot 292 \\ &= 145 \cdot 437 - 217(729 - 437) \\ &\equiv 362 \cdot 437 \; mod \; 729. \end{aligned}$$

Thus, $a' \equiv 362 \cdot 142 \equiv 374 \ mod \ 729$, and then $b' \equiv 134 - 675 \cdot 374 \equiv 647 \ mod \ 729$. Now applying the deciphering transformation to the digraphs "ND", "XB" and "HO" of our message — they correspond to the integers 354, 622 and 203, respectively — we obtain the integers 365, 724 and 24. Writing $365 = 13 \cdot 27 + 14$, $724 = 26 \cdot 27 + 22$, $24 = 0 \cdot 27 + 24$, we put together the plaintext digraphs into the message "NO WAY". Finally, to find the enciphering key we compute $a \equiv a'^{-1} \equiv 374^{-1} \equiv 614 \ mod \ 729$ (again using the Euclidean algorithm) and $b \equiv -a'^{-1}b' \equiv -614 \cdot 647 \equiv 47 \ mod \ 729$.

**Remark.** Although affine cryptosystems with digraphs (i.e., modulo $N^2$) are better than the ones using single letters (i.e., modulo $N$), they also have drawbacks. Notice that the second letter of each ciphertext digraph depends only on the second letter of the plaintext digraph. This is because that second letter depends on the mod-$N$ value of $C \equiv aP + b \ mod \ N^2$, which depends only on $P$ modulo $N$, i.e., only on the second letter of the plaintext digraph. Thus, one could obtain a lot of information (namely, $a$ and $b$ modulo $N$) from a frequency analysis of the even-numbered letters of the ciphertext message. A similar remark applies to mod-$N^k$ affine transformations of $k$-letter blocks.

## *Exercises*

1. In certain computer bulletin-board systems it is customary, if you want to post a message that may offend some people (e.g., a dirty joke), to encipher the letters (but not the blanks or punctuation) by a translation $C \equiv P + b \ mod \ 26$. It is then easy to decipher the text if one wants to, but no one is forced to see a message that jars on the nerves. Decipher the punchline of the following story (use frequency analysis to find $b$): At an international convention of surgeons, representatives of different countries were comparing notes on recent advances in reattaching severed parts of the body. The French, Americans and Russians were being especially boastful. The French surgeon said, "We sewed a leg on an injured runner, and a year later he placed in a national 1000-meter race." "Using the most advanced surgical procedures," the Russian surgeon chimed in, "we were able to put back an athlete's entire arm, and a year later with the same arm he established a new world record for the shot put." But they all fell silent when the American, not to be outdone, announced that "Jr frjrq n fzvyr ba n ubefr'f nff, naq n lrne yngre vg jnf ryrpgrq Cerfvqrag!" (Note: We are using a 26-letter alphabet, but we have inserted blanks and punctuation for ease of reading.)
2. Using frequency analysis, cryptanalyze and decipher the following message, which you know was enciphered using a shift transformation of single-letter plaintext message units in the 26-letter alphabet:

   PXPXKXENVDRUXVTNLXHYMXGMAXYKXJN

XGVRFXMAHWGXXWLEHGZXKVBIAXKMXQM.

3. In the 27-letter alphabet (with blank=26), use the affine enciphering transformation with key $a = 13$, $b = 9$ to encipher the message "HELP ME."
4. In a long string of ciphertext which was encrypted by means of an affine map on single-letter message units in the 26-letter alphabet, you observe that the most frequently occurring letters are "Y" and "V", in that order. Assuming that those ciphertext message units are the encryption of "E" and "T", respectively, read the message "QAOOYQQEVHEQV".
5. You are trying to cryptanalyze an affine enciphering transformation of single-letter message units in a 37-letter alphabet. This alphabet includes the numerals 0–9, which are labeled by themselves (i.e., by the integers 0–9). The letters A—Z have numerical equivalents 10—35, respectively, and blank=36. You intercept the ciphertext "OH7F86BB46R3627O266BB9" (here the O's are the letter "oh", not the numeral zero). You know that the plaintext ends with the signature "007" (zero zero seven). What is the message?
6. You intercept the ciphertext "OFJDFOHFXOL", which was enciphered using an affine transformation of single-letter plaintext units in the 27-letter alphabet (with blank=26). You know that the first word is "I " ("I" followed by blank). Determine the enciphering key, and read the message.
7. (a) How many different shift transformations are there with an $N$-letter alphabet?

   (b) Find a formula for the number of different affine enciphering transformations there are with an $N$-letter alphabet.

   (c) How many affine transformations are there when $N = 26, 27, 29, 30$?
8. A plaintext message unit $P$ is said to be *fixed* for a given enciphering transformation $f$ if $f(P) = P$. Suppose we are using an affine enciphering transformation on single-letter message units in an $N$-letter alphabet. In this problem we also assume that the affine map is *not* a shift, i.e., that $a \neq 1$.

   (a) Prove that if $N$ is a prime number, then there is always exactly one fixed letter.

   (b) Prove (for any $N$) that if our affine transformation is linear, i.e., if $b = 0$, then it has at least one fixed letter; and that, if $N$ is even, then a linear enciphering transformation has at least two fixed letters.

   (c) Give an example for some $N$ of an affine enciphering transformation which has no fixed letter.
9. Now suppose that our message units are digraphs in an $N$-letter alphabet. Find a formula for the number of different affine enciphering transformations there are. How many are there when $N = 26, 27, 29, 30$?

10. You intercept the ciphertext message "PWULPZTQAWHF", which you know was encrypted using an affine map on digraphs in the 26-letter alphabet, where, as in the text, a digraph whose two letters have numerical equivalents $x$ and $y$ corresponds to the integer $26x + y$. An extensive statistical analysis of earlier ciphertexts which had been coded by the same enciphering map shows that the most frequently occurring digraphs in all of that ciphertext are "IX" and "TQ", in that order. It is known that the most common digraphs in the English language are "TH" and "HE", in that order.
(a) Find the deciphering key, and read the message.
(b) You decide to have the intended recipient of the message incapacitated, but you don't want the sender to know that anything is amiss. So you want to impersonate the sender's accomplice and reply "GOODWORK". Find the enciphering key, and determine the appropriate ciphertext.
11. You intercept the coded message "DXM SCE DCCUVGX ", which was enciphered using an affine map on digraphs in a 30-letter alphabet, in which A—Z have numerical equivalents 0—25, blank=26, ?=27, !=28, '=29. A frequency analysis shows that the most common digraphs in earlier ciphertexts are "M ", "U ", and "IH", in that order. Suppose that in the English language the most frequently occurring digraphs (in this particular 30-letter alphabet) are "E ", "S ", and " T", in that order.
(a) Find the deciphering key, and read the message.
(b) Find the enciphering key, and encrypt the message "YES I'M JOKING!"
12. The same techniques apply, of course, if one is using some other alphabet besides the Latin alphabet. For example, this exercise uses the Russian alphabet (it is not necessary, or even helpful, to know Russian or the Cyrillic alphabet in order to do this exercise). Use the following numerical equivalents for the Cyrillic alphabet:

| | | | | | | | | | | |
|---|---|---|---|---|---|---|---|---|---|---|
| А | Б | В | Г | Д | Е | Ё | Ж | З | И | Й |
| 0 | 1 | 2 | 3 | 4 | 5 | 6 | 7 | 8 | 9 | 10 |
| К | Л | М | Н | О | П | Р | С | Т | У | Ф |
| 11 | 12 | 13 | 14 | 15 | 16 | 17 | 18 | 19 | 20 | 21 |
| Х | Ц | Ч | Ш | Щ | Ъ | Ы | Ь | Э | Ю | Я |
| 22 | 23 | 24 | 25 | 26 | 27 | 28 | 29 | 30 | 31 | 32 |

Suppose that you intercept the coded message "ЦНТИ", which was enciphered using an affine map on digraphs in the above 33-letter alphabet. A frequency analysis of earlier ciphertext shows that the most frequently occurring ciphertext digraphs are "ЦЯ" and "ЫТ", in that order. Suppose it is known that the two most frequently occurring digraphs in the Russian language are "HO" and "ET". Find the deciphering key, and write out the plaintext message.

13. Recall from Exercise 8 that a *fixed* plaintext message unit is one that the given enciphering transformation keeps the same. Find all fixed digraphs for the enciphering transformation in Exercise 11.

14. By the *product* (or *composition*) of two cryptosystems, we mean the cryptosystem that results from enciphering a plaintext using the first cryptosystem and then treating the resulting ciphertext as plaintext for the second cryptosystem, i.e., encrypting a second time using the second system. More precisely, we must assume that the set $\mathcal{C}_1$ of ciphertext message units for the first cryptosystem is contained in the set of plaintext message units for the second system. Let $f_1$ and $f_2$ be the enciphering functions; then the product cryptosystem is given by the enciphering function $f = f_2 \circ f_1$. If we let $I$ (for "intermediate text") denote a ciphertext message unit for the first system, and let $\mathcal{I} = \mathcal{C}_1$ denote the set of intermediate texts, then the product cryptosystem can be represented schematically by the composite diagram:

$$\mathcal{P} \xrightarrow{f_1} \mathcal{I} \xrightarrow{f_2} \mathcal{C}.$$

Prove that:
(a) The product of two shift enciphering transformations is also a shift enciphering transformation.
(b) The product of two linear enciphering transformations is a linear enciphering transformation.
(c) The product of two affine enciphering transformations is an affine enciphering transformation.

15. Here is a slightly more complicated cryptosystem, in which the plaintexts and ciphertexts are written in different alphabets. We choose an $N$-letter alphabet for plaintexts and an $M$-letter alphabet for ciphertexts, where $M > N$. As usual, we regard digraphs in the $N$-letter alphabet as two-digit integers written to the base $N$, i.e., as integers between 0 and $N^2 - 1$; and we similarly regard digraphs in the $M$-letter alphabet as integers between 0 and $M^2 - 1$. Now choose any integer $L$ between $N^2$ and $M^2$: $N^2 \leq L \leq M^2$. Also choose integers $a$ and $b$ with $g.c.d.(a, L) = 1$. We encipher a plaintext digraph $P$ using the rule $C \equiv aP + b \ mod\ L$ (in which $C$ is taken to be the least nonnegative residue modulo $L$ which satisfies the congruence). (Here the set $\mathcal{P}$ of all possible digraphs $P$ consists of all integers from 0 to $N^2 - 1$; but the set $\mathcal{C}$ of all possible ciphertext digraphs $C$ in the larger alphabet is only part of the integers from 0 to $M^2 - 1$, in fact, it is the subset of the integers less than $L$ that arises from applying the enciphering rule to all possible plaintext digraphs.) Suppose that the plaintext alphabet is the 27-letter alphabet (as in Exercise 3), and the ciphertext alphabet is the 30-letter alphabet in Exercise 11. Suppose

that $L = 853$. Further suppose that you know that the two most frequently occurring plaintext digraphs "E " and "S " have encryptions "FQ" and "LE", respectively. Find the deciphering key, and read the message "YAVAOCH'D!"

16. Continuing along the lines of Exercise 15, here is an example of how one can, without too much extra work, create a cryptosystem that is much harder to break. Let $f_1$ be one cryptosystem of the type described in Exercise 15, i.e., given by the rule $f_1(P) \equiv a_1P + b_1 \ mod \ L_1$, and let $f_2$ be a second cryptosystem of the same type. Here the $N$ and $M$ are the same, but the $a$'s, $b$'s and $L$'s are different. We suppose that $L_2 > L_1$. We then construct the *product* of the two cryptosystems (see Exercise 14), i.e., we encrypt a plaintext message unit $P$ by successively applying the two rules:

$$I \equiv a_1P + b_1 \ mod \ L_1,$$
$$C \equiv a_2I + b_2 \ mod \ L_2.$$

(In the first rule $I$ is the nonnegative integer less than $L_1$ that satisfies the congruence, and in the second rule $C$ is less than $L_2$.) Because the moduli $L_1$ and $L_2$ are different, Exercise 14(c) does not apply, and this product cryptosystem is not generally an affine system. Here we suppose that the two alphabets of $M$ and $N$ letters are always the same, but we are free to frequently change our choice of the parameters $a_1$, $b_1$, $L_1$, $a_2$, $b_2$, $L_2$, subject, of course, to the conditions: $N^2 \leq L_1 < L_2 \leq M^2$, $g.c.d.(a_1, L_1) = 1$, $g.c.d.(a_2, L_2) = 1$. Thus, the enciphering key consists of the six-tuple of parameter values $\{a_1, b_1, L_1, a_2, b_2, L_2\}$. Let the plaintext and ciphertext alphabets be as in Exercise 15, consisting of 27 and 30 letters, respectively. If the enciphering key is $\{247, 109, 757, 675, 402, 881\}$, explain how to *de*cipher, and decipher the message "D!RAJ'KCTN".

# 2 Enciphering Matrices

Suppose we have an $N$-letter alphabet and want to send digraphs (two-letter blocks) as our message units. In §1 we saw how we can let each digraph correspond to an integer considered modulo $N^2$, i.e., to an element of $\mathbf{Z}/N^2\mathbf{Z}$. An alternate possibility is to let each digraph correspond to a *vector*, i.e., to a pair of integers $\binom{x}{y}$ with $x$ and $y$ each considered modulo $N$. For example, if we're using the 26-letter alphabet A—Z with numerical equivalents 0—25, respectively, then the digraph NO corresponds to the vector $\binom{13}{14}$. See the diagram at the top of the next page.

We picture each digraph $P$ as a point on an $N \times N$ square array. That is, we have an "$xy$-plane," except that each axis, rather than being a copy

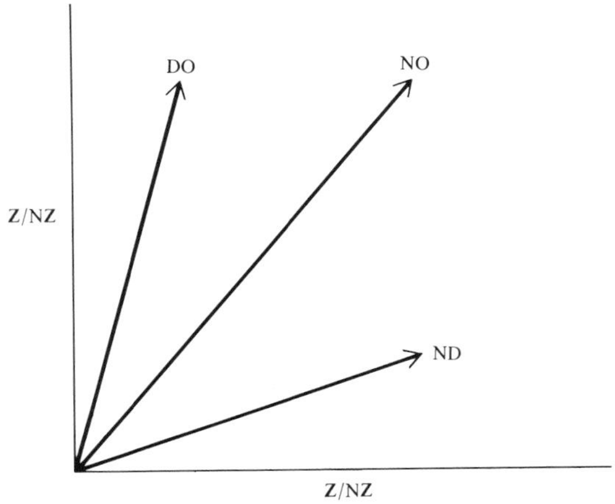


of the real number line, is now a copy of $\mathbf{Z}/N\mathbf{Z}$. Just as the real $xy$-plane is often denoted $\mathbf{R}^2$, this $N \times N$ array is denoted $(\mathbf{Z}/N\mathbf{Z})^2$.

Once we visualize digraphs as vectors (points in the plane), we then interpret an "enciphering transformation" as a rearrangement of the $N \times N$ array of points. More precisely, an enciphering map is a 1-to-1 function from $(\mathbf{Z}/N\mathbf{Z})^2$ to itself.

**Remark.** For several centuries one of the most popular methods of encryption was the so-called "Vigenère cipher." This can be described as follows. For some fixed $k$, regard blocks of $k$ letters as vectors in $(\mathbf{Z}/N\mathbf{Z})^k$. Choose some fixed vector $b \in (\mathbf{Z}/N\mathbf{Z})^k$ (usually $b$ was the vector corresponding to some easily remembered "key–word"), and encipher by means of the vector translation $C = P + b$ (where the ciphertext message unit $C$ and the plaintext message unit $P$ are $k$-tuples of integers modulo $N$). This cryptosystem, unfortunately, is almost as easy to break as a single-letter translation (see Example 1 of the last section). Namely, if one knows (or can guess) $N$ and $k$, then one simply breaks up the ciphertext in blocks of $k$ letters and performs a frequency analysis on the first letter in each block to determine the first component of $b$, then the same for the second letter in each block, and so on.

**Review of linear algebra.** We now review how one works with vectors in the real $xy$-plane and with $2 \times 2$–matrices with real entries. Recall that, given a $2 \times 2$ array of numbers

$$\begin{pmatrix} a & b \\ c & d \end{pmatrix} \qquad \text{and a vector in the plane} \qquad \begin{pmatrix} x \\ y \end{pmatrix}$$

(we shall write vectors as columns), one can *apply the matrix to the vector* to obtain a new vector, as follows:

$$\begin{pmatrix} a & b \\ c & d \end{pmatrix} \begin{pmatrix} x \\ y \end{pmatrix} =_{def} \begin{pmatrix} ax + by \\ cx + dy \end{pmatrix}.$$

For a fixed matrix, this function from one vector to another vector is called a *linear transformation*, meaning that it preserves sums and constant multiples of vectors. Using this notation, we can view any set of simultaneous equations of the form $ax + by = e, \quad cx + dy = f$ as equivalent to a single matrix equation $AX = B$, where $A$ denotes the matrix

$$\begin{pmatrix} a & b \\ c & d \end{pmatrix},$$

$X$ denotes the vector of unknowns $\binom{x}{y}$, and $B$ denotes the vector of constants $\binom{e}{f}$. Stated in words, the simultaneous equations can thus be interpreted as asking to find a vector which when "multiplied" by a certain known matrix gives a certain known vector. Thus, it is analogous to the simple equation $ax = b$, which is solved by multiplying both sides by $a^{-1}$ (assuming $a \neq 0$). Similarly, one way to solve the matrix equation $AX = B$ is to find the inverse of the matrix $A$, and then apply $A^{-1}$ to both sides to obtain the unique vector solution $X = A^{-1}B$.

By the inverse of the matrix $A$ we mean the matrix which multiplies by it to give the identity matrix

$$\begin{pmatrix} 1 & 0 \\ 0 & 1 \end{pmatrix}$$

(the matrix which, when applied to any vector, keeps that vector the same). But not all matrices have inverses. It is not hard to prove that a matrix

$$A = \begin{pmatrix} a & b \\ c & d \end{pmatrix}$$

has an inverse if and only if its *determinant* $D =_{def} ad - bc$ is nonzero, and that its inverse in that case is

$$\frac{1}{D} \begin{pmatrix} d & -b \\ -c & a \end{pmatrix} = \begin{pmatrix} D^{-1}d & -D^{-1}b \\ -D^{-1}c & D^{-1}a \end{pmatrix}.$$

There are three possibilities for the solutions of the system of simultaneous equations $AX = B$. First, if the determinant $D$ is nonzero, then there is precisely one solution $X = \binom{x}{y}$. If $D = 0$, then either there are no solutions or there are infinitely many. The three possibilities have a simple geometric interpretation. The two equations give straight lines in the $xy$-plane. If $D \neq 0$, then they intersect in exactly one point $(x, y)$. Otherwise, they are parallel lines, which means either that they don't meet at all (the simultaneous equations have no common solution) or else that they are really the same line (the equations have infinitely many common solutions).

Next, let us suppose that we have a bunch of vectors $X_1 = \binom{x_1}{y_1}, \ldots,$ $X_k = \binom{x_k}{y_k}$, arranged as the columns of a $2 \times k$–matrix. Then we define the matrix product

$$AX = \begin{pmatrix} a & b \\ c & d \end{pmatrix} \begin{pmatrix} x_1 & \ldots & x_k \\ y_1 & \ldots & y_k \end{pmatrix} =_{def} \begin{pmatrix} ax_1 + by_1 & \ldots & ax_k + by_k \\ cx_1 + dy_1 & \ldots & cx_k + dy_k \end{pmatrix},$$

i.e., we simply apply the matrix $A$ to each column vector in order, obtaining new column vectors. For example, the product of two $2 \times 2$–matrices is:

$$\begin{pmatrix} a & b \\ c & d \end{pmatrix} \begin{pmatrix} a' & b' \\ c' & d' \end{pmatrix} = \begin{pmatrix} aa' + bc' & ab' + bd' \\ ca' + dc' & cb' + dd' \end{pmatrix}.$$

Similar facts hold for $3 \times 3$–matrices, which can be applied to 3-dimensional column-vectors, and so on. However, the formulas for the determinant and inverse matrix are more complicated. This concludes our brief review of linear algebra over the real numbers.

**Linear algebra modulo $N$.** In §1, when we were dealing with single characters and enciphering maps of $\mathbf{Z}/N\mathbf{Z}$, we found that two easy types of maps to work with were:
(a) "linear" maps $C = aP$, where $a$ is invertible in $\mathbf{Z}/N\mathbf{Z}$;
(b) "affine" maps $C = aP + b$, where $a$ is invertible in $\mathbf{Z}/N\mathbf{Z}$.
We have a similar situation when our message units are digraph-vectors. We first consider linear maps. The difference when we work with $(\mathbf{Z}/N\mathbf{Z})^2$ rather than $\mathbf{Z}/N\mathbf{Z}$ is that now instead of an integer $a$ we need a $2 \times 2$–matrix, which we shall denote $A$. We start by giving a systematic explanation of the type of matrices we need.

Let $R$ be any commutative ring, i.e., a set with multiplication and addition satisfying the same rules as in a field, except that we do *not* require that any nonzero element have a multiplicative inverse. For example, $\mathbf{Z}/N\mathbf{Z}$ is always a ring, but it is not a field unless $N$ is prime. We let $R^*$ denote the subset of invertible elements of $R$. For example, $(\mathbf{Z}/N\mathbf{Z})^* = \{0 < j < N \mid g.c.d.(j, N) = 1\}$.

If $R$ is a commutative ring, we let $M_2(R)$ denote the set of all $2 \times 2$–matrices with entries in $R$, with addition and multiplication defined in the usual way for matrices. We call $M_2(R)$ a "matrix ring over $R$"; $M_2(R)$ itself is a ring, but it is *not* a commutative ring, i.e., in matrix multiplication the order of the factors makes a difference.

Earlier in this section, the matrices considered were the case when $R = \mathbf{R}$ is the ring (actually, field) of real numbers. Recall that a matrix

$$\begin{pmatrix} a & b \\ c & d \end{pmatrix}$$

with real numbers $a$, $b$, $c$, $d$ has a multiplicative inverse if and only if the determinant $D = ad - bc$ is nonzero, and in that case the inverse matrix is

$$\begin{pmatrix} D^{-1}d & -D^{-1}b \\ -D^{-1}c & D^{-1}a \end{pmatrix}.$$

We have a similar situation when we work over an arbitrary ring $R$.

Namely, suppose that

$$A = \begin{pmatrix} a & b \\ c & d \end{pmatrix} \in M_2(R)$$

and $D = det(A) =_{def} ad - bc$ is in $R^*$. Let $D^{-1}$ denote the multiplicative inverse of $D$ in $R$. Then

$$\begin{pmatrix} D^{-1}d & -D^{-1}b \\ -D^{-1}c & D^{-1}a \end{pmatrix} \begin{pmatrix} a & b \\ c & d \end{pmatrix} = \begin{pmatrix} D^{-1}(da - bc) & 0 \\ 0 & D^{-1}(-cb + ad) \end{pmatrix}$$
$$= \begin{pmatrix} 1 & 0 \\ 0 & 1 \end{pmatrix},$$

and we obtain the same result

$$\begin{pmatrix} 1 & 0 \\ 0 & 1 \end{pmatrix}$$

if we multiply in the opposite order. Thus, $A$ has an inverse matrix given by the same formula as in the real number case:

$$A^{-1} = \begin{pmatrix} D^{-1}d & -D^{-1}b \\ -D^{-1}c & D^{-1}a \end{pmatrix}.$$

**Example 1.** Find the inverse of

$$A = \begin{pmatrix} 2 & 3 \\ 7 & 8 \end{pmatrix} \in M_2(\mathbf{Z}/26\mathbf{Z}).$$

**Solution.** Here $D = 2 \cdot 8 - 3 \cdot 7 = -5 = 21$ in $\mathbf{Z}/26\mathbf{Z}$. Since $g.c.d.(21, 26) = 1$, the determinant $D$ has an inverse, namely $21^{-1} = 5$. Thus,

$$A^{-1} = \begin{pmatrix} 5 \cdot 8 & -5 \cdot 3 \\ -5 \cdot 7 & 5 \cdot 2 \end{pmatrix} = \begin{pmatrix} 40 & -15 \\ -35 & 10 \end{pmatrix} = \begin{pmatrix} 14 & 11 \\ 17 & 10 \end{pmatrix}.$$

We check that $\begin{pmatrix} 14 & 11 \\ 17 & 10 \end{pmatrix} \begin{pmatrix} 2 & 3 \\ 7 & 8 \end{pmatrix} = \begin{pmatrix} 105 & 130 \\ 104 & 131 \end{pmatrix} = \begin{pmatrix} 1 & 0 \\ 0 & 1 \end{pmatrix}$. Here, since we are working in $\mathbf{Z}/26\mathbf{Z}$, we are using "=" to mean that the entries are congruent modulo 26.

Just as in the real number case, a $2 \times 2$–matrix

$$\begin{pmatrix} a & b \\ c & d \end{pmatrix}$$

with entries in a ring $R$ can be multiplied by a column-vector $\binom{x}{y}$ with $x,\ y \in R$ to get a new vector $\binom{x'}{y'}$:

$$\begin{pmatrix} x' \\ y' \end{pmatrix} = \begin{pmatrix} a & b \\ c & d \end{pmatrix} \begin{pmatrix} x \\ y \end{pmatrix} = \begin{pmatrix} ax+by \\ cx+dy \end{pmatrix}.$$

This gives a "linear map" from vectors to vectors, meaning that a linear combination $\binom{k_1x_1+k_2x_2}{k_1y_1+k_2y_2}$, where $k_1$ and $k_2$ are in the ring $R$, is taken to $\binom{k_1x_1'+k_2x_2'}{k_1y_1'+k_2y_2'}$. The only difference with the situation earlier in our review of linear algebra is that now everything is in our ring $R$ rather than in the real numbers.

We shall want to apply all of this when our ring is $R = \mathbf{Z}/N\mathbf{Z}$. The next proposition will be stated in that case, although the analogous proposition is true for any $R$.

**Proposition III.2.1.** *Let*

$$A = \begin{pmatrix} a & b \\ c & d \end{pmatrix} \in M_2(\mathbf{Z}/N\mathbf{Z}) \quad \text{and set} \quad D = ad - bc.$$

*The following are equivalent:*
(a) *g.c.d.$(D,N)=1$;*
(b) *$A$ has an inverse matrix;*
(c) *if $x$ and $y$ are not both $0$ in $\mathbf{Z}/N\mathbf{Z}$, then $A\binom{x}{y} \neq \binom{0}{0}$;*
(d) *$A$ gives a 1-to-1 correspondence of $(\mathbf{Z}/N\mathbf{Z})^2$ with itself.*

**Proof.** We already showed that (a)$\Longrightarrow$(b). It suffices now to prove that (b)$\Longrightarrow$(d)$\Longrightarrow$(c)$\Longrightarrow$(a).

Suppose that (b) holds. Then part (d) also holds, because $A^{-1}$ gives the inverse map from $\binom{x'}{y'}$ to $\binom{x}{y}$. Next, if we have (d), then $\binom{x}{y} \neq \binom{0}{0}$ implies that $A\binom{x}{y} \neq A\binom{0}{0} = \binom{0}{0}$, and so (c) holds. Finally, we prove (c)$\Longrightarrow$(a) by showing that (a) false $\Longrightarrow$ (c) false. So suppose that (a) is false, and set $m = g.c.d.(D,N) > 1$ and let $m' = N/m$. Three cases are possible.

*Case (i).* If all four entries of $A$ are divisible by $m$, set $\binom{x}{y} = \binom{m'}{m'}$, to get a contradiction to (c).

*Case (ii).* If $a$ and $b$ are not both divisible by $m$, set $\binom{x}{y} = \binom{-bm'}{am'}$. Then

$$A\begin{pmatrix} x \\ y \end{pmatrix} = \begin{pmatrix} a & b \\ c & d \end{pmatrix} \begin{pmatrix} -bm' \\ am' \end{pmatrix} = \begin{pmatrix} -abm' + bam' \\ -cbm' + dam' \end{pmatrix} = \begin{pmatrix} 0 \\ Dm' \end{pmatrix} = \begin{pmatrix} 0 \\ 0 \end{pmatrix},$$

because $m|D$ and so $N = mm'|Dm'$.

*Case (iii).* If $c$ and $d$ are not both divisible by $m$, set $\binom{x}{y} = \binom{dm'}{-cm'}$, and proceed as in case (ii). These three cases exhaust all possibilities. Thus, (a) false implies (c) false. This completes the proof of Proposition III.2.1.

**Example 2.** Solve the following systems of simultaneous congruences:

(a)

$$2x + 3y \equiv 1 \ mod \ 26,$$
$$7x + 8y \equiv 2 \ mod \ 26;$$

(b)

$$x + 3y \equiv 1 \ mod \ 26,$$
$$7x + 9y \equiv 2 \ mod \ 26;$$

(c)

$$x + 3y \equiv 1 \ mod \ 26,$$
$$7x + 9y \equiv 1 \ mod \ 26.$$

**Solution.** The matrix form of the system (a) is $AX \equiv B \ mod \ 26$, where $A$ is the matrix in Example 1, $X = \binom{x}{y}$, and $B = \binom{1}{2}$. We obtain the unique solution

$$X \equiv A^{-1}B \equiv \begin{pmatrix} 14 & 11 \\ 17 & 10 \end{pmatrix} \begin{pmatrix} 1 \\ 2 \end{pmatrix} \equiv \begin{pmatrix} 10 \\ 11 \end{pmatrix} \ mod \ 26.$$

The matrix of the systems (b)–(c) does not have an inverse modulo 26, since its determinant is 14, which has a common factor of 2 with 26. However, we can work modulo 13, i.e., we can find the solution to the same congruence mod 13 and see if it gives a solution which works modulo 26. Modulo 13 we obtain

$$\begin{pmatrix} x \\ y \end{pmatrix} \equiv \begin{pmatrix} 9 & 10 \\ 6 & 1 \end{pmatrix} \begin{pmatrix} e \\ f \end{pmatrix}$$

(where $\binom{e}{f} = \binom{1}{2}$ in part (b) and $\binom{1}{1}$ in part (c)). This gives $\binom{x}{y} \equiv \binom{3}{8}$ and $\binom{6}{7}$ *mod* 13, respectively. Testing the possibilities modulo 26, we find that in part (b) there are *no* solutions, and in part (c) there are *two* solutions: $x = 6$, $y = 7$ and $x = 19$, $y = 20$.

Another way to solve systems of equations (preferable sometimes, especially when the matrix is not invertible) is to eliminate one of the variables (e.g., in parts (b) and (c), one could subtract 7 times the first congruence from the second).

To return to cryptography, we see from Proposition III.2.1 that we can get enciphering transformations of our digraph-vectors by using matrices $A \in M_2(\mathbf{Z}/N\mathbf{Z})$ whose determinant has no common factor with $N$:

$$A = \begin{pmatrix} a & b \\ c & d \end{pmatrix}, \qquad D = ad - bc, \qquad g.c.d.(D, N) = 1.$$

Namely, each plaintext message unit $P = \binom{x}{y}$ is taken to a ciphertext $C = \binom{x'}{y'}$ by the rule

$$C = AP, \qquad \text{i.e.,} \qquad \begin{pmatrix} x' \\ y' \end{pmatrix} = \begin{pmatrix} a & b \\ c & d \end{pmatrix} \begin{pmatrix} x \\ y \end{pmatrix}.$$

To decipher a message, we simply apply the inverse matrix:

$$P = A^{-1}AP = A^{-1}C, \qquad \text{i.e.,} \qquad \begin{pmatrix} x \\ y \end{pmatrix} = \begin{pmatrix} D^{-1}d & -D^{-1}b \\ -D^{-1}c & D^{-1}a \end{pmatrix} \begin{pmatrix} x' \\ y' \end{pmatrix}.$$

**Example 3.** Working in the 26-letter alphabet, use the matrix $A$ in Example 1 to encipher the message unit "NO."

**Solution.** We have:

$$AP = \begin{pmatrix} 2 & 3 \\ 7 & 8 \end{pmatrix} \begin{pmatrix} 13 \\ 14 \end{pmatrix} = \begin{pmatrix} 68 \\ 203 \end{pmatrix} = \begin{pmatrix} 16 \\ 21 \end{pmatrix},$$

and so $C = AP$ is "QV."

**Remark.** To encipher a plaintext sequence of $k$ digraphs $P = P_1P_2P_3 \cdots P_k$, we can write the $k$ vectors as columns of a $2 \times k$–matrix, which we also denote $P$, and then multiply the $2 \times 2$–matrix $A$ by the $2 \times k$-matrix $P$ to get a $2 \times k$–matrix $C = AP$ of coded digraph-vectors.

**Example 4.** Continue as in Example 3 to encipher the plaintext "NOANSWER."

**Solution.** The numerical equivalent of "NOANSWER" is the sequence of vectors $\binom{13}{14}\binom{0}{13}\binom{18}{22}\binom{4}{17}$. We have

$$C = AP = \begin{pmatrix} 2 & 3 \\ 7 & 8 \end{pmatrix} \begin{pmatrix} 13 & 0 & 18 & 4 \\ 14 & 13 & 22 & 17 \end{pmatrix} = \begin{pmatrix} 68 & 39 & 102 & 59 \\ 203 & 104 & 302 & 164 \end{pmatrix}$$
$$= \begin{pmatrix} 16 & 13 & 24 & 7 \\ 21 & 0 & 16 & 8 \end{pmatrix},$$

i.e., the coded message is "QVNAYQHI."

**Example 5.** In the situation of Examples 3–4, decipher the ciphertext "FWMDIQ."

**Solution.** We have:

$$P = A^{-1}C = \begin{pmatrix} 14 & 11 \\ 17 & 10 \end{pmatrix} \begin{pmatrix} 5 & 12 & 8 \\ 22 & 3 & 16 \end{pmatrix}$$
$$= \begin{pmatrix} 0 & 19 & 2 \\ 19 & 0 & 10 \end{pmatrix} = \text{"ATTACK."}$$

As in §1, suppose that we have some limited information from which we want to analyze how to decipher a string of ciphertext. We know that the "enemy" is using digraph-vectors in an $N$-letter alphabet and a linear enciphering transformation $C = AP$. However, we do not have the enciphering "key" — the matrix $A$ — or the deciphering "key" — the matrix $A^{-1}$. But suppose we are able to determine two pairs of plaintext and ciphertext digraphs: $C_1 = AP_1$ and $C_2 = AP_2$. Perhaps we learned this information from an analysis of the frequency of occurrence of digraphs in a long string of ciphertext. Or perhaps we know from some outside source that a certain 4-letter plaintext segment corresponds to a certain 4-letter ciphertext. In that case we can proceed as follows to determine $A$ and $A^{-1}$. We put the two columns $P_1$ and $P_2$ together into a $2 \times 2$–matrix $P$, and similarly for the ciphertext columns. We obtain an equation of $2 \times 2$–matrices: $C = AP$, in which $C$ and $P$ are known to us, and $A$ is the unknown. We can solve for $A$ by multiplying both sides by $P^{-1}$:

$$A = APP^{-1} = CP^{-1}.$$

Similarly, from the equation $P = A^{-1}C$ we can solve for $A^{-1}$:

$$A^{-1} = PC^{-1}.$$

**Example 6.** Suppose that we know that our adversary is using a $2 \times 2$ enciphering matrix with a 29-letter alphabet, where A—Z have the usual numerical equivalents, blank=26, ?=27, !=28. We receive the message

"GFPYJP X?UYXSTLADPLW,"

and we suppose that we know that the last five letters of plaintext are our adversary's signature "KARLA." Since we don't know the sixth letter from the end of the plaintext, we can only use the last four letters to make two digraphs of plaintext. Thus, the ciphertext digraphs DP and LW correspond to the plaintext digraphs AR and LA, respectively. That is, the matrix $P$ made up from AR and LA is the result of applying the unknown deciphering matrix $A^{-1}$ to the matrix $C$ made up from DP and LW:

$$\begin{pmatrix} 0 & 11 \\ 17 & 0 \end{pmatrix} = A^{-1}\begin{pmatrix} 3 & 11 \\ 15 & 22 \end{pmatrix}.$$

Thus,

$$A^{-1} = \begin{pmatrix} 0 & 11 \\ 17 & 0 \end{pmatrix}\begin{pmatrix} 3 & 11 \\ 15 & 22 \end{pmatrix}^{-1} = \begin{pmatrix} 0 & 11 \\ 17 & 0 \end{pmatrix}\begin{pmatrix} 3 & 13 \\ 23 & 7 \end{pmatrix} = \begin{pmatrix} 21 & 19 \\ 22 & 18 \end{pmatrix},$$

and the full plaintext message is

$$\begin{aligned} &\begin{pmatrix} 21 & 19 \\ 22 & 18 \end{pmatrix}\begin{pmatrix} 6 & 15 & 9 & 26 & 27 & 24 & 18 & 11 & 3 & 11 \\ 5 & 24 & 15 & 23 & 20 & 23 & 19 & 0 & 15 & 22 \end{pmatrix} \\ &\qquad = \begin{pmatrix} 18 & 17 & 10 & 26 & 19 & 13 & 14 & 28 & 0 & 11 \\ 19 & 8 & 4 & 0 & 26 & 14 & 13 & 10 & 17 & 0 \end{pmatrix} \\ &\qquad = \text{"STRIKE AT NOON! KARLA."} \end{aligned}$$

**Remark.** In order for this to work, notice that the matrix $P$ formed by the two known plaintext digraphs must be invertible, i.e., its determinant $D$ must have no common factor with the number of letters $N$. What if we are not so fortunate? If we happen to know another ciphertext-plaintext pair, then we could try to use that pair of columns in place of either the first or second columns of $P$ and $C$, hoping to obtain then an invertible matrix. But suppose we have no further information, or that none of the known plaintext digraphs give us an invertible matrix $P$. Then we cannot find $A^{-1}$ exactly. However, we might be able to get enough information about $A^{-1}$ to cut down drastically the number of possibilities for the deciphering matrix. We now illustrate this with an example. (For more on this, see the exercises at the end of the section.)

**Example 7.** Suppose we know than our adversary is using an enciphering matrix $A$ in the 26-letter alphabet. We intercept the ciphertext "WKNCCHSSJH," and we know that the first word is "GIVE." We want to find the deciphering matrix $A^{-1}$ and read the message.

**Solution.** If we try to proceed as in Example 6, writing

$$P = \text{“}GIVE\text{”} = \begin{pmatrix} 6 & 21 \\ 8 & 4 \end{pmatrix},$$

$$C = \text{“}WKNC\text{”} = \begin{pmatrix} 22 & 13 \\ 10 & 2 \end{pmatrix}, \qquad \text{and} \qquad A^{-1} = PC^{-1};$$

we immediately run into a problem, since $det(C) = 18$ and $g.c.d.(18, 26) = 2$. We can proceed as follows. Let $\overline{A}$ denote the reduction modulo 13 of the matrix $A$, and similarly for $\overline{P}$ and $\overline{C}$. If we consider these matrices in $M_2(\mathbf{Z}/13\mathbf{Z})$, we can take $C^{-1}$ (more precisely, $\overline{C}^{-1}$), because $g.c.d.(det(C), 13) = 1$. Thus, from $\overline{P} = \overline{A}^{-1}\overline{C}$ we can compute

$$\overline{A}^{-1} = \overline{P}\,\overline{C}^{-1} = \begin{pmatrix} 6 & 8 \\ 8 & 4 \end{pmatrix} \begin{pmatrix} 9 & 0 \\ 10 & 2 \end{pmatrix}^{-1} = \begin{pmatrix} 2 & 4 \\ 3 & 2 \end{pmatrix}.$$

Since the entries of $A^{-1}$; which are integers mod 26, must reduce to

$$\begin{pmatrix} 2 & 4 \\ 3 & 2 \end{pmatrix}$$

modulo 13, it follows that there are two possibilities for each entry in the matrix $A^{-1}$. More precisely,

$$A^{-1} = \begin{pmatrix} 2 & 4 \\ 3 & 2 \end{pmatrix} + 13A_1,$$

where $A_1 \in M_2(\mathbf{Z}/2\mathbf{Z})$ is a $2 \times 2$–matrix of 0's and 1's. That leaves $2^4 = 16$ possibilities. However, in the first place, since $A^{-1}$ is invertible, its determinant must be prime to 26, and hence also prime to 2 (i.e., odd). This consideration rules out all but 6 possibilities for $A_1$. In the second place, when we substitute

$$\begin{pmatrix} 2 & 4 \\ 3 & 2 \end{pmatrix} + 13A_1$$

for $A^{-1}$ in the equation

$$A^{-1}\begin{pmatrix} 22 & 13 \\ 10 & 2 \end{pmatrix} = \begin{pmatrix} 6 & 21 \\ 8 & 4 \end{pmatrix}$$

(this means entry-by-entry congruence mod 26), we eliminate all but 2 possibilities, namely,

$$A_1 = \begin{pmatrix} 1 & 0 \\ 1 & 1 \end{pmatrix} \quad \text{or} \quad \begin{pmatrix} 1 & 1 \\ 1 & 1 \end{pmatrix},$$

i.e.,

$$A^{-1} = \begin{pmatrix} 15 & 4 \\ 16 & 15 \end{pmatrix} \quad \text{or} \quad \begin{pmatrix} 15 & 17 \\ 16 & 15 \end{pmatrix}.$$

Attempting to decipher with the first matrix yields "GIVEGHEMHP," which must be wrong. Deciphering with the second matrix

$$A^{-1} = \begin{pmatrix} 15 & 17 \\ 16 & 15 \end{pmatrix}$$

leads to "GIVETHEMUP." So that must be correct. Although a certain amount of trial and error is involved, it's better than running through all 157,248 possibilities for a deciphering matrix $A^{-1} \in M_2(\mathbf{Z}/26\mathbf{Z})^*$.

**Remark.** In Example 7 it would perhaps be more efficient to adjust the entries in $\overline{A}^{-1}$ by multiples of 13 so that they become divisible by 2, i.e., to define $A_1$ by writing:

$$A^{-1} = \begin{pmatrix} 2 & 4 \\ 16 & 2 \end{pmatrix} + 13A_1.$$

Then one can obtain information on $A_1$ by working modulo 2, since we now have $A_1C \equiv P \ mod \ 2$.

**Affine enciphering transformations.** A more general way to encipher a digraph-vector $P = \binom{x}{y}$ is to apply a $2 \times 2$–matrix $A = \left(\begin{smallmatrix} a & b \\ c & d \end{smallmatrix}\right) \in M_2(\mathbf{Z}/N\mathbf{Z})$ and then add a constant vector $B = \binom{e}{f}$:

$$C = AP + B,$$

i.e.,

$$\begin{pmatrix} x' \\ y' \end{pmatrix} = \begin{pmatrix} a & b \\ c & d \end{pmatrix}\begin{pmatrix} x \\ y \end{pmatrix} + \begin{pmatrix} e \\ f \end{pmatrix} = \begin{pmatrix} ax + by + e \\ cx + dy + f \end{pmatrix}.$$

This is called an "affine" map, and is analogous to the enciphering function $C = aP + b$ that we studied in §1 when we were using single-letter message units. Of course, as before, we are using "=" to mean the corresponding entries are congruent mod N.

The inverse transformation that expresses $P$ in terms of $C$ can be found by subtracting $B$ from both sides and then applying $A^{-1}$ to both sides:

$$P = A^{-1}C - A^{-1}B.$$

This is also an affine transformation $P = A'C + B'$, where $A' = A^{-1}$ and $B' = -A^{-1}B$. Notice that we must assume that $A$ is an invertible matrix in order to be able to decipher uniquely.

Suppose we know that our adversary is using an affine enciphering transformation of digraph-vectors with an $N$-letter alphabet. To determine $A$ and $B$ (or to determine $A' = A^{-1}$ and $B' = -A^{-1}B$), we need at least *three* digraph pairs. Suppose we know that the ciphertext digraphs $C_1$, $C_2$, $C_3$ correspond to the plaintext digraphs $P_1$, $P_2$, $P_3$:

$$P_1 = A'C_1 + B'$$
$$P_2 = A'C_2 + B'$$
$$P_3 = A'C_3 + B'.$$

To find $A'$ and $B'$ we can proceed as follows. Subtract the last equation from the first two, and then make a $2 \times 2$–matrix $P$ from the two columns $P_1 - P_3$ and $P_2 - P_3$ and a $2 \times 2$–matrix $C$ from the two columns $C_1 - C_3$ and $C_2 - C_3$. We obtain the matrix equation $P = A'C$, which can be solved for $A'$ (provided that $C$ is invertible) as we did in the case of linear enciphering transformations. Finally, once we find $A' = A^{-1}$, we can determine $B'$ from any of the above three equations, e.g., $B' = P_1 - A'C_1$.

## *Exercises*

1. Use frequency analysis to decrypt the following message, which was encoded in the 26-letter alphabet using a Vigenère cipher with a 3-letter key-word. Do this in the following way. To find the first letter of the key-word, work with the sequence consisting of every third letter starting with the first. Do not assume that the most frequently occurring letter is necessarily the ciphertext for "E". List the four most frequently occurring letters, and try out the possibility that each one in turn is the encryption of "E". If one of the other three frequently occurring letters would then have to be the encryption, say, of "Z" or "Q", then you know that you made a wrong choice for "E". By an elimination process, find the letter that must be "E" and then the key-word letter which produces that translation. In this way find the key-word and decipher the message:

AWYVPQCTBLWYLPASQJWUPGBUSHFACELDLLDLWLBWAFAHS
EBYJXXACELWCJTQMARKDDLWCSXBUDLKDPLXSEQCJTNWPR
WSRGBCLWPGJEZIFWIMJDLLDAGCQMAYLTGLPPJXTWSGFRM
VTLGUYUXJAIGWHCPXQLTBXDPVTAGSGFVRZTWTGMMVFLXR
LDKWPRLWCSXPHDPLPKSHQGULMBZWGQAPQCTBAURZTWSHQ
MBCVXAGJJVGCSSGLIFWNQSXBFDGSHIWSFGLRZTWEPLSVC
VIFWNQSXBOWCFHMETRZXLYPPJXTWSGFRMVTRZTWHWMFTB
OPQZXLYIMFPLVWYVIFWDPAVGFPJETQKPEWGCSSRGIFWB

2. Find the inverses of the following matrices mod $N$. Write the entries in the inverse matrix as nonnegative integers less than $N$.

(a) $\begin{pmatrix} 1 & 3 \\ 4 & 3 \end{pmatrix}$ *mod* 5 (b) $\begin{pmatrix} 1 & 3 \\ 4 & 3 \end{pmatrix}$ *mod* 29 (c) $\begin{pmatrix} 15 & 17 \\ 4 & 9 \end{pmatrix}$ *mod* 26

(d) $\begin{pmatrix} 40 & 0 \\ 0 & 21 \end{pmatrix}$ *mod* 841 (e) $\begin{pmatrix} 197 & 62 \\ 603 & 271 \end{pmatrix}$ *mod* 841

In Exercises 3—5, find all solutions $\binom{x}{y}$ modulo $N$, writing $x$ and $y$ as nonnegative integers less than $N$.

3.

(a) $\begin{aligned} x + 4y &\equiv 1 \ mod\ 9 \\ 5x + 7y &\equiv 1 \ mod\ 9 \end{aligned}$ (b) $\begin{aligned} x + 4y &\equiv 1 \ mod\ 9 \\ 5x + 8y &\equiv 1 \ mod\ 9 \end{aligned}$

(c) $\begin{aligned} x + 4y &\equiv 1 \ mod\ 9 \\ 5x + 8y &\equiv 2 \ mod\ 9 \end{aligned}$ (d) $\begin{aligned} x + 4y &\equiv 0 \ mod\ 9 \\ 5x + 8y &\equiv 0 \ mod\ 9 \end{aligned}$

4.

(a) $\begin{aligned} 17x + 11y &\equiv 7 \ mod\ 29 \\ 13x + 10y &\equiv 8 \ mod\ 29 \end{aligned}$ (b) $\begin{aligned} 17x + 11y &\equiv 0 \ mod\ 29 \\ 13x + 10y &\equiv 0 \ mod\ 29 \end{aligned}$

(c) $\begin{aligned} 9x + 20y &\equiv 0 \ mod\ 29 \\ 16x + 13y &\equiv 0 \ mod\ 29 \end{aligned}$ (d) $\begin{aligned} 9x + 20y &\equiv 10 \ mod\ 29 \\ 16x + 13y &\equiv 21 \ mod\ 29 \end{aligned}$

(e) $\begin{aligned} 9x + 20y &\equiv 1 \ mod\ 29 \\ 16x + 13y &\equiv 2 \ mod\ 29 \end{aligned}$

5.

(a) $\begin{aligned} 480x + 971y &\equiv 416 \ mod\ 1111 \\ 297x + 398y &\equiv 319 \ mod\ 1111 \end{aligned}$ (b) $\begin{aligned} 480x + 971y &\equiv 109 \ mod\ 1111 \\ 297x + 398y &\equiv 906 \ mod\ 1111 \end{aligned}$

(c) $\begin{aligned} 480x + 971y &\equiv 0 \ mod\ 1111 \\ 297x + 398y &\equiv 0 \ mod\ 1111 \end{aligned}$ (d) $\begin{aligned} 480x + 971y &\equiv 0 \ mod\ 1111 \\ 298x + 398y &\equiv 0 \ mod\ 1111 \end{aligned}$

(e) $\begin{aligned} 480x + 971y &\equiv 648 \ mod\ 1111 \\ 298x + 398y &\equiv 1004 \ mod\ 1111 \end{aligned}$

6. The *Fibonacci* numbers can be defined by the rule $f_1 = 1$, $f_2 = 1$, $f_3 = 2$, $f_{n+1} = f_n + f_{n-1}$ for $n > 1$, or, equivalently, by means of the matrix equation

$$\begin{pmatrix} f_{n+1} & f_n \\ f_n & f_{n-1} \end{pmatrix} = \begin{pmatrix} 1 & 1 \\ 1 & 0 \end{pmatrix}^n$$

(see Exercise 10 of §I.2). Using the matrix form of the definition, prove that $f_n$ is even if and only if $n$ is divisible by 3. More generally, prove that $f_n$ is divisible by $a$ if and only if $n$ is divisible by $b$ for the following $a$ and $b$: (a) $a = 2$, $b = 3$; (b) $a = 3$, $b = 4$; (c) $a = 5$, $b = 5$; (d) $a = 7$, $b = 8$; (e) $a = 8$, $b = 6$; (f) $a = 11$, $b = 10$.

7. You intercept the message "SONAFQCHMWPTVEVY," which you know resulted from a *linear* enciphering transformation of digraph-vectors, where the sender used the usual 26-letter alphabet A—Z with numerical equivalents 0—25, respectively. An earlier statistical analysis of a long string of intercepted ciphertext revealed that the most frequently occurring ciphertext digraphs were "KH" and "XW" in that order. You take a guess that those digraphs correspond to "TH" and "HE," respectively, since those are the most frequently occurring digraphs in most long plaintext messages on the subject you think is being discussed. Find the deciphering matrix, and read the message.
8. You intercept the message "ZRIXXYVBMNPO," which you know resulted from a linear enciphering transformation of digraph-vectors in a 27-letter alphabet, in which A—Z have numerical equivalents 0—25, and blank=26. You have found that the most frequently occurring ciphertext digraphs are "PK" and "RZ." You guess that they correspond to the most frequently occurring plaintext digraphs in the 27-letter alphabet, namely, "E " (E followed by blank) and "S ." Find the deciphering matrix, and read the message.
9. You intercept the message "!IWGVIEX!ZRADRYD," which was sent using a linear enciphering transformation of digraph-vectors in a 29-letter alphabet, in which A—Z have numerical equivalents 0—25, blank=26, ?=27, !=28. You know that the last five letters of plaintext are the sender's signature "MARIA."
   (a) Find the deciphering matrix, and read the message.
   (b) Find the enciphering matrix, and, impersonating Maria's friend Jo, send the following reply in code: "DAMN FOG! JO."
10. In this exercise we are again working with the Cyrillic alphabet (see Exercise 12 of the last section). We use a 34-letter alphabet, where in addition to the numerical equivalents listed before we have blank=33. Suppose that still the two most frequently occurring digraphs in Russian are taken to be "HO" and "ET." Meanwhile, we find that in a long string of ciphertext the most frequently occurring digraphs are "ЮТ" and "ЧМ." We know that the encryption uses a linear enciphering transformation of digraph-vectors in the 34-letter alphabet. Read the intercepted message "СХНСЪШОНЩЗ."
11. Prove that the *product* (see Exercise 14 of the last section) of a cryptosystem with enciphering matrix $A_1 \in M_2(\mathbf{Z}/N\mathbf{Z})^*$ and a cryptosys-

tem with enciphering matrix $A_2 \in M_2(\mathbf{Z}/N\mathbf{Z})^*$ is also a linear enciphering transformation.

12. In order to increase the difficulty of breaking your cryptosystem, you decide to encipher a digraph-vector in the 26-letter alphabet by first applying the matrix

$$\begin{pmatrix} 3 & 11 \\ 4 & 15 \end{pmatrix},$$

working modulo 26, and then applying the matrix

$$\begin{pmatrix} 10 & 15 \\ 5 & 9 \end{pmatrix},$$

working modulo 29. (Note that applying two matrices in succession while working with the same modulus is equivalent to applying a single matrix, as shown in Exercise 11; but if you change modulus the two-step encryption is much more complicated.) Thus, while your plaintexts are in the 26-letter alphabet, your ciphertexts will be in the 29-letter alphabet we used in Exercise 9.
(a) Encipher the message "SEND."
(b) Describe how to decipher a ciphertext by applying two matrices in succession, and decipher "ZMOY."

13. Prove that if a non-invertible $A \in M_2(\mathbf{Z}/N\mathbf{Z})$ is used to encipher digraph vectors by means of the formula $C = AP$, then every ciphertext one sends can be deciphered as coming from at least two different possible plaintexts.

14. You intercept the message "S GNLIKD?KOZQLLIOMKUL.VY" (here the blank after the S is part of the message). Suppose that a linear enciphering transformation $C = AP$ is being used with a 30-letter alphabet, in which A—Z have the usual numerical equivalents 0—25, blank=26, .=27, ,=28, ?=29. You also know that the last six letters of the plaintext are the signature KARLA followed by a period. Find the deciphering matrix $A^{-1}$ and the full plaintext message.

15. You intercept the message "KVW? TA!KJB?FVR ." (The blanks after ? and R are part of the message, but the final . is not.) You know that a linear enciphering transformation is being used with a 30-letter alphabet, in which A—Z have numerical equivalents 0—25, blank=26, ?=27, !=28, .=29. You further know that the first six letters of the plaintext are "C.I.A." Find the deciphering matrix $A^{-1}$ and the full plaintext message.

16. Suppose that $N = mn$, where $g.c.d.(m, n) = 1$. Any $A \in M_2(\mathbf{Z}/N\mathbf{Z})$ can be considered in $M_2(\mathbf{Z}/m\mathbf{Z})$ or $M_2(\mathbf{Z}/n\mathbf{Z})$ by simply reducing the entries modulo $m$ or $n$. Let $\overline{A}$ and $\tilde{A}$ denote the corresponding matrices in $M_2(\mathbf{Z}/m\mathbf{Z})$ and $M_2(\mathbf{Z}/n\mathbf{Z})$, respectively.
(a) Prove that the map that takes $A$ to the pair $(\overline{A}, \tilde{A})$ is a 1-to-1 correspondence between $M_2(\mathbf{Z}/N\mathbf{Z})$ and the set $M_2(\mathbf{Z}/m\mathbf{Z}) \times M_2(\mathbf{Z}/n\mathbf{Z})$ of all pairs of matrices, one modulo $m$ and one modulo $n$.

(b) Prove that the map in part (a) gives a 1-to-1 correspondence between the set $M_2(\mathbf{Z}/N\mathbf{Z})^*$ of invertible matrices mod $N$ and the set $M_2(\mathbf{Z}/m\mathbf{Z})^* \times M_2(\mathbf{Z}/n\mathbf{Z})^*$.

17. For $p$ a prime, find the number of elements in $M_2(\mathbf{Z}/p\mathbf{Z})^*$ in two ways, and check that your answers agree:
(a) Count the number of solutions in $\mathbf{F}_p$ of the equation $ad - bc = 0$, and subtract this from the number of elements in $M_2(\mathbf{Z}/p\mathbf{Z})$.
(b) Any $A \in M_2(\mathbf{Z}/p\mathbf{Z})^*$ must take $\binom{1}{0}$ and $\binom{0}{1}$ to two linearly independent vectors, i.e., the first can be any nonzero vector, and then the second can be any vector not a multiple of the first. Count the number of possibilities.

18. Prove that a matrix in $M_2(\mathbf{Z}/p^\alpha\mathbf{Z})$ is invertible if and only if its reduction mod $p$ in $M_2(\mathbf{Z}/p\mathbf{Z})$ is invertible. Then find the number of elements in $M_2(\mathbf{Z}/p^\alpha\mathbf{Z})^*$.

19. Using Exercises 16–18, find a formula for the number of elements in $M_2(\mathbf{Z}/N\mathbf{Z})^*$. Call this number $\varphi_2(N)$. Recall the formula for the number $\varphi(N)$ of elements in $(\mathbf{Z}/N\mathbf{Z})^*$: $\varphi(N) = N\prod_{p|n}(1-\frac{1}{p})$. Write your formula for $\varphi_2(N)$ in a similar form. How many possible $2 \times 2$ enciphering matrices $A$ are there when $N = 26$, 29, 30?

20. Let $\varphi_k(N)$ denote the number of invertible $k \times k$–matrices with entries in $\mathbf{Z}/N\mathbf{Z}$. *Guess* a formula for $\varphi_k(N)$. This formula is not hard to prove by the method in Exercise 16(b).

**Remark.** The approach in Exercises 16–20 is typical of many proofs and computations modulo $N$. Using a multiplicativity property, one first reduces to the case of a prime power. Then, using a "lifting argument" (see Exercise 20 of § II.2 for another example of this), one reduces to the case of a prime, i.e., we can then work in a *field* $\mathbf{F}_p$. Once we are working with a field, we can more easily use our geometric intuition, as in Exercise 17(b) above. All of linear algebra that we first learn over the real numbers goes through word–for–word over any field. For example, a congruence of the form $ax + by \equiv c \bmod p$ can be depicted by a "line" in the "plane" over the field $\mathbf{F}_p$; a second such congruence will either meet the first line in a single point, be parallel to the first line, or else coincide with the first line. In the case of congruences with a composite modulus $N$, on the other hand, there are other possibilities, which occur when the determinant of the coefficient matrix has a nontrivial common factor with $N$.

21. How many possible affine enciphering transformations are there for digraphs in an $N$-letter alphabet? How many are there when $N = 26$, 29, 30?

22. Suppose that you want to find a deciphering matrix $A^{-1} \in M_2(\mathbf{Z}/N\mathbf{Z})^*$ from the equation $P = A^{-1}C$, where $P$ and $C$ are made up from two known pairs of plaintext–ciphertext digraphs. Suppose that *g.c.d.* $(det(C), N) = p$, where $p$ is a prime dividing $N$ only to the first power. Let $n = N/p$.
(a) Find the number of possibilities for $A^{-1}$ you will be left with after

solving the congruence $P \equiv A^{-1}C\ mod\ n$ and after taking into account that $p \not| det(A^{-1})$.
(b) Suppose that $p$ does *not* divide all of the entries in $C$. Describe how to use the congruence $P \equiv A^{-1}C\ mod\ p$ to further reduce the number of possibilities for $A^{-1}$. How many possibilities are you now left with? Example 8 and Exercise 15 illustrate this in the case $p = 2$.

23. You want to find a $2 \times 2$ enciphering matrix $A$ modulo 30. You have two plaintext/ciphertext digraph pairs (in a 30-letter alphabet), which enables you to write $AP \equiv C\ mod\ 30$, where

$$P = \begin{pmatrix} 3 & 3 \\ 2 & 5 \end{pmatrix}, \qquad C = \begin{pmatrix} 17 & 8 \\ 8 & 29 \end{pmatrix}.$$

(a) Working modulo 10, write $A$ in the form $A \equiv A_0 + 10A_1\ mod\ 30$, where $A_1$ is an unknown matrix modulo 3 (whose entries are 0, 1 or 2) and $A_0$ is a matrix you know from your mod 10 computations. Choose $A_0$ so that all of its entries are between 0 and 29 and are divisible by 3.
(b) Working modulo 3, find the second column of the matrix $A_1$.
(c) How many possibilities are there for the original matrix $A$? List them all.

24. Let

$$A = \begin{pmatrix} a & b \\ c & d \end{pmatrix} \in M_2(\mathbf{Z}/N\mathbf{Z})^*$$

be the matrix of a linear enciphering transformation of digraphs in an $N$-letter alphabet. By a *fixed digraph* of $A$ we mean a digraph vector $P$ whose corresponding ciphertext vector $C$ is the same as $P$, i.e., $AP = P$. In this problem we suppose that $A$ is not the identity matrix. (After all, there's no point in considering the enciphering transformation that doesn't even make a half-hearted attempt to disguise anything.)
(a) Show that the digraph "AA"$= \binom{0}{0}$ is always fixed, and find a condition on

$$\begin{pmatrix} a & b \\ c & d \end{pmatrix}$$

which is equivalent to "AA" being the *only* fixed digraph.
(b) If $N$ is a prime number and if "AA" is not the only fixed digraph, prove that there are exactly $N$ fixed digraphs.

25. You intercept the message

"WUXHURWZNQR XVUEXU!JHALGQGJ?"

which you know was encoded using an **affine** transformation of vectors $\binom{x}{y}$ in an 841-letter alphabet. Here the numerical equivalent of a digraph is the number $x = 29x_1 + x_2$, where $x_1$ is the number of the first letter and $x_2$ is the number of the second letter in the digraph (the 29 letters are numbered as in Exercise 9). Thus, each block of four letters

gives a column $\binom{x}{y}$: the first two letters give the integer $x$ and the next two letters give $y$. You also know that the last 12 letters of the above ciphertext correspond to the signature "HEADQUARTERS."
(a) Find the deciphering transformation and read the message.
(b) Find the enciphering transformation and make a coded message that inpersonates headquarters and says "CANCEL LAST ORDER!" followed by two blanks and the signature "HEADQUARTERS."

26. How many possible affine enciphering transformations are there in the situation of Exercise 25 (with an 841-letter digraph alphabet)?
27. How many possible affine enciphering transformations are there for **tri**-graphs (3-component vectors) in a 26-letter alphabet?
28. You intercept the message

"FBRTLWUGAJQINZTHHXTEPHBNXSW,"

which you know was encoded using a linear enciphering transformation of trigraphs in the 26-letter alphabet A—Z with numerical equivalents 0—25. You also know that the last three trigraphs are the sender's signature "JAMESBOND." Find the deciphering matrix and read the message.

# References for Chapter III

1. L. S. Hill, "Concerning certain linear transformation apparatus of cryptography," *American Math. Monthly*, **38** (1931), 135–154.
2. D. Kahn, *The Codebreakers, the Story of Secret Writing*, Macmillan, 1967.
3. K. H. Rosen, *Elementary Number Theory and Its Applications*, 3rd ed., Addison–Wesley, 1993.

# IV
# Public Key

## 1 The idea of public key cryptography

Recall that a cryptosystem consists of a 1-to-1 enciphering transformation $f$ from a set $\mathcal{P}$ of all possible plaintext message units to a set $\mathcal{C}$ of all possible ciphertext message units. Actually, the term "cryptosystem" is more often used to refer to a whole family of such transformations, each corresponding to a choice of *parameters* (the sets $\mathcal{P}$ and $\mathcal{C}$, as well as the map $f$, may depend upon the values of the parameters). For example, for a fixed $N$-letter alphabet (with numerical equivalents also fixed once and for all), we might consider the affine cryptosystem (or "family of cryptosystems") which for each $a \in (\mathbf{Z}/N\mathbf{Z})^*$ and $b \in \mathbf{Z}/N\mathbf{Z}$ is the map from $\mathcal{P} = \mathbf{Z}/N\mathbf{Z}$ to $\mathcal{C} = \mathbf{Z}/N\mathbf{Z}$ defined by $C \equiv aP + b \ mod\ N$. In this example, the sets $\mathcal{P}$ and $\mathcal{C}$ are fixed (because $N$ is fixed), but the enciphering transformation $f$ depends upon the choice of parameters $a$, $b$. The enciphering transformation can then be described by (i) an algorithm, which is the same for the whole family, and (ii) the values of the parameters. The values of the parameters are called the *enciphering key* $K_E$. In our example, $K_E$ is the pair $(a, b)$. In practice, we shall suppose that the algorithm is publicly known, i.e., the general procedure used to encipher cannot be kept secret. However, the keys can easily be changed periodically and, if one wants, kept secret.

One also needs an algorithm and a key in order to decipher, i.e., compute $f^{-1}$. The key is called the *deciphering key* $K_D$. In our example of the affine cryptosystem family, deciphering is also accomplished by an affine map, namely $P \equiv a^{-1}C - a^{-1}b \ mod\ N$, and so the deciphering transformation uses the same algorithm as the enciphering transformation, except with a different key, namely, the pair $(a^{-1}, -a^{-1}b)$. (In some cryptosystems, the deciphering algorithm, as well as the key, is different from the enciphering algorithm.) We shall always suppose that the deciphering and enciphering algorithms are publicly known, and that it is the keys $K_E$ and $K_D$ which can be concealed.

Let us suppose that someone wishes to communicate secretly using the above affine cryptosystem $C \equiv aP + b$. We saw in § III.1 that it is not hard to break the system if one uses single-letter message units in an $N$-letter alphabet. It is a little more difficult to break the system if one uses digraphs, which can be regarded as symbols in an $N^2$-letter alphabet. It would be safer to use blocks of $k$ letters, which have numerical equivalents in $\mathbf{Z}/N^k\mathbf{Z}$. At least for $k > 3$ it is not easy to use frequency analysis, since the number of possible $k$-letter blocks is very large, and one will find many that are close contenders for the title of most frequently occurring $k$-graph. If we want to increase $k$, we must be concerned about the length of time it takes to do various arithmetic tasks (the most important one being finding $a^{-1}$ by the Euclidean algorithm) involved in setting up our keys and carrying out the necessary transformations every time we send a message or our friend at the other end deciphers a message from us. That is, it is useful to have big-$O$ estimates for the order of magnitude of time (as the parameters increase, i.e., as the cryptosystem becomes "larger") that it takes to: encipher (knowing $K_E$), decipher (knowing $K_D$), or break the code by enciphering without knowledge of $K_E$ or deciphering without knowledge of $K_D$.

In all of the examples in Chapter III — and in all of the cryptosystems used historically until about fifteen years ago — it is not really necessary to specify the deciphering key once the enciphering key (and the general algorithms) are known. Even if we are working with large numbers — such as $N^k$ with $k$ fairly large — it is possible to determine the deciphering key from the enciphering key using an order of magnitude of time which is roughly the same as that needed to implement the various algorithms. For example, in the case of an affine enciphering transformation of $\mathbf{Z}/N^k\mathbf{Z}$, once we know the enciphering key $K_E = (a, b)$ we can compute the deciphering key $K_D = (a^{-1} \mathit{mod}\ N^k, -a^{-1}b \mathit{mod}\ N^k)$ by the Euclidean algorithm in $O(\mathit{log}^3(N^k))$ bit operations.

Thus, with a traditional cryptosystem anyone who knew enough to decipher messages could, with little or no extra effort, determine the enciphering key. Indeed, it was considered naive or foolish to think that someone who had broken a cipher might nevertheless not know the enciphering key. We see this in the following passage from the autobiography of a well-known historical personality:

> Five or six weeks later, she [Madame d'Urfé] asked me if I had deciphered the manuscript which had the transmutation procedure. I told her that I had.

"Without the key, sir, excuse me if I believe the thing impossible."

"Do you wish me to name your key, madame?"

"If you please."

I then told her the key-word, which belonged to no language, and I saw her surprise. She told me that it was impossible, for she believed herself the only possessor of that word which she kept in her memory and which she had never written down.

I could have told her the truth — that the same calculation which had served me for deciphering the manuscript had enabled me to learn the word — but on a caprice it struck me to tell her that a genie had revealed it to me. This false disclosure fettered Madame d'Urfé to me. That day I became the master of her soul, and I abused my power. Every time I think of it, I am distressed and ashamed, and I do penance now in the obligation under which I place myself of telling the truth in writing my memoirs.

— Casanova, 1757, quoted in D. Kahn's *The Codebreakers*

The situation persisted for another 220 years after this encounter between Casanova and Madame d'Urfé: knowledge of how to encipher and knowledge of how to decipher were regarded as essentially equivalent in any cryptosystem. However, in 1976 W. Diffie and M. Hellman discovered an entirely different type of cryptosystem and invented "public key cryptography."

By definition, a public key cryptosystem has the property that someone who knows only how to encipher cannot use the enciphering key to find the deciphering key without a prohibitively lengthy computation. In other words the enciphering function $f: \mathcal{P} \longrightarrow \mathcal{C}$ is easy to compute once the enciphering key $K_E$ is known, but it is very hard in practice to compute the inverse function $f^{-1}: \mathcal{C} \longrightarrow \mathcal{P}$. That is, from the standpoint of realistic computability, the function $f$ is not invertible (without some additional information — the deciphering key $K_D$). Such a function $f$ is called a *trapdoor function.* That is, a trapdoor function $f$ is a function which is easy to compute but whose inverse $f^{-1}$ is hard to compute without having some additional auxiliary information beyond what is necessary to compute $f$. The inverse $f^{-1}$ is easy to compute, however, for someone who has this information $K_D$ (the "deciphering key").

There is a closely related concept of a *one-way* function. This is a function $f$ which is easy to compute but for which $f^{-1}$ is hard to compute and cannot be made easy to compute even by acquiring some additional information. While the notion of a trapdoor function apparently appeared for the first time in 1978 along with the invention of the RSA public-key cryptosystem, the notion of a one-way function is somewhat older. What seems to have been the first use of one-way functions for cryptography was described in Wilkes' book about time-sharing systems that was published in 1968. The author describes a new *one-way cipher* used by R. M. Needham in order to make it possible for a computer to verify passwords without storing information that could be used by an intruder to impersonate a legitimate user.

> In Needham's system, when the user first sets his password, or whenever he changes it, it is immediately subjected to the enciphering process, and it is the enciphered form that is stored in the computer. Whenever the password is typed in response to a demand from the supervisor for the user's identity to be established, it is again enciphered and the result compared with the stored version. It would be of no immediate use to a would-be malefactor to obtain a copy of the list of enciphered passwords, since he would have to decipher them before he could use them. For this purpose, he would need access to a computer and even if full details of the enciphering algorithm were available, the deciphering process would take a long time.

In 1974, G. Purdy published the first detailed description of such a one-way function. The original passwords and their enciphered forms are regarded as integers modulo a large prime $p$, and the "one-way" map $\mathbf{F}_p \longrightarrow \mathbf{F}_p$ is given by a polynomial $f(x)$ which is not hard to evaluate by computer but which takes an unreasonably long time to invert. Purdy used $p = 2^{64} - 59$, $f(x) = x^{2^{24}+17} + a_1 x^{2^{24}+3} + a_2 x^3 + a_3 x^2 + a_4 x + a_5$, where the coefficients $a_i$ were arbitrary 19-digit integers.

The above definitions of a public key cryptosystem and a one-way or trapdoor function are not precise from a rigorous mathematical standpoint. The notion of "realistic computability" plays a basic role. But that is an empirical concept that is affected by advances in computer technology (e.g., parallel processor techniques) and the discovery of new algorithms which speed up the performance of arithmetic tasks (sometimes by a large factor). Thus, it is possible that an enciphering transformation that can safely be regarded as a one-way or trapdoor function in 1994 might lose its one-way or trapdoor status in 2004 or in the year 2994.

It is conceivable that some transformation could be *proved* to be trapdoor. That is, there could be a theorem that provides a nontrivial lower bound for the number of bit operations that would be required ("on the average," i.e., for random values of the key parameters) in order to figure out and implement a deciphering algorithm without the deciphering key. Here one would have to allow the possibility of examining a large number of corresponding plaintext–ciphertext message units (as in our frequency analysis of the simple systems in Chapter III), because, by the definition of a public key system, any user can generate an arbitrary number of plaintext–ciphertext pairs. One would also have to allow the use of "probabilistic" methods which, while not guaranteed to break the code at once, would be likely to work if repeated many times. (Examples of probabilistic algorithms will be given in the next chapter.) Unfortunately, no such theorems have been proved for any of the functions that have been used as enciphering maps. Thus, while there are now many cryptosystems which empirically seem to earn the right to be called "public key," there is no cryptosystem in existence which is *provably* public key.

The reason for the name "public key" is that the information needed to send secret messages — the enciphering key $K_E$ — can be made public information without enabling anyone to read the secret messages. That is, suppose we have some population of users of the cryptosystem, each one of whom wants to be able to receive confidential communications from any of the other users without a third party (either another user or an outsider) being able to decipher the message. Some central office can collect the enciphering key $K_{E,A}$ from each user $A$ and publish all of the keys in a "telephone book" having the form

| | |
|---|---|
| AAA Banking Company | (9974398087453939, 2975290017591012) |
| Aardvark, Aaron | (8870004228331, 7234752637937) |
| ⋮ | ⋮ |

Someone wanting to send a message merely has to look up the enciphering key in this "telephone book" and then use the general enciphering algorithm with the key parameters corresponding to the intended recipient. Only the intended recipient has the matching deciphering key needed to read the message.

In earlier ages this type of system would not have seemed to have any particularly striking advantages. Traditionally, cryptography was used mainly for military and diplomatic purposes. Usually there was a small, well-defined group of users who could all share a system of keys, and new keys could be distributed periodically (using couriers) so as to keep the enemy guessing.

However, in recent years the actual and potential applications of cryptography have expanded to include many other areas where communication systems play a vital role — collecting and keeping records of confidential data, electronic financial transactions, and so on. Often one has a large network of users, any two of whom should be able to keep their communications secret from all other users as well as intruders from outside the network. Two parties may share a secret communication on one occasion, and then a little later one of them may want to send a confidential message to a third party. That is, the "alliances" — who is sharing a secret with whom — may be continually shifting. It might be impractical always to be exchanging keys with all possible confidential correspondents.

Notice that with a public key system it is possible for two parties to initiate secret communications without ever having had any prior contact, without having established any prior trust for one another, without exchanging any preliminary information. All of the information necessary to send an enciphered message is publicly available.

**Classical vesus public key.** By a *classical* cryptosystem (also called a *private key* cryptosystem or a *symmetrical* cryptosystem), we mean a cryptosystem in which, once the enciphering information is known, the deciphering transformation can be implemented in approximately the same order of magnitude of time as the enciphering transformation. All of the cryptosystems in Chapter III are classical. Occasionally, it takes a little longer for the deciphering — because one needs to apply the Euclidean algorithm to find an inverse modulo $N$ or one must invert a matrix (and this can take a fairly long time if we work with $k \times k$ -matrices for $k$ larger than 2) — nevertheless, the additional time required is not prohibitive. (Moreover, usually the additional time is required only once — to find $K_D$ — after which it takes no longer to decipher than to encipher.) For example, we might need only $O(log^2 B)$ to encipher a message unit, and $O(log^3 B)$ bit operations to decipher one by finding $K_D$ from $K_E$, where $B$ is a bound on the size of the key parameters. Notice the role of big-O estimates here.

If, on the other hand, the enciphering time were polynomial in $log\ B$ and the deciphering time (based on knowledge of $K_E$ but not $K_D$) were, say, polynomial in $B$ but not in $\log B$, then we would have a *public key* rather than a classical cryptosystem.

**Authentication.** Often, one of the most important parts of a message is the *signature*. A person's signature — hopefully, written with an idiosyncratic flourish of the pen which is hard to duplicate — lets the recipient know that the message really is from the person whose name is typed below. If the message is particularly important, it might be necessary to use additional methods to *authenticate* the communication. And in electronic communication, where one does not have a physical signature, one has to rely entirely on other methods. For example, when an officer of a corporation wants to withdraw money from the corporate account by telephone, he/she is often asked to give some personal information (e.g., mother's maiden name) which the corporate officer knows and the bank knows (from data submitted when the account was opened) but which an imposter would not be likely to know.

In public key cryptography there is an especially easy way to identify oneself in such a way that no one could be simply pretending to be you. Let A (Alice) and B (Bob) be two users of the system. Let $f_A$ be the enciphering transformation with which any user of the system sends a message to Alice, and let $f_B$ be the same for Bob. For simplicity, we shall assume that the set $\mathcal{P}$ of all possible plaintext message units and the set $\mathcal{C}$ of all possible ciphertext message units are equal, and are the same for all users. Let $P$ be Alice's "signature" (perhaps including an identification number, a statement of the time the message was sent, etc.). It would not be enough for Alice to send Bob the encoded message $f_B(P)$, since *everyone* knows how to do that, so there would be no way of knowing that the signature was not forged. Rather, at the beginning (or end) of the message Alice transmits $f_B f_A^{-1}(P)$. Then, when Bob deciphers the whole message, including this part, by applying $f_B^{-1}$, he finds that everything has become plaintext except for a small section of jibberish, which is $f_A^{-1}(P)$. Since Bob knows that the message is claimed to be from Alice, he applies $f_A$ (which he knows, since Alice's enciphering key is public), and obtains $P$. Since no one other than Alice could have applied the function $f_A^{-1}$ which is inverted by $f_A$, he knows that the message was from Alice.

**Hash functions.** A common way to sign a document is with the help of a *hash function.* Roughly speaking, a hash function is an easily computable map $f: x \mapsto h$ from a very long input $x$ to a much shorter output $h$ (for example, from strings of about $10^6$ bits to strings of 150 or 200 bits) that has the following property: *it is not computationally feasible to find two different inputs $x$ and $x'$ such that $f(x') = f(x)$.* If part of Alice's "signature" consists of the hash value $h = f(x)$, where $x$ is the entire text of her message, then Bob can verify not only that the message was really sent by Alice, but also that it wasn't tampered with during transmission. Namely, Bob applies the hash function $f$ to his deciphered plaintext from Alice, and checks that the result agrees with the value $h$ in Alice's signature. By assumption, no tamperer would have been able to change $x$ without changing the value $h = f(x)$.

**Key exchange.** In practice, the public key cryptosystems for sending messages tend to be slower to implement than the classical systems that are in current use. The number of plaintext message units per second that can be transmitted is less. However, even if a network of users feels attached to the traditional type of cryptosystem, they may want to use a public key cryptosystem in an auxiliary capacity to send one another their keys $K = (K_E, K_D)$ for the classical system. Thus, the ground rules for the classical cryptosystem can be agreed upon, and keys can be periodically exchanged, using the slower public key cryptography; while the large volume of messages would then be sent by the faster, older methods.

**Probabilistic Encryption.** Most of the number theory based cryptosystems for message transmission are *deterministic*, in the sense that a given plaintext will always be encrypted into the same ciphertext any time it is sent. However, deterministic encryption has two disadvantages: (1) if an eavesdropper knows that the plaintext message belongs to a small set (for example, the message is either "yes" or "no"), then she can simply encrypt all possibilities in order to determine which is the supposedly secret message; and (2) it seems to be very difficult to *prove* anything about the security of a system if the encryption is deterministic. For these reasons, *probabilistic encryption* was introduced. We will not discuss this further or give examples in this book. For more information, see the fundamental papers on the subject by Goldwasser and Micali (*Proc. 14th ACM Symp. Theory of Computing*, 1982, 365–377, and *J. Comput. System Sci.* **28** (1984), 270–299).

## *Exercises*

1. Suppose that $m$ users want to be able to communicate with one another using a classical cryptosystem. Each user insists on being able to communicate with each other user without the remaining $m-2$ users eavesdropping. How many keys $K = (K_E, K_D)$ must be developed? How many keys are needed if they are using a public key cryptosystem? How many keys are needed for each type of cryptosystem if $m = 1000$?
2. Suppose that a network of investors and stockbrokers is using public key cryptography. The investors fear that their stockbrokers will buy stock without authorization (in order to receive the commission) and then, when the investor's money is lost, claim that they had received instructions (producing as evidence an enciphered message to buy the stock, claiming that it came from the investor). The stockbrokers, on the other hand, fear that in cases when they buy according to the investor's instructions and the stock loses money, the investor will claim that he never sent the instruction, and that it was sent by an imposter or by the stockbroker himself. Explain how this problem can be solved by public key cryptography, so that when all of these sleazy people end up in court suing one another, there is proof of who is to blame for the reckless investing and consequent loss of money. (Suppose that, in the case of a lawsuit between investor A and stockbroker B, the judge is given all of the relevant enciphering/deciphering information — the keys $K_A = (K_{E,A}, K_{D,A})$ and $K_B = (K_{E,B}, K_{D,B})$ and the software necessary to encipher and decipher.)
3. Suppose that two countries A and B want to reach an agreement to ban underground nuclear tests. Neither country trusts the other, in both cases for good reason. Nevertheless, they must agree on a system of verification devices to be implanted at various locations on the territory of the two countries. Each verification device consists of a sophisticated seismograph, a small computer for interpreting the seismograph reading and generating a message, and a radio transmitter. Explain how public key cryptography can be used to enable all of the following (at first glance seemingly contradictory) conditions to be met:
   a. Country A insists on knowing the plaintext content of all messages emanating from its territory, in order to be sure that the devices are not used in coordination with espionage activities by Country B.
   b. Country B insists that Country A cannot fabricate a message from the devices which broadcast from its territory (i.e., a message saying that everything's OK, when in fact the seismograph has detected a treaty violation).
   c. Country A insists that, if Country B falsely claims to have received notification from the device of a treaty violation, then any interested third country will be able to determine that, in fact, no such message was sent.

d. Same as conditions a–c with the roles of the two countries reversed.
e. The verification devices in both countries must be identical, and must be constructed jointly by scientists from both countries.

4. The purpose of this problem is to construct a long–distance coin flip using any two-to-one trapdoor function. For example, suppose that two chess players at distant parts of the world are playing chess by mail or telephone and want a fair way to determine who plays white. Or suppose that when making preparations for an international ice–hockey match, representatives of the two teams decide to flip a coin to see which country hosts the match, without having to arrange a meeting (or trust a third party) to "flip the coin."
By a system of two-to-one trapdoor functions, we mean an algorithm which, given a key $K_E$ of a suitable type, constructs a function $f: \mathcal{P} \longrightarrow \mathcal{C}$ such that every element $c$ in the image of $f$ has exactly two preimages $p_1, p_2 \in \mathcal{P}$ such that $f(p_j) = c$; and an algorithm which, given a key $K_D$ which "reverses $K_E$," can find both preimages of any $c$ in the image of $f$. Here we assume that it is computationally infeasible to find $K_D$ knowing only $K_E$. Given an element $p_1 \in \mathcal{P}$, notice that one can find the other element $p_2$ having the same image if one knows both $K_E$ and $K_D$ (namely, find both inverses of $f(p_1)$); but we assume that, knowing only $K_E$, one cannot feasibly compute the companion element $p_2$ for *any* $p_1$ at all.
Suppose that Player A (Aniuta) and Player B (Björn) want to use this set-up to flip a coin. Aniuta generates a pair of keys $K_E$ and $K_D$ and sends $K_E$ (but *not* $K_D$) to Björn. Explain a procedure that has a 50%–50% chance of each player "winning" (give a suitable definition of "winning"), and that has adequate safeguards against cheating.

## References for § IV.1


1. M. Blum, "Coin-flipping by telephone — a protocol for solving impossible problems," *IEEE Proc., Spring Compcon.*, 133–137.
2. W. Diffie and M. E. Hellman, "New directions in cryptography," *IEEE Transactions on Information Theory IT-22* (1976), 644–654.
3. D. Chaum, "Achieving electronic privacy," *Scientific American*, **267** (1992), 96–101.
4. S. Goldwasser, "The search for provably secure cryptosystems," *Cryptology and Computational Number Theory, Proc. Symp. Appl. Math.* **42** (1990), 89–113.
5. M. E. Hellman, "The mathematics of public-key cryptography," *Scientific American*, **241** (1979), 146–157.
6. E. Kranakis, *Primality and Cryptography*, John Wiley & Sons, 1986.
7. R. Rivest, "Cryptography," *Handbook of Theoretical Computer Science*, Vol. A, Elsevier, 1990, 717–755.

8. G. Ruggiu, "Cryptology and complexity theories," *Advances in Cryptology, Proceedings of Eurocrypt 84,* Springer-Verlag, 1985, 3–9.

## 2 RSA

In looking for a trapdoor function $f$ to use for a public key cryptosystem, one wants to use an idea which is fairly simple conceptually and lends itself to easy implementation. On the other hand, one wants to have very strong empirical evidence — based on a long history of attempts to find algorithms for $f^{-1}$ — that decryption cannot feasibly be accomplished without knowledge of the secret deciphering key. For this reason it is natural to look at an ancient problem of number theory: the problem of finding the complete factorization of a large composite integer whose prime factors are not known in advance. The success of the so-called "RSA" cryptosystem (from the last names of the inventors Rivest, Shamir, and Adleman), which is one of the oldest (16 years old) and most popular public key cryptosystems, is based on the tremendous difficulty of factoring.

We now describe how RSA works. Each user first chooses two extremely large prime numbers $p$ and $q$ (say, of about 100 decimal digits each), and sets $n = pq$. Knowing the factorization of $n$, it is easy to compute $\varphi(n) = (p-1)(q-1) = n + 1 - p - q$. Next, the user randomly chooses an integer $e$ between 1 and $\varphi(n)$ which is prime to $\varphi(n)$.

**Remark.** Whenever we say "random" we mean that the number was chosen with the help of a random-number generator (or "pseudo-random" number generator), i.e., a computer program that generates a sequence of digits in a way that no one could duplicate or predict, and which is likely to have all of the statistical properties of a truly random sequence. A lot has been written concerning efficient and secure ways to generate random numbers, but we shall not concern ourselves with this question here. In the RSA cryptosystem we need a random number generator not only to choose $e$, but also to choose the large primes $p$ and $q$ (so that no one could guess our choices by looking at tables of special types of primes, for example, Mersenne primes or factors of $b^k \pm 1$ for small $b$ and relatively small $k$). What does a "randomly generated" prime number mean? Well, first generate a large random integer $m$. If $m$ is even, replace $m$ by $m + 1$. Then apply suitable *primality tests* to see if the odd number $m$ is prime (primality tests will be examined systematically in the next chapter). If $m$ is not prime, try $m+2$, then $m+4$, and so on, until you reach the first prime number $\geq m$, which is what you take as your "random" prime. According to the Prime Number Theorem (for the statement see Exercise 13 of § I.1), the frequency of primes among the numbers near $m$ is about $1/log(m)$, so you can expect to test $O(log\, m)$ numbers for primality before reaching the first prime $\geq m$.

Similarly, the "random" number $e$ prime to $\varphi(n)$ can be chosen by first generating a random (odd) integer with an appropriate number of bits, and then successively incrementing it until one finds an $e$ with $g.c.d.(e, \varphi(n)) = 1$. (Alternately, one can perform primality tests until one finds a prime $e$, say between $max(p, q)$ and $\varphi(n)$; such a prime must necessarily satisfy $g.c.d.(e, \varphi(n)) = 1$.)

Thus, each user A chooses two primes $p_A$ and $q_A$ and a random number $e_A$ which has no common factor with $(p_A - 1)(q_A - 1)$. Next, $A$ computes $n_A = p_A q_A$, $\varphi(n_A) = n_A + 1 - p_A - q_A$, and also the multiplicative inverse of $e_A$ modulo $\varphi(n_A)$: $d_A \underset{\text{def}}{=} e_A^{-1} \; mod \; \varphi(n_A)$. She makes public the enciphering key $K_{E,A} = (n_A, \, e_A)$ and conceals the deciphering key $K_{D,A} = (n_A, \, d_A)$. The enciphering transformation is the map from $\mathbf{Z}/n_A\mathbf{Z}$ to itself given by $f(P) \equiv P^{e_A} \; mod \; n_A$. The deciphering transformation is the map from $\mathbf{Z}/n_A\mathbf{Z}$ to itself given by $f^{-1}(C) \equiv C^{d_A} \; mod \; n_A$. It is not hard to see that these two maps are inverse to one another, because of our choice of $d_A$. Namely, performing $f$ followed by $f^{-1}$ or $f^{-1}$ followed by $f$ means raising to the $d_A e_A$-th power. But, because $d_A e_A$ leaves a remainder of 1 when divided by $\varphi(n_A)$, this is the same as raising to the 1-st power (see the corollary of Proposition I.3.5, which gives this in the case when $P$ has no common factor with $n_A$; if $g.c.d.(P, n_A) > 1$, see Exercise 6 below).

From the description in the last paragraph, it seems that we are working with sets $\mathcal{P} = \mathcal{C}$ of plaintext and ciphertext message units that vary from one user to another. In practice, we would probably want to choose $\mathcal{P}$ and $\mathcal{C}$ uniformly throughout the system. For example, suppose we are working in an $N$-letter alphabet. Then let $k < \ell$ be suitably chosen positive integers, such that, for example, $N^k$ and $N^\ell$ have approximately 200 decimal digits. We take as our plaintext message units all blocks of $k$ letters, which we regard as $k$-digit base-$N$ integers, i.e., we assign them numerical equivalents between 0 and $N^k$. We similarly take ciphertext message units to be blocks of $\ell$ letters in our $N$-letter alphabet. Then each user must choose his/her large primes $p_A$ and $q_A$ so that $n_A = p_A q_A$ satisfies $N^k < n_A < N^\ell$. Then any plaintext message unit, i.e., integer less than $N^k$, corresponds to an element in $\mathbf{Z}/n_A\mathbf{Z}$ (for any user's $n_A$); and, since $n_A < N^\ell$, the image $f(P) \in \mathbf{Z}/n_A\mathbf{Z}$ can be uniquely written as an $\ell$-letter block. (Not all $\ell$-letter blocks can arise — only those corresponding to integers less than $n_A$ for the particular user's $n_A$.)

**Example 1.** For the benefit of a reader who doesn't have a computer handy (or does not have good multiple precision software), we shall sacrifice realism and choose most of our examples so as to involve relatively small integers. Choose $N = 26$, $k = 3$, $\ell = 4$. That is, the plaintext consists of trigraphs and the ciphertext consists of four-graphs in the usual 26-letter alphabet. To send the message "YES" to a user A with enciphering key $(n_A, e_A) = (46927, \, 39423)$, we first find the numerical equivalent of "YES," namely: $24 \cdot 26^2 + 4 \cdot 26 + 18 = 16346$, and then compute $16346^{39423} \; mod \; 46927$, which is $21166 = 1 \cdot 26^3 + 5 \cdot 26^2 + 8 \cdot 26 + 2 =$"BFIC."

The recipient A knows the deciphering key $(n_A, d_A) = (46927, 26767)$, and so computes $21166^{26767}$ *mod* $46927 = 16346$ ="YES." How did user A generate her keys? First, she multiplied the primes $p_A = 281$ and $q_A = 167$ to get $n_A$; then she chose $e_A$ at random (but subject to the condition that $g.c.d.(e_A, 280) = g.c.d.(e_A, 166) = 1$). Then she found $d_A = e_A^{-1}$ *mod* $280 \cdot 166$. The numbers $p_A$, $q_A$, $d_A$ remain secret.

In Example 1, how cumbersome are the computations? The most time-consuming step is modular exponentiation, e.g., $16346^{39423}$ *mod* 46927. But this can be done by the repeated squaring method (see § I.3) in $O(k^3)$ bit operations, where $k$ is the number of bits in our integers. Actually, if we were working with much larger integers, potentially the most time-consuming step would be for each user A to find two very large primes $p_A$ and $q_A$. In order to quickly choose suitable very large primes, one must use an efficient primality test. Such tests will be described in the next chapter.

**Remarks.** **1.** In choosing $p$ and $q$, user A should take care to see that certain conditions hold. The most important are: that the two primes not be too close together (for example, one should be a few decimal digits longer than the other); and that $p-1$ and $q-1$ have a fairly small g.c.d. and both have at least one large prime factor. Some of the reasons for these conditions are indicated in the exercises below. Of course, if someone discovers a factorization method that works quickly under certain other conditions on $p$ and $q$, then future users of RSA would have to take care to avoid those conditions as well.

**2.** In § I.3 we saw that, when $n$ is a product of two primes $p$ and $q$, knowledge of $\varphi(n)$ is equivalent to knowledge of the factorization. Let's suppose now that we manage to break an RSA system by determining a positive integer $d$ such that $a^{ed} \equiv a$ *mod* $n$ for all $a$ prime to $n$. This is equivalent to $ed-1$ being a multiple of the least common multiple of $p-1$ and $q-1$. Knowing this integer $m = ed - 1$ is weaker than actually knowing $\varphi(n)$. But we now give a procedure that with a high probability is nevertheless able to use the integer $m$ to factor $n$.

So suppose we know $n$ — which is a product of two unknown primes — and also an integer $m$ such that $a^m \equiv 1$ *mod* $n$ for all $a$ prime to $n$. Notice that any such $m$ must be even (as we see by taking $a = -1$). We first check whether $m/2$ has the same property, in which case we can replace $m$ by $m/2$. If $a^{m/2}$ is *not* $\equiv 1$ *mod* $n$ for all $a$ prime to $n$, then we must have $a^{m/2} \not\equiv 1$ *mod* $n$ for at least 50% of the $a$'s in $(\mathbf{Z}/n\mathbf{Z})^*$ (this statement is proved in exactly the same way as part (a) of Exercise 21 in § II.2). Thus, if we test several dozen randomly chosen $a$'s and find that in all cases $a^{m/2} \equiv 1$ *mod* $n$, then with very high probability we have this congruence for all $a$ prime to $n$, and so may replace $m$ by $m/2$. We keep on doing this until we no longer have the congruence when we take half of the exponent. There are now two possibilities:

(i) $m/2$ is a multiple of one of the two numbers $p-1$, $q-1$ (say, $p-1$) but not both. In this case $a^{m/2}$ is always $\equiv 1$ *mod* $p$ but exactly 50%

of the time is congruent to $-1$ rather than $+1$ modulo $q$.

(ii) $m/2$ is not a multiple of either $p-1$ or $q-1$. In this case $a^{m/2}$ is $\equiv 1$ modulo both $p$ and $q$ (and hence modulo $n$) exactly 25% of the time, it is $\equiv -1$ modulo both $p$ and $q$ exactly 25% of the time, and for the remaining 50% of the values of $a$ it is $\equiv 1$ modulo one of the primes and $\equiv -1$ modulo the other prime.

Thus, by trying $a$'s at random with high probability we will soon find an $a$ for which $a^{m/2}-1$ is divisible by one of the two primes (say, $p$) but not the other. (Each randomly selected $a$ has a 50% chance of satisfying this statement.) Once we find such an $a$ we can immediately factor $n$, because $g.c.d.(n,\ a^{m/2}-1)=p$.

The above procedure is an example of a *probabilistic algorithm*. We shall encounter other probabilistic algorithms in the next chapter.

**3.** How do we send a signature in RSA? When discussing authentication in the last section, we assumed for simplicity that $\mathcal{P}=\mathcal{C}$. We have a slightly more complicated set-up in RSA. Here is one way to avoid the problem of different $n_A$'s and different block sizes ($k$, the number of letters in a plaintext message unit, being less than $\ell$, the number of letters in a ciphertext message unit). Suppose that, as in the last section, Alice is sending her signature (some plaintext $P$) to Bob. She knows Bob's enciphering key $K_{E,B}=(n_B,e_B)$ and her own deciphering key $K_{D,A}=(n_A,d_A)$. What she does is send $f_Bf_A^{-1}(P)$ if $n_A<n_B$, or else $f_A^{-1}f_B(P)$ if $n_A>n_B$. That is, in the former case she takes the least positive residue of $P^{d_A}$ modulo $n_A$; then, regarding that number modulo $n_B$, she computes $(P^{d_A}\ mod\ n_A)^{e_B}\ mod\ n_B$, which she sends as a ciphertext message unit. In the case $n_A>n_B$, she first computes $P^{e_B}\ mod\ n_B$ and then, working modulo $n_A$, she raises this to the $d_A$-th power. Clearly, Bob can verify the authenticity of the message in the first case by raising to the $d_B$-th power modulo $n_B$ and then to the $e_A$-th power modulo $n_A$; in the second case he does these two operations in the reverse order.

## *Exercises*

1. Suppose that the following 40-letter alphabet is used for all plaintexts and ciphertexts: A—Z with numerical equivalents 0—25, blank=26, .=27, ?=28, \$=29, the numerals 0—9 with numerical equivalents 30—39. Suppose that plaintext message units are digraphs and ciphertext message units are trigraphs (i.e., $k=2$, $\ell=3$, $40^2<n_A<40^3$ for all $n_A$).
(a) Send the message "SEND \$7500" to a user whose enciphering key is $(n_A,e_A)=(2047,179)$.
(b) Break the code by factoring $n_A$ and then computing the deciphering key $(n_A,d_A)$.
(c) Explain why, even without factoring $n_A$, a codebreaker could find the deciphering key rather quickly. In other words, why (in addition to its small size) is 2047 a particularly bad choice for $n_A$?

2. Try to break the code whose enciphering key is $(n_A, e_A) = (536813567, 3602561)$. Use a computer to factor $n_A$ by the stupidest known algorithm, i.e., dividing by all odd numbers 3, 5, 7,.... If you don't have a computer available, try to guess a prime factor of $n_A$ by trying special classes of prime numbers. After factoring $n_A$, find the deciphering key. Then decipher the message BNBPPKZAVQZLBJ, under the assumption that the plaintext consists of 6-letter blocks in the usual 26-letter alphabet (converted to an integer between 0 and $26^6 - 1$ in the usual way) and the ciphertext consists of 7-letter blocks in the same alphabet. It should be clear from this exercise that even a 29-bit choice of $n_A$ is far too small.
3. Suppose that both plaintexts and ciphertexts consist of trigraph message units, but while plaintexts are written in the 27-letter alphabet (consisting of A—Z and blank=26), ciphertexts are written in the 28-letter alphabet obtained by adding the symbol "/" (with numerical equivalent 27) to the 27-letter alphabet. We require that each user A choose $n_A$ between $27^3 = 19683$ and $28^3 = 21952$, so that a plaintext trigraph in the 27-letter alphabet corresponds to a residue $P$ modulo $n_A$, and then $C = P^{e_A}\ mod\ n_A$ corresponds to a ciphertext trigraph in the 28-letter alphabet.
(a) If your deciphering key is $K_D = (n, d) = (21583, 20787)$, decipher the message "YSNAUOZHXXH " (one blank at the end).
(b) If in part (a) you know that $\varphi(n) = 21280$, find (i) $e = d^{-1} mod\ \varphi(n)$, and (ii) the factorization of $n$.
4. Show why the 35-bit integer 23360947609 is a particularly bad choice for $n = pq$, because the two prime factors are too close to one another; that is, show that $n$ can easily be factored by "Fermat factorization" as follows. Note that if $n = pq$ (say $p > q$), then $n = (\frac{p+q}{2})^2 - (\frac{p-q}{2})^2$. If $p$ and $q$ are close together, then $s = (p-q)/2$ is small and $t = (p+q)/2$ is an integer only slightly larger than $\sqrt{n}$ having the property that $t^2 - n$ is a perfect square. If you test the successive integers $t > \sqrt{n}$, you'll soon find one such that $n = t^2 - s^2$, at which point you have $p = t + s$, $q = t - s$. (See Exercise 3 of § I.2 and also §3 of Chapter V.)
5. Suppose that you have a quick algorithm (a probabilistic algorithm) for solving the equation $x^2 \equiv a\ mod\ p$ for any prime $p$ and any quadratic residue $a$. For example, by trying random integers and computing the Legendre symbol, with high probability we can find a nonresidue; then we can apply the algorithm described in § II.2. Suppose, however, that there is no good algorithm for solving $x^2 \equiv a\ mod\ n$ for $a$ a square modulo $n$ and $n = pq$ a product of two large primes, unless one knows the factorization of $n$ (in which case one can find a square root modulo $p$ and modulo $q$ and then use the Chinese Remainder Theorem to find a square root modulo $n$). Suppose that $p$ and $q$ are not both $\equiv 1\ mod\ 4$. Let $K_E = n$, and let $K_D = \{p, q\}$ be its factorization. Let $\mathcal{P} = \mathcal{C} = (\mathbf{Z}/n\mathbf{Z})^*/\pm 1$, which is the set of pairs $(x, -x)$ of residues

modulo $n$ prime to $n$, where negatives are grouped with one another. Let $f: \mathcal{P} \longrightarrow \mathcal{C}$ be the map $x \mapsto x^2 \; mod \; n$. Show that this set-up is an example of Exercise 4 in the last section. This gives us a way to implement long-distance coin flips.

6. Let $n$ be any squarefree integer (i.e., product of distinct primes). Let $d$ and $e$ be positive integers such that $de-1$ is divisible by $p-1$ for every prime divisor $p$ of $n$. (For example, this is the case if $de \equiv 1 \; mod \; \varphi(n)$.) Prove that $a^{de} \equiv a \; mod \; n$ for *any* integer $a$ (whether or not it has a common factor with $n$).
7. Prove the statements in Remark 2 about the percent of the time the different congruences for $a^{m/2}$ occur in cases (i) and (ii).

## References for § IV.2


1. L. M. Adleman, R. L. Rivest and A. Shamir, "A method for obtaining digital signatures and public-key cryptosystems," *Communications of the ACM*, **21** (1978), 120–126.
2. R. L. Rivest, "RSA chips (past/present/future)," *Advances in Cryptology, Proceedings of Eurocrypt 84,* Springer, 1985, 159–165.
3. J. A. Gordon, "Strong primes are easy to find," *Advances in Cryptology, Proceedings of Eurocrypt 84,* Springer, 1985, 216–223.


# 3 Discrete log

The RSA system discussed in the last section is based on the fact that finding two large primes and multiplying them together to get $n$ is far easier than going in the other direction (given $n$, finding the two primes). There are other fundamental processes in number theory which apparently also have this "trapdoor" or "one-way" property. One of the most important is raising to a power in a large finite field.

When working with the real numbers, exponentiation (finding $b^x$ to a prescribed accuracy) is not significantly easier than the inverse operation (finding $log_b x$ to a prescribed accuracy). But now suppose we have a finite group, such as $(\mathbf{Z}/n\mathbf{Z})^*$ or $\mathbf{F}_q^*$ (with the group operation of multiplication). Because of the repeated–squaring method (see § I.3), one can compute $b^x$ for large $x$ rather rapidly (in time which is polynomial in $log\, x$). But, if we're given an element $y$ which we know to be of the form $b^x$ (we suppose that the "base" $b$ is fixed), how can we find the power of $b$ that gives $y$, i.e., how can we compute $x = log_b y$ (where here "log" has a different but analogous meaning than before)? This question is called the "discrete logarithm problem." The word "discrete" distinguishes the finite group situation from the classical (continuous) situation.

**Definition.** If $G$ is a finite group, $b$ is an element of $G$, and $y$ is an element of $G$ which is a power of $b$, then the *discrete logarithm* of $y$ to the base $b$ is any integer $x$ such that $b^x = y$.

**Example 1.** If we take $G = \mathbf{F}_{19}^* = (\mathbf{Z}/19\mathbf{Z})^*$ and let $b$ be the generator 2 (see Example 1 of § II.1), then the discrete logarithm of 7 to the base 2 is 6.

**Example 2.** In $\mathbf{F}_9^*$ with $\alpha$ a root of $X^2 - X - 1$ (see Example 2 of § II.1), the discrete logarithm of $-1$ to the base $\alpha$ is 4.

At the end of this section we shall briefly discuss the present state of algorithms to solve the discrete logarithm problem in finite fields. First we describe several public key cryptosystems or special purpose public key arrangements that are based on the computational difficulty of solving the discrete logarithm problem in finite fields.

**The Diffie–Hellman key exchange system.** Because public key cryptosystems are relatively slow compared to classical cryptosystems (at least at our present stage of technology and theoretical knowledge), it is often more realistic to use them in a limited role in conjunction with a classical cryptosystem in which the actual messages are transmitted. In particular, the process of agreeing on a key for a classical cryptosystem can be accomplished fairly efficiently using a public key system. The first detailed proposal for doing this, due to W. Diffie and M. E. Hellman, was based on the discrete logarithm problem.

We suppose that the key for the classical cryptosystem is a large randomly chosen positive integer (or a collection of such integers). For example, suppose we want to use an affine matrix transformation of pairs of digraphs (see § III.2)

$$C \equiv \begin{pmatrix} a & b \\ c & d \end{pmatrix} P + \begin{pmatrix} e \\ f \end{pmatrix} \quad mod\ N^2,$$

where $0 \leq a,\ b,\ c,\ d,\ e,\ f < N^2$ and $P$ is a column vector consisting of the numerical equivalents of two successive plaintext digraphs (i.e., altogether a four-letter block) in an $N$-letter alphabet. Once we have a randomly selected integer $k$ between 0 and $N^{12}$, we can take $a,\ b,\ c,\ d,\ e,\ f$ to be the six digits in $k$ written to the base $N^2$. (We must check that $ad - bc$ is invertible modulo $N^2$, i.e., that it has no common factor with $N$; otherwise we choose another random integer $k$.)

We observe that choosing a random integer in some interval is equivalent to choosing a random element of a large finite field of roughly the same size. Let us suppose, for example, that we want to choose a random positive $k < N^{12}$. If our finite field is a prime field of $p$ elements, we simply let an element of $\mathbf{F}_p$ correspond to an integer from 0 to $p-1$ in the usual way; if the resulting integer is larger than $N^{12}$, we reduce it modulo $N^{12}$.

If our finite field is $\mathbf{F}_{p^f}$, we first choose an $\mathbf{F}_p$-basis of this field, so that every element corresponds to an $f$-tuple of elements of $\mathbf{F}_p$; then such an $f$-tuple gives an integer less than $p^f$ if we consider the coordinates as digits of an integer written to the base $p$. **Warning:** This gives a 1-to-1 correspondence between $\mathbf{F}_{p^f}$ and $\mathbf{Z}/p^f\mathbf{Z} = \{0, 1, 2, \ldots, p^f - 1\}$. But these two sets have a very different structure under addition and multiplication. The first is a *field*, i.e., all of the $p^f - 1$ nonzero elements have inverses, while the second is a *ring* in which $p^{f-1}$ of the $p^f$ elements (the multiples of $p$) fail to have inverses.

We now describe the Diffie–Hellman method for generating a random element of a large finite field $\mathbf{F}_q$. We suppose that $q$ is public knowledge: everyone knows what finite field our key will be in. We also suppose that $g$ is some fixed element of $\mathbf{F}_q$, which is also not kept secret. Ideally, $g$ should be a generator of $\mathbf{F}_q^*$; however, this is not absolutely necessary. The method described below for generating a key will lead only to elements of $\mathbf{F}_q$ which are powers of $g$; thus, if we really want our random element of $\mathbf{F}_q^*$ to have a chance of being any element, $g$ must be a generator.

Suppose that two users $A$ (Aïda) and $B$ (Bernardo) want to agree upon a key — a random element of $\mathbf{F}_q^*$ — which they will use to encrypt their subsequent messages to one another. Aïda chooses a random integer $a$ between 1 and $q - 1$, which she keeps secret, and computes $g^a \in \mathbf{F}_q$, which she makes public. Bernardo does the same: he chooses a random $b$ and makes public $g^b$. The secret key they use is then $g^{ab}$. Both users can compute this key. For example, Aïda knows $g^b$ (which is public knowledge) and her own secret $a$. However, a third party knows only $g^a$ and $g^b$. If the following assumption holds for the multiplicative group $\mathbf{F}_q^*$, then an unauthorized third party will be unable to determine the key.

**Diffie–Hellman assumption.** It is computationally infeasible to compute $g^{ab}$ knowing only $g^a$ and $g^b$.

The Diffie–Hellman assumption is *a priori* at least as strong as the assumption that discrete logarithms cannot be feasibly computed in the group. That is, if discrete logarithms can be computed, then obviously the Diffie–Hellman assumption fails. Some people would conjecture that the converse implication also holds, but that is still an open question. In other words, no one can imagine a way of passing from $g^a$ and $g^b$ to $g^{ab}$ without first being able to determine $a$ or $b$; but it is conceivable that such a way might exist.

**Example 3.** Suppose we're using a shift encryption of single–letter message units in the 26-letter alphabet (see Example 1 of § III.1): $C \equiv P + B \bmod 26$. (We're using $B$ rather than $b$ to denote the shift key so as not to confuse it with the $b$ in the last paragraph.) To choose $B$, take the least nonnegative residue modulo 26 of a random element in $\mathbf{F}_{53}$. Let $g = 2$ (which is a generator of $\mathbf{F}_{53}$). Suppose Aïda picked at random $a = 29$, and looked up Bernardo's public $2^b$, which is, say, $12 \in \mathbf{F}_{53}$. She then knows that the enciphering key is $12^{29} = 21 \in \mathbf{F}_{53}$, i.e., $B = 21$. Meanwhile, she has made public $2^{29} = 45$, and so Bernardo can also find the key $B = 21$ by raising 45 to the $b$-th power (his secret exponent is $b = 19$). Of course, there is no security in working with such a small field; an outsider could easily find the discrete logarithm to the base 2 of 12 or 45 modulo 53. And in any case there is no security in using a shift encryption of single-letter message units. But this example illustrates the mechanics of the Diffie–Hellman key exchange system.

**The Massey–Omura cryptosystem for message transmission.** We suppose that everyone has agreed upon a finite field $\mathbf{F}_q$, which is fixed and publicly known. Each user of the system secretly selects a random integer $e$ between 0 and $q-1$ such that $g.c.d.(e, q-1) = 1$ and, using the Euclidean algorithm, computes its inverse $d = e^{-1}\ mod\ q-1$, i.e., $de \equiv 1\ mod\ q-1$. If user $A$ (Alice) wants to send a message $P$ to Bob, first she sends him the element $P^{e_A}$. This means nothing to Bob, who, not knowing $d_A$ (or $e_A$, for that matter), cannot recover $P$. But, without attempting to make sense of it, he raises it to *his* $e_B$, and sends $P^{e_A e_B}$ back to Alice. The third step is for Alice to unravel the message part of the way by raising to the $d_A$-th power; because $P^{d_A e_A} = P$ (by Proposition II.1.1), this means that she returns $P^{e_B}$ to Bob, who can read the message by raising this to the $d_B$-th power.

The idea behind this system is rather simple, and it can be generalized to settings where one is using other processes besides exponentiation in finite fields. However, some words of caution are in order. First of all, notice that it is absolutely necessary to use a good signature scheme along with the Massey–Omura system. Otherwise, any person $C$ who is not supposed to know the message $P$ could pretend to be Bob, returning to Alice $P^{e_A e_C}$; not knowing that an intruder was using his own $e_C$, she would proceed to raise to the $d_A$ and make it possible for $C$ to read the message. Thus, the message $P^{e_A e_B}$ from Bob to Alice must be accompanied by some authentification, i.e., some message in some signature scheme which only Bob could have sent.

In the second place, it is important that, after a user such as $B$ or $C$ has deciphered various messages $P$, and so knows various pairs ($P$, $P^{e_A}$), he cannot use that information to determine $e_A$. That is, suppose Bob could solve the discrete log problem in $\mathbf{F}_q^*$, thereby determining from $P$ and $P^{e_A}$ what $e_A$ must be. In that case he could quickly compute $d_A = e_A^{-1}\ mod\ q-1$ and then intercept and read all future messages from Alice, whether intended for him or not.

**The ElGamal cryptosystem.** We start by fixing a very large finite field $\mathbf{F}_q$ and an element $g \in \mathbf{F}_q^*$ (preferably, but not necessarily, a generator). We suppose that we are using plaintext message units with numerical equivalents $P$ in $\mathbf{F}_q$. Each user $A$ randomly chooses an integer $a = a_A$, say in the range $0 < a < q-1$. This integer $a$ is the secret deciphering key. The public enciphering key is the element $g^a \in \mathbf{F}_q$.

To send a message $P$ to the user $A$, we choose an integer $k$ at random, and then send $A$ the following pair of elements of $\mathbf{F}_q$:

$$(g^k,\ Pg^{ak}).$$

Notice that we can compute $g^{ak}$ without knowing $a$, simply by raising $g^a$ to the $k$-th power. Now $A$, who knows $a$, can recover $P$ from this pair by raising the first element $g^k$ to the $a$-th power and dividing the result into the second element (or, equivalently, raising $g^k$ to the $(q-1-a)$-th power and multiplying by the second element). In other words, what we send $A$ consists of a disguised form of the message — $P$ is "wearing a mask" $g^{ak}$ — along with a "clue," namely $g^k$, which can be used to take off the mask (but the clue can be used only by someone who knows $a$).

Someone who can solve the discrete log problem in $\mathbf{F}_q$ breaks the cryptosystem by finding the secret deciphering key $a$ from the public enciphering key $g^a$. In theory, there could be a way to use knowledge of $g^k$ and $g^a$ to find $g^{ak}$ — and hence break the cipher — without solving the discrete log problem. However, as we mentioned in our discussion of the Diffie–Hellman key exchange system, it is conjectured that there is no way to go from $g^k$ and $g^a$ to $g^{ak}$ without essentially solving the discrete logarithm problem.

**The Digital Signature Standard.** In 1991 the U.S. government's National Institute of Standards and Technology (NIST) proposed a Digital Signature Standard (DSS). The role of DSS is expected to be analogous to that of the much older Data Encryption Standard (DES), i.e., it is supposed to provide a standard digital signature method for use by government and commercial organizations. But while DES is a classical ("private key") cryptosystem, in order to construct digital signatures it is necessary to use public key cryptography. NIST chose to base their signature scheme on the discrete log problem in a prime finite field. The DSS is very similar to a signature scheme that was originally proposed by Schnorr (see the references below). It is also similar to a signature scheme of ElGamal (see Exercise 9 below). We now describe how the DSS works.

To set up the scheme (in order later to be able to sign messages), each user Alice proceeds as follows: (1) she chooses a prime $q$ of about 160 bits (to do this, she uses a random number generator and a primality test); (2) she then chooses a second prime $p$ that is $\equiv 1\ (mod\ q)$ and has about 512 bits; (3) she chooses a generator of the unique cyclic subgroup of $\mathbf{F}_p^*$ of order $q$ (by computing $g_0^{(p-1)/q}\ (mod\ p)$ for a random integer $g_0$; if this number is $\neq 1$, it will be a generator); (4) she takes a random integer $x$ in the range $0 < x < q$ as her secret key, and sets her public key equal to $y = g^x\ (mod\ p)$.

Now suppose that Alice wants to sign a message. She first applies a hash function to her plaintext (see §1), obtaining an integer $h$ in the range $0 < h < q$. She next picks a random integer $k$ in the same range, computes $g^k\ (mod\ p)$, and sets $r$ equal to the least nonnegative residue modulo $q$ of the latter number (i.e., $g^k$ is first computed modulo $p$, and the result is then reduced modulo the smaller prime $q$). Finally, Alice finds an integer $s$ such that $sk \equiv h + xr \ (mod\ q)$. Her signature is then the pair $(r, s)$ of integers modulo $q$.

To verify the signature, the recipient Bob computes $u_1 = s^{-1}h \ (mod\ q)$ and $u_2 = s^{-1}r \ (mod\ q)$. He then computes $g^{u_1}y^{u_2} \ (mod\ p)$. If the result agrees modulo $q$ with $r$, he is satisfied. (Note that $g^{u_1}y^{u_2} = g^{s^{-1}(h+xr)} = g^k \ (mod\ p)$.)

This signature scheme has the advantage that signatures are fairly short, consisting of two 160-bit numbers (the magnitude of $q$). On the other hand, the security of the system seems to depend upon intractability of the discrete log problem in the multiplicative group of the rather large field $\mathbf{F}_p$. Although to break the system it would suffice to find discrete logs in the smaller subgroup generated by $g$, in practice this seems to be no easier than finding arbitrary discrete logarithms in $\mathbf{F}_p^*$. Thus, the DSS seems to have attained a fairly high level of security without sacrificing small signature storage and implementation time.

**Algorithms for finding discrete logs in finite fields.** We first suppose that all of the prime factors of $q-1$ are small. In this case we sometimes say that $q-1$ is "smooth." With this assumption there is a fast algorithm for finding the discrete log of an element $y \in \mathbf{F}_q^*$ to the base $b$. For simplicity, we shall suppose that $b$ is a generator of $\mathbf{F}_q^*$. We now describe this algorithm, which is due to Silver, Pohlig and Hellman.

First, for each prime $p$ dividing $q-1$, we compute the $p$-th roots of unity $r_{p,j} = b^{j(q-1)/p}$ for $j = 0, 1, \ldots, p-1$. (As usual, we use the repeated squaring method to raise $b$ to a large power.) With our table of $\{r_{p,j}\}$ we are ready to compute the discrete log of any $y \in \mathbf{F}_q^*$. (Note that, if $b$ is fixed, this first computation needs only be done once, after which the same table is used for any $y$.)

Our object is to find $x$, $0 \le x < q-1$, such that $b^x = y$. If $q-1 = \prod_p p^\alpha$ is the prime factorization of $q-1$, then it suffices to find $x \ mod\ p^\alpha$ for each $p$ dividing $q-1$; from this $x$ is uniquely determined using the algorithm in the proof of the Chinese Remainder Theorem (Proposition I.3.3). So we now fix a prime $p$ dividing $q-1$, and show how to determine $x \ mod\ p^\alpha$.

Suppose that $x \equiv x_0 + x_1 p + \cdots + x_{\alpha-1}p^{\alpha-1} \ (mod\ p^\alpha)$ with $0 \le x_i < p$. To find $x_0$ we compute $y^{(q-1)/p}$. We get a $p$-th root of 1, since $y^{q-1} = 1$. Since $y = b^x$, it follows that $y^{(q-1)/p} = b^{x(q-1)/p} = b^{x_0(q-1)/p} = r_{p,x_0}$. Thus, we compare $y^{(q-1)/p}$ with the $\{r_{p,j}\}_{0 \le j < p}$ and set $x_0$ equal to the value of $j$ for which $y^{(q-1)/p} = r_{p,j}$.

Next, to find $x_1$, we replace $y$ by $y_1 = y/b^{x_0}$. Then $y_1$ has discrete log $x - x_0 \equiv x_1 p + \cdots x_{\alpha-1}p^{\alpha-1} \ (mod\ p^\alpha)$. Since $y_1$ is a $p$-th power, we have $y_1^{(q-1)/p} = 1$ and $y_1^{(q-1)/p^2} = b^{(x-x_0)(q-1)/p^2} = b^{(x_1+x_2p+\cdots)(q-1)/p} = b^{x_1(q-1)/p} = r_{p,x_1}$. So we can compare $y_1^{(q-1)/p^2}$ with $\{r_{p,j}\}$ and set $x_1$ equal to the value of $j$ for which $y_1^{(q-1)/p^2} = r_{p,j}$.

It should now be clear how we can proceed inductively to find all $x_0, x_1,$

$\ldots, x_{\alpha-1}$. Namely, for each $i = 1, 2, \ldots, \alpha - 1$ set

$$y_i = y/b^{x_0 + x_1 p + \cdots + x_{i-1} p^{i-1}},$$

which has discrete log congruent *mod* $p^\alpha$ to $x_i p^i + \cdots + x_{\alpha-1} p^{\alpha-1}$. Since $y_i$ is a $p^i$-th power, we have $y_i^{(q-1)/p^i} = 1$ and $y_i^{(q-1)/p^{i+1}} = b^{(x_i + x_{i+1}p + \cdots)(q-1)/p}$ $= b^{x_i(q-1)/p} = r_{p,x_i}$. So we set $x_i$ equal to the value of $j$ for which $y_i^{(q-1)/p^{i+1}}$ $= r_{p,j}$.

When we are done we will have $x$ *mod* $p^\alpha$. After doing this for each $p|q-1$, we finally use the Chinese Remainder Theorem to find $x$.

This algorithm works well when all of the primes dividing $q-1$ are small. But clearly the computation of the table of $\{r_{p,j}\}$ and the comparison of the $y_i^{(q-1)/p^{i+1}}$ with this table will take a long time if $q-1$ is divisible by a large prime. (By "large" we mean of at least about 20 digits. If $p|q-1$ is smaller than about $10^{20}$, then one can combine the Silver–Pohlig–Hellman algorithm with Shanks' "giant step — baby step" method; see pp. 9, 575–576 of Knuth, Vol. 2.)

**Example 4.** Find the discrete log of 28 to the base 2 in $\mathbf{F}_{37}^*$ using the Silver–Pohlig–Hellman algorithm. (2 is a generator of $\mathbf{F}_{37}^*$.)

**Solution.** Here $37 - 1 = 2^2 \cdot 3^2$. We compute $2^{18} \equiv 1$ (*mod* 37), and so $r_{2,0} = 1$, $r_{2,1} = -1$. (For $p = 2$, always $\{r_{2,j}\} = \{\pm 1\}$.) Next, $2^{36/3} \equiv 26$, $2^{2\cdot 36/3} \equiv 10$ (*mod* 37), and so $\{r_{3,j}\} = \{1,\ 26,\ 10\}$. Now let $28 \equiv 2^x$ (*mod* 37). We first take $p = 2$ and find $x$ *mod* 4, which we write as $x_0 + 2x_1$. We compute $28^{36/2} \equiv 1$ (*mod* 37), and hence $x_0 = 0$. We then compute $28^{36/4} \equiv -1$ (*mod* 37), and hence $x_1 = 1$, i.e., $x \equiv 2$ (*mod* 4). Next we take $p = 3$ and find $x$ *mod* 9, which we write as $x_0 + 3x_1$. (Of course, for each $p$ the $x_i$ are defined differently.) To find $x_0$, we compute $28^{36/3} \equiv 26$ (*mod* 37), and so $x_0 = 1$. We then compute $(28/2)^{36/9} = 14^4 \equiv 10$ (*mod* 37); thus, $x_1 = 2$, and so $x \equiv 1 + 2 \cdot 3 = 7$ (*mod* 9). It remains to find the unique $x$ *mod* 36 such that $x \equiv 2$ (*mod* 4) and $x \equiv 7$ (*mod* 9). This is $x = 34$. Thus, $28 = 2^{34}$ in $\mathbf{F}_{37}^*$.

**The index–calculus algorithm for discrete logs.** The reader may want to skip this subsection for now, or read it lightly, and come back to it for a closer examination while reading §V.3, since the index–calculus algorithm for computing discrete logs in finite fields has much in common with the factor–base method for factoring large integers.

Here we shall suppose that $q = p^n$ is a fairly large power of a small prime $p$, and $b$ is a generator of $\mathbf{F}_q^*$. The index–calculus algorithm finds for any $y \in \mathbf{F}_q^*$ the value of $x$ *mod* $q-1$ such that $y = b^x$.

Let $f(X) \in \mathbf{F}_p[X]$ be any irreducible polynomial of degree $n$; then $\mathbf{F}_q$ is isomorphic to the residue ring $\mathbf{F}_p[X]/f(X)$. Any element $a \in \mathbf{F}_q = \mathbf{F}_p[X]/f(X)$ can be written (uniquely) as a polynomial $a(X) \in \mathbf{F}_p[X]$ of degree at most $n-1$. In particular, our base $b = b(X)$ is such a polynomial. The "constants" are the elements of $\mathbf{F}_p \subset \mathbf{F}_q$.

We first note that $b' = b^{(q-1)/(p-1)}$ is a generator of $\mathbf{F}_p^*$ (see Exercise 17 of §II.1). Thus, we immediately know the discrete logs to the base $b$ of these constants once we solve the discrete log problem in $\mathbf{F}_p^*$ (to the base $b'$). But we have assumed that $p$ is small, and so a table of such discrete logs can easily be constructed. In the important special case $p = 2$, in fact, the only nonzero constant is 1, whose discrete log to any base is 0. In what follows we shall suppose that we can easily find the discrete log of a constant.

For the rest of this section we shall let $ind(a(X))$ (from the word "index") denote the discrete log of $a(X) \in \mathbf{F}_q^*$ to the base $b(X)$. The base $b(X)$ is fixed throughout the discussion, and so will not be indicated in the notation.

There are two basic stages of the index–calculus algorithm. The first stage is called a "precomputation," because it does not depend on the element $y(X) \in \mathbf{F}_q^*$ whose discrete log we ultimately want to determine. It has only to be carried out once, and can then be used for many computations of various discrete logs to the fixed base $b(X)$. (Recall that there was also an analogous precomputation stage in the Silver–Pohlig–Hellman algorithm, namely, the compilation of the table of $\{r_{p,j}\}$.)

We first choose a subset $B \subset \mathbf{F}_q$ which will serve as our "basis." Usually $B$ consists of all monic irreducible polynomials over $\mathbf{F}_p$ of degree $\leq m$, where $m < n$ is determined in some optimal way so that the set $B$ has a suitable size $h = \#(B)$ of intermediate magnitude between $p = \#(\mathbf{F}_p)$ and $q = p^n = \#(\mathbf{F}_q)$. The precomputation stage consists in determining the discrete logs of all $a(X) \in B$, as follows.

Choose a random integer $t$ between 1 and $q-2$, and compute $b^t \in \mathbf{F}_q$, i.e., compute the polynomial $c(X) \in \mathbf{F}_p[X]$ of degree $< n$ such that

$$c(X) \equiv b(X)^t \mod f(X).$$

(Here one uses the repeated squaring method, at each step reducing modulo $f(X)$.) Factor out the leading coefficient $c_0$ from $c(x)$, and determine whether or not the resulting monic polynomial can be written as a product of the $a(X) \in B$, i.e., whether or not $c(X)$ can be written in the form

$$c(X) = c_0 \prod_{a \in B} a(X)^{\alpha_{c,a}}.$$

One way to determine this is to run through all $a(X) \in B$ and divide $c(X)$ successively by $a(X)^{\alpha_{c,a}}$ (where $\alpha_{c,a}$ is the highest power of $a(X)$ which divides $c(X)$ in $\mathbf{F}_p[X]$). If the constant $c_0$ is all that remains after dividing by powers of all of the $a(X) \in B$, then $c(X)$ has the above form; otherwise, start over again at the beginning of this paragraph with a different random integer $t$. (A second way — in some cases quicker — to determine whether $c(X)$ factors into a product of $a(X) \in B$ is simply to factor $c(X)$ using an algorithm for factoring elements of $\mathbf{F}_p[X]$. For a description of a good algorithm for this purpose (due to Berlekamp), see Volume II of Knuth, §4.6.2.)

Now suppose that we have found a $c(X) \equiv b(X)^t \ \ mod \ f(X)$ which has the desired type of factorization. Taking the discrete log of both sides of the above equality, we obtain

$$ind(c(X)) - ind(c_0) = \sum_{a \in B} \alpha_{c,a} ind(a(X)),$$

where equality here should be interpreted as congruence modulo $q-1$ (since the discrete log is defined only modulo $q-1$). The left side of this equality is known, since $ind(c(X)) = t$ and the discrete logs of constants are assumed to be known. The coefficients $\alpha_{c,a}$ on the right are also known. The unknowns are the $h$ values $ind(a(X))$, $a(X) \in B$, on the right.

Thus, we have obtained a linear equation in $\mathbf{Z}/(q-1)\mathbf{Z}$ with $h$ unknowns. Now suppose we continue to choose random integers $t$ until we obtain a large number of different $c(X)$'s which factor into a product of $a(X)$'s. As soon as we obtain $h$ independent congruences of the type

$$t - ind(c_0) \equiv \sum_{a \in B} \alpha_{c,a} ind(a(X)) \ \ mod \ q-1$$

(here "independent" means that the determinant of the coefficient matrix $\{\alpha_{c,a}\}$ is prime to $q-1$), then we can solve the system for the unknowns modulo $q-1$. (See §III.2 for a discussion of linear algebra modulo $N = q-1$.) This completes the first stage of the index–calculus algorithm. The precomputation has given us a large "data–base," namely the discrete logs of all $a(X) \in B$, from which to compute any discrete log we are interested in.

Before proceeding to a description of the second stage of the index–calculus algorithm, we should comment on the choice of $m$, which was not specified when we described $B \subset \mathbf{F}_p[X]$ as the set of all monic irreducible polynomials of degree $\le m$. The size $h$ of the set $B$ grows rapidly as $m$ increases. For example, if $m$ is prime, then we saw (Corollary to Proposition II.1.8) that in degree $m$ alone there are $(p^m - p)/m$ monic irreducible polynomials. Since we are required to find at least $h$ different $c(X)$'s which give us the $h \times h$ system of independent linear congruences in the $h$ unknowns $ind(a(X))$, and then we have to solve the system, it would be helpful if $h$ were not too large, i.e., if $m$ were not too large. On the other hand, if $m$ is small, then a "typical" monic polynomial $c_0^{-1}c(X)$ of degree $\le n-1$ is not likely to factor into a product of $a(X)$ of degree $\le m$; it is more likely to have at least one irreducible factor of degree $> m$. That is, if $m$ is small, it will take us an inordinate amount of time to make even a single lucky random choice of $t$ for which $c(X) \equiv b(X)^t \ mod \ f(X)$ has the desired type of factorization. Thus, $m$ must be not too small, though quite a bit smaller than $n$. The optimal choice of $m$ — depending, of course, on $p$ and $n$ — requires a lengthy analysis of probabilities and time estimates, which go beyond the scope of this book. For example, when $p = 2$ and $n = 127$, the best choice turns out to be $m = 17$ (in which case $h = 16510$). The value $q = 2^{127}$ is a popular choice, because $\#(\mathbf{F}_{2^{127}}^*) = 2^{127} - 1$ is a Mersenne prime.

We now return to the index–calculus algorithm, and describe the final stage. Here we suppose that $y(X) \in \mathbf{F}_q^*$ is the element whose discrete log we wish to compute, and that stage one has already given us the values of $ind(a(X))$ for all $a(X) \in B$. We again choose a random $t$ between 1 and $q-2$, and compute $y_1 = yb^t$, i.e., the unique polynomial $y_1(X) \in \mathbf{F}_p[X]$ of degree $< n$ satisfying $y_1(X) \equiv y(X)b(X)^t \mod f(X)$. As in the first stage of the algorithm, we test whether $y_1(X)$ factors into a constant $y_0$ times a product of powers of $a(X)$, $a(X) \in B$. If not, we choose another random $t$, and so on, until we finally have an integer $t$ such that $y_1(X) \equiv y_0 \prod_{a \in B} a(X)^{\alpha_a}$. As soon as this happens, we are done, because $ind(y) = ind(y_1) - t$, by the definition of $y_1$; and $ind(y_1) = ind(y_0) + \sum \alpha_a ind(a(X))$, in which we know all of the terms on the right. This completes the description of the index–calculus algorithm.

It should be mentioned that in the popular case $p = 2$, an improved method due to D. Coppersmith has significantly speeded up the process of finding discrete logs. For this reason, a discrete log cryptosystem using $\mathbf{F}_{2^n}^*$ is no longer regarded as secure unless $n$ is of the order of 1000. Despite this, these fields $\mathbf{F}_{2^n}$ remain popular because they lend themselves to efficient programming. For a good survey (covering what was known as of 1985), the reader is referred to A. Odlyzko's article (see References below).

If $q = p^n$ is an odd prime power which is $k$ bits long, it turns out that, roughly speaking, the order of magnitude of time needed to solve the discrete log problem in $\mathbf{F}_q^*$ is comparable to what is needed to factor a $k$-bit integer. That is, from an empirical point of view, the discrete log problem seems to be about as difficult as factoring (though no one has been able to prove a theorem to this effect). In fact, when we discuss factoring algorithms and time estimates for them in the next chapter, we will see that one of the fundamental methods of factoring large integers bears a striking resemblance to the index–calculus algorithm for finding discrete logs.

Thus, at this point it is too early to say whether the public key cryptosystems of the RSA type (based on the difficulty of factoring integers) or the discrete log cryptosystems will eventually prove to be the more secure.

## *Exercises*

**Note:** Exercises 4, 6, 7(c) and 8 should be attempted only if you have the use of a computer with multiple precision arithmetic programs. (All that is really needed is a program for computing $a^b \mod m$ for very large integers $a$, $b$ and $m$; recall that $a^{-1} \mod p$ can be computed by taking $a^{p-2}$.)

1. If one has occasion to do a lot of arithmetic in a fixed finite field $\mathbf{F}_q$ which is not too large, it can save time first to compose a complete "table of logarithms." In other words, choose a generator $g$ of $\mathbf{F}_q$ and

make a 2-column list of all pairs $n$, $g^n$ as $n$ goes from 1 to $q-1$; then make third and fourth columns listing all pairs $a$, $log_g a$. That is, list the elements $a$ of $\mathbf{F}_q^*$ in some convenient order in the third column, and then run down the first two columns, putting each $n$ in the fourth column next to the $a$ which is $g^n$. For example, to do this for $\mathbf{F}_9$ (see Example 2 in § II.1), we choose $g=\alpha$ to be a root of $X^2-X-1$, and make the following table:

| $n$ | $g^n$ | $a$ | $log_g a$ |
|---|---|---|---|
| 1 | $\alpha$ | 1 | 8 |
| 2 | $\alpha+1$ | $-1$ | 4 |
| 3 | $-\alpha+1$ | $\alpha$ | 1 |
| 4 | $-1$ | $\alpha+1$ | 2 |
| 5 | $-\alpha$ | $\alpha-1$ | 7 |
| 6 | $-\alpha-1$ | $-\alpha$ | 5 |
| 7 | $\alpha-1$ | $-\alpha+1$ | 3 |
| 8 | 1 | $-\alpha-1$ | 6 |

Then multiplication or division involves nothing more than addition or subtraction modulo $q-1$ and looking at the table. For example, to multiply $\alpha-1$ by $-\alpha-1$, we find the two numbers in the third column, add the two corresponding logarithms: $7+6\equiv 5\ mod\ 8$, and then find the answer $-\alpha$ in the second column next to 5.

(a) Make a log table for $\mathbf{F}_{31}^*$, and use it to compute $16\cdot 17$, $19\cdot 13$, $1/17$, $20/23$.

(b) Make a log table for $\mathbf{F}_8^*$, and use it to compute the following (where $\alpha$ is a root of $X^3+X+1$; your answers should not involve any higher power of $\alpha$ than $\alpha^2$): $(\alpha+1)(\alpha^2+\alpha)$, $(\alpha^2+\alpha+1)(\alpha^2+1)$, $1/(\alpha^2+1)$, $\alpha/(\alpha^2+\alpha+1)$.

2. At first glance, it may seem that we could use the cyclic group $(\mathbf{Z}/p^\alpha\mathbf{Z})^*$ (see Exercise 2(a) in § II.1) instead of $\mathbf{F}_q^*$ as a setting for the discrete logarithm problem. However, the discrete log problem for $(\mathbf{Z}/p^\alpha\mathbf{Z})$ for $\alpha>1$ turns out to be essentially no more time-consuming (even if $\alpha$ is fairly large) than for $\alpha=1$ (i.e., $\mathbf{F}_p$). More precisely, using the same technique that is given below in this exercise, one can prove that, once one solves the discrete log problem modulo $p$, going the rest of the way (i.e., solving it modulo $p^\alpha$) takes polynomial time in $log(p^\alpha)=\alpha\, log\, p$. (Recall that no algorithm is known which solves the discrete log problem modulo $p$ for large $p$ in polynomial time in $log\, p$; and experts doubt that such an algorithm exists.) In this exercise, we show that in the case $p=3$ there's a straightforward algorithm which solves the discrete log problem modulo $3^\alpha$ in time which is polynomial in $\alpha$.

Thus, suppose we take $g=2$ (it is easy to show that 2 is a generator of $(\mathbf{Z}/3^\alpha\mathbf{Z})^*$ for any $\alpha$), we have some integer $a$ not divisible by 3, and we want to solve the congruence $2^x\equiv a\ mod\ 3^\alpha$. Prove that the following algorithm always finds $x$ and takes polynomial time in $\alpha$, and estimate (using the $O$-notation) the number of bit operations required to find $x$:

(i) Show that the discrete log problem is equivalent to the congruence with $a$ moved to the left (i.e., $2^x a \equiv 1$). Next, show that without loss of generality we may assume that $a \equiv 1 \; mod \; 3$ and $x$ is even. Thus, we can replace our original congruence with the congruence $4^x a \equiv 1 \; mod \; 3^\alpha$.

(ii) Write $x = x_0 + 3x_1 + \cdots + 3^{\alpha-2}x_{\alpha-2}$, where the $x_j$ are base-3 digits. Take $x_{-1} = 0$. Then the congruence

$$4^{x_0+3x_1+\cdots+3^{j-2}x_{j-2}} a \equiv 1 \; mod \; 3^j \qquad (*)_j$$

holds for $j = 1$. Set $g_1 = 4$. In the course of the algorithm as a by-product we will compute $g_j \underset{\text{def}}{=} 4^{3^{j-1}} \; mod \; 3^\alpha$. Set $a_1 = a$, and for $j > 1$ define $a_j$ to be the least positive residue mod $3^\alpha$ of $4^{x_0+3x_1+\cdots+3^{j-2}x_{j-2}} a$; we will compute $a_j$ below as we go along.

(iii) Suppose that $j > 1$ and we have found $x_0, \ldots, x_{j-3}$ such that the congruence $(*)_{j-1}$ holds (i.e., $(*)$ with $j-1$ in place of $j$). Further suppose that we have computed $g_{j-1} = 4^{3^{j-2}} \; mod \; 3^\alpha$ and also $a_{j-1}$. First set $x_{j-2}$ equal to $(1 - a_{j-1})/3^{j-1}$ modulo 3. (Notice that $a_{j-1} \equiv 1 \; mod \; 3^{j-1}$ because of $(*)_{j-1}$.) Next, compute $a_j = g_{j-1}^{x_{j-2}} a_{j-1} \; mod \; 3^\alpha$. Finally, if $j < \alpha$, compute $g_j$ by raising $g_{j-1}$ to the 3-rd power, working modulo $3^\alpha$.

(iv) When you reach $j = \alpha$, you're done.

3. You and your friend agree to communicate using affine enciphering transformations $C \equiv AP + B \; mod \; N$ (see Examples 3 and 4 in § III.1, where lowercase letters $a$ and $b$ were used for the coefficients of the transformation). Your message units are single letters in the 31-letter alphabet with A—Z corresponding to 0—25, blank=26, .=27, ?=28, !=29, '=30. You regard the key $K_E = (A, B)$ as an element $A + Bi$ in the field of $31^2$ elements (where $i$ denotes a square root of $-1$ in that field). You also agree to exchange keys using the Diffie–Hellman system, and to choose $g = 4 + i$. Then you randomly choose a secret integer $a = 209$. Your friend sends you her $g^b = 1 + 19i$.
(a) Find the enciphering key.
(b) What element of $\mathbf{F}_{961}$ must you send your friend in order that she can also find the key?
(c) Find the deciphering transformation.
(d) Read the message "BUVCFIWOUJTZ!H."

4. You receive the ciphertext "VHNHDOAM," which was sent to you using a $2 \times 2$ enciphering matrix

$$\begin{pmatrix} a & b \\ c & d \end{pmatrix}$$

applied to digraphs in the usual 26-letter alphabet. The enciphering matrix was determined using the Diffie–Hellman key exchange method, as follows. Working in the prime field of 3602561 elements, your correspondent sent you $g^b = 983776$. Your randomly chosen Diffie-Hellman exponent $a$ is 1082389. Finally, you agree to get a matrix from a key number $K_E \in \mathbf{F}_{3602561}$ by writing the least nonnegative residue of $K_E$ modulo $26^4$ in the form $a \cdot 26^3 + b \cdot 26^2 + c \cdot 26 + d$ (where $a$, $b$, $c$, $d$ are digits in the base 26). If the resulting matrix is not invertible modulo 26, replace $K_E$ by $K_E + 1$ and try again. Take as the enciphering matrix the first invertible matrix that arises from the successive integers starting with $K_E$.

(a) Use this information to find the enciphering matrix.

(b) Find the deciphering matrix, and read the message.

5. Suppose that each user $A$ has a secret pair of transformations $f_A$ and $f_A^{-1}$ from $\mathcal{P}$ to $\mathcal{P}$, where $\mathcal{P}$ is a fixed set of plaintext message units. They want to transmit information securely using the Massey–Omura technique, i.e., Alice sends $f_A(P)$ to Bob, who then sends $f_B(f_A(P))$ back to her, and so on. Give the conditions that the system of $f_A$'s must satisfy in order for this to work.

6. Let $p$ be the Fermat prime 65537, and let $g = 5$. You receive the message (29095, 23846), which your friend composed using the ElGamal cryptosystem in $\mathbf{F}_p^*$, using your public key $g^a$. Your secret key, needed for deciphering, is $a = 13908$. You have agreed to convert integers in $\mathbf{F}_p$ to trigraphs in the 31-letter alphabet of Exercise 3 by writing them to the base 31, the digits in the $31^2-$, the $31-$ and $1-$ place being the numerical equivalents of the three letters in the trigraph. Decipher the message.

7. (a) Show that choosing $\mathbf{F}_p$ with $p = 2^{2^k} + 1$ a Fermat prime is an astoundingly bad idea, by constructing a polynomial time algorithm for solving the discrete log problem in $\mathbf{F}_p^*$ (i.e., an algorithm which is polynomial in $log\, p$). To do this, suppose that $g$ is a generator (e.g., 5 or 7, as shown in Exercise 15 of § II.2) and for a given $a$ you want to find $x$, where $0 \le x < p - 1 = 2^{2^k}$, such that $g^x \equiv a\ mod\ p$. Write $x$ in binary, and pattern your algorithm after the algorithm for extracting square roots modulo $p$ that was described at the end of § II.2.

   (b) Find a big-$O$ estimate (in terms of $p$) for the number of bit operations required to find the integer $x$ by means of the algorithm in part (a).

   (c) Use the algorithm in part (a) to find the value of $k$ in Exercise 6.

8. Suppose that your plaintext message units are 18-letter blocks written in the usual 26-letter alphabet, where the numerical equivalent of such a block is an 18-digit base-26 integer (written in order of decreasing powers of 26). You receive the message

$$(82746592004375034872957717,\ 164063768437915425954819351),$$

which was enciphered using the ElGamal cryptosystem in the prime field of 297262705009139006771611927 elements, using your public key $g^a$. Your secret key is $a = 10384756843984756438549809$. Decipher the message.

9. Here is a scheme (also due to ElGamal) for sending a signature using a large prime finite field $\mathbf{F}_p$. Explain why Alice can do all the steps required to send her signature (in time polynomial in $log\, p$), why Bob can verify that Alice must have sent the signature, and why the system would fail if an imposter could solve the discrete logarithm problem in $\mathbf{F}_p^*$.
We suppose that a fixed $p$ and a fixed $g \in \mathbf{F}_p^*$ are publicly known. Each user $A$ also chooses a random integer $a_A$, $0 < a_A < p-1$, which is kept secret, and publishes $y_A = g^{a_A}$.
To send her signature — which is composed of message units with numerical equivalents $S$ in the range $0 \le S < p-1$ — Alice first chooses a random integer $k$ prime to $p-1$. She computes $r = g^k \, mod\, p$, and then solves the following congruence for the unknown $x$: $g^S \equiv y^r r^x \, mod\, p$. She sends Bob the pair $(r, x)$ along with her signature $S$. Bob verifies that $g^S$ is in fact $\equiv y^r r^x \, mod\, p$, and he is happy, secure in his confidence that Alice did send the message $S$.
10. Using the Silver–Pohlig–Hellman algorithm, find the discrete log of 153 to the base 2 in $\mathbf{F}_{181}^*$. (2 is a generator of $\mathbf{F}_{181}^*$.)
11. (a) What is the percent likelihood that a random polynomial over $\mathbf{F}_2$ of degree exactly 10 factors into a product of polynomials of degree $\le 2$? What is the likelihood that a random nonzero polynomial of degree at most 10 factors into such a product?
(b) What is the probability that a random monic polynomial over $\mathbf{F}_3$ of degree exactly 10 factors into a product of polynomials of degree $\le 2$? What is the probability that a random monic polynomial of degree at most 10 factors into such a product?
12. For $n > m \ge 1$, let $P_p(n, m)$ denote the probability that a random monic polynomial over $\mathbf{F}_p$ of degree at most $n$ is a product of irreducible factors all of degree $\le m$.
(a) Prove that for any fixed $n$ and $m$, $P(n, m) = \lim_{p\to\infty} P_p(n, m)$ exists and is strictly between 0 and 1.
(b) Find an explicit expression for $P(n, 2)$.
(c) Compute $P(n, 2)$ exactly for all $n \le 7$.

## References for § IV.3

1. L. M. Adleman, "A subexponential algorithm for the discrete logarithm problem with applications to cryptography," *Proc. 20th Annual Symposium on the Foundations of Computer Science* (1979), 55–60.

2. L. M. Adleman and J. DeMarrais, "A subexponential algorithm for discrete logarithms over all finite fields," *Math. Comp.* **61** (1993), 1–15.
3. D. Coppersmith, "Fast evaluation of logarithms in fields of characteristic two," *IEEE Transactions on Information Theory IT-30* (1984), 587–594.
4. D. Coppersmith, A. Odlyzko, and R. Schroeppel, "Discrete logarithms in $GF(p)$," *Algorithmica* **1** (1986), 1–15.
5. W. Diffie and M. E. Hellman, "New directions in cryptography," *IEEE Transactions on Information Theory IT-22* (1976), 644–654.
6. T. ElGamal, "A public key cryptosystem and a signature scheme based on discrete logarithms," *IEEE Transactions on Information Theory IT-31*, (1985), 469–472.
7. T. ElGamal, "A subexponential-time algorithm for computing discrete logarithms over $GF(p^2)$," *IEEE Transactions on Information Theory IT-31* (1985), 473–481.
8. M. Fellows and N. Koblitz, "Fixed-parameter complexity and cryptography," *Proc. Tenth Intern. Symp. Appl. Algebra, Algebraic Algorithms and Error Correcting Codes* (San Juan, Puerto Rico), 1993.
9. D. Gordon, "Discrete logarithms in $GF(p)$ using the number field sieve," *SIAM J. Discrete Math.* **6** (1993), 124–138.
10. D. Gordon and K. McCurley, "Massively parallel computation of discrete logarithms," *Advances in Cryptology — Crypto '92*, Springer-Verlag, 1993.
11. D. E. Knuth, *The Art of Computer Programming*, Vol. II, Addison–Wesley, 1973.
12. B. LaMacchia and A. Odlyzko, "Computation of discrete logarithms in prime fields," *Designs, Codes and Cryptography* **1** (1991), 47–62.
13. J. L. Massey, "Logarithms in finite cyclic groups — cryptographic issues," *Proc. 4th Benelux Symposium on Information Theory* (1983), 17–25.
14. K. McCurley, "The discrete logarithm problem," *Cryptology and Computational Number Theory, Proc. Symp. Appl. Math.* **42** (1990), 49–74.
15. A. M. Odlyzko, "Discrete logarithms in finite fields and their cryptographic significance," *Advances in Cryptology, Proc. Eurocrypt 84*, Springer, 1985, 224–314.
16. P. K. S. Wah and M. Z. Wang, "Realization and application of the Massey–Omura lock," *Proc. International Zürich Seminar* (1984), 175–182.

# 4 Knapsack

In this section we describe another type of public key cryptosystem, which is based on the so-called "knapsack problem." Suppose you have a large knapsack which you are packing in preparation for a long hike in the wilderness. You have a large number of items (say, $k$ items) of volume $v_i$, $i = 0, \ldots, k-1$, to fit into the knapsack, which holds a total volume $V$. Suppose that you are an experienced knapsack packer, and can always fit items in with no wasted space. You want to take the biggest load possible, so you want to find some subset of the $k$ items that exactly fills the knapsack. In other words, you want to find some subset $I \subset \{1, \ldots, k\}$ such that $\sum_{i\in I} v_i = V$, if such a subset exists. This is the *general knapsack problem*. We shall further assume that $V$ and all of the $v_i$ are positive integers. An equivalent way to state the problem is then as follows:

**The knapsack problem.** Given a set $\{v_i\}$ of $k$ positive integers and an integer $V$, find a $k$-bit integer $n = (\epsilon_{k-1}\epsilon_{k-2}\cdots\epsilon_1\epsilon_0)_2$ (where the $\epsilon_i \in \{0,1\}$ are the binary digits of $n$) such $\sum_{i=0}^{k-1} \epsilon_i v_i = V$, if such an $n$ exists.

Note that there may be no solution $n$ or many solutions, or there might be a unique solution, depending on the $k$-tuple $\{v_i\}$ and the integer $V$.

A special case of the knapsack problem is the *superincreasing knapsack problem*. This is the case when the $v_i$, arranged in increasing order, have the property that each one is greater than the *sum* of all of the earlier $v_i$.

**Example 1.** The 5-tuple $(2, 3, 7, 15, 31)$ is a superincreasing sequence.

It is known that the general knapsack problem is in a very difficult class of problems, called "NP-complete" problems. This means that it is equivalent in difficulty to the notorious "traveling salesman problem." In particular, if the central conjecture in complexity theory is true, as most everyone believes it is, then there does not exist an algorithm which solves an arbitrary knapsack problem in time polynomial in $k$ and $\log B$, where $B$ is a bound on the size of $V$ and the $v_i$.

However, the superincreasing knapsack problem is much, much easier to solve. Namely, we look down the $v_i$, starting with the largest, until we get to the first one that is $\leq V$. We include the corresponding $i$ in our subset $I$ (i.e., we take $\epsilon_i = 1$), replace $V$ by $V - v_i$, and then continue down the list of $v_i$ until we find one that is less than or equal to this difference. Continuing in this way, we eventually either obtain a subset of $\{v_i\}$ which sums to $V$, or else we exhaust all of $\{v_i\}$ without getting $V - \sum_{i\in I} v_i$ equal to 0, in which case there is no solution. We now write the algorithm in a more formal way that could be easily converted to a computer program.

The following polynomial time algorithm solves the knapsack problem for a given superincreasing $k$-tuple $\{v_i\}$ and integer $V$:

1. Set $W$ equal to $V$, and set $j = k$.
2. Starting with $\epsilon_{j-1}$ and decreasing the index of $\epsilon$, choose all of the $\epsilon_i$ equal to 0 until you get to the first $i$ — call it $i_0$ — such that $v_{i_0} \leq W$. Set $\epsilon_{i_0} = 1$.
3. Replace $W$ by $W - v_{i_0}$, set $j = i_0$, and, if $W > 0$, go back to step 2.
4. If $W = 0$, you're done. If $W > 0$, and all of the remaining $v_i$ are $> W$, then you know there is no solution $n = (\epsilon_{k-1}\cdots\epsilon_0)_2$ to the problem. Notice that the solution (if there is one) is unique.

**Example 2.** Let the $v_i$ be as in Example 1, and take $V = 24$. Then, working from right to left in our 5-tuple $\{2, 3, 7, 15, 31\}$, we see that $\epsilon_4 = 0$, $\epsilon_3 = 1$ (at which point we replace 24 by $24 - 15 = 9$), $\epsilon_2 = 1$ (at which point we replace 9 by $9 - 7 = 2$), $\epsilon_1 = 0$, $\epsilon_0 = 1$. Thus, $n = (01101)_2 = 13$.

We now describe how to construct the knapsack cryptosystem (also called the Merkle–Hellman system). We first suppose that our plaintext message units have $k$-bit integers $P$ as their numerical equivalents. For example, if we're working with single letters in the 26-letter alphabet, then every letter corresponds to one of the 5-bit integers from $0 = (00000)_2$ to $25 = (11001)_2$ in the usual way.

Next, each user chooses a superincreasing $k$-tuple $\{v_0, \ldots, v_{k-1}\}$, an integer $m$ which is greater than $\sum_{i=0}^{k-1} v_i$, and an integer $a$ prime to $m$, $0 < a < m$. This is done by some random process. For example, we could choose an arbitrary sequence of $k + 1$ positive integers $z_i$, $i = 0, 1, \ldots, k$, less than some convenient bound; set $v_0 = z_0$, $v_i = z_i + v_{i-1} + v_{i-2} + \cdots + v_0$ for $i = 1, \ldots, k - 1$; and set $m$ equal to $z_k + \sum_{i=0}^{k-1} v_i$. Then one can choose a random positive $a_0 < m$ and take $a$ to be the first integer $\geq a_0$ that is prime to $m$. After that, one computes $b = a^{-1}\ mod\ m$ (i.e., $b$ is the least positive integer such that $ab \equiv 1\ mod\ m$), and also computes the $k$-tuple $\{w_i\}$ defined by $w_i = av_i\ mod\ m$ (i.e., $w_i$ is the least positive residue of $av_i$ modulo $m$). The user keeps the numbers $v_i$, $m$, $a$, and $b$ all secret, but publishes the $k$-tuple of $w_i$. That is, the enciphering key is $K_E = \{w_0, \ldots, w_{k-1}\}$. The deciphering key is $K_D = (b, m)$ (which, along with the enciphering key, enables one to determine $\{v_0, \ldots, v_{k-1}\}$).

Someone who wants to send a plaintext $k$-bit message $P = (\epsilon_{k-1}\epsilon_{k-2} \cdots \epsilon_1\epsilon_0)_2$ to a user with enciphering key $\{w_i\}$ computes $C = f(P) = \sum_{i=0}^{k-1} \epsilon_i w_i$, and transmits that integer.

To read the message, the user first finds the least positive residue $V$ of $bC$ modulo $m$. Since $bC \equiv \sum \epsilon_i b w_i \equiv \sum \epsilon_i v_i\ mod\ m$ (because $bw_i \equiv bav_i \equiv v_i\ mod\ m$), it follows that $V = \sum \epsilon_i v_i$. (Here we are using the fact that both $V < m$ and $\sum \epsilon_i v_i \leq \sum v_i < m$ to convert the congruence modulo $m$ to equality.) It is then possible to use the above algorithm for superincreasing knapsack problems to find the unique solution $(\epsilon_{k-1} \cdots \epsilon_0)_2 = P$ of the problem of finding a subset of the $\{v_i\}$ which sums exactly to $V$. In this way we recover the message $P$.

Note that an eavesdropper who knows only $\{w_i\}$ is faced with the knapsack problem $C = \sum \epsilon_i w_i$, which is *not* a superincreasing problem, because the superincreasing property of the $k$-tuple of $v_i$ is destroyed when $v_i$ is replaced by the least positive residue of $av_i$ modulo $m$. Thus, the above algorithm cannot be used, and, at first glance, the unauthorized person seems to be faced with a much more difficult problem. We shall return to this point later.

**Example 3.** Suppose that our plaintext message units are single letters with 5-bit numerical equivalents from $(00000)_2$ to $(11001)_2$, as above. Suppose that our secret deciphering key is the superincreasing 5-tuple in Example 1. Let us choose $m = 61$, $a = 17$; then $b = 18$ and the enciphering key is $(34, 51, 58, 11, 39)$. To send the message 'WHY' our correspondent would compute 'W'$= (10110)_2 \mapsto 51 + 58 + 39 = 148$, 'H'$= (00111)_2 \mapsto 34 + 51 + 58 = 143$, 'Y'$= (11000)_2 \mapsto 11 + 39 = 50$. To read the message $148, 143, 50$, we first multiply by 18 modulo 61, obtaining $41, 12, 46$. Proceeding as in Example 2 with $V = 41$, $V = 12$, and $V = 46$, we recover the plaintext $(10110)_2$, $(00111)_2$, $(11000)_2$.

Of course, as usual there is no security using single-letter message units with such a small value of $k = 5$; Example 3 is meant only to illustrate the mechanics of the system.

For a while, many people were optimistic about the possibilities for knapsack cryptosystems. Since the problem of breaking the system is in a very difficult class of problems (NP-complete problems), they reasoned, the system should be secure.

However, there was a fallacy in that reasoning. The type of knapsack problem $C = \sum \epsilon_i w_i$ that must be solved, while not a superincreasing knapsack problem, is nevertheless of a very special type, namely, it is obtained from a superincreasing problem by a simple transformation, i.e., multiplying everything by $a$ and reducing modulo $m$. In 1982, Shamir found an algorithm to solve this type of knapsack problem that is polynomial in $k$. Thus, the original Merkle–Hellman cryptosystem cannot be regarded as a secure public key cryptosystem.

One way around Shamir's algorithm is to make the knapsack system a little more complicated by using a sequence of transformations of the form $x \mapsto ax \ mod \ m$ for different $a$ and $m$. For example, we might simply use two transformations corresponding to $(a_1, m_1)$ and $(a_2, m_2)$. That is, we first replace our superincreasing sequence $\{v_i\}$ by $\{w_i\}$, where $w_i$ is the least positive residue of $a_1 v_i \ mod \ m_1$, and then obtain a third sequence $\{u_i\}$ by taking the least positive residue $u_i = a_2 w_i \ mod \ m_2$. Here we choose random $m_1$, $m_2$, $a_1$ and $a_2$ subject to the conditions $m_1 > \sum v_i$, $m_2 > km_1$, and $g.c.d.(a_1, m_1) = g.c.d.(a_2, m_2) = 1$. The public key is then the $k$-tuple of $u_i$, and the enciphering function is $C = f(P) = \sum_{i=0}^{k-1} \epsilon_i u_i$, where $P = (\epsilon_{k-1} \cdots \epsilon_1)_2$. To decipher the ciphertext using the key $K_D = (b_1, m_1, b_2, m_2)$ (where $b_1 = a_1^{-1} \ mod \ m_1$ and $b_2 = a_2^{-1} \ mod \ m_2$), we first compute the least positive residue of $b_2 C$ modulo $m_2$, and then take the result, multiply it by $b_1$, and reduce modulo $m_1$. Since $b_2 C \equiv \sum \epsilon_i w_i \ mod \ m_2$, and since $m_2 > km_1 > \sum w_i$, it follows that the result of reducing $b_2 C \ mod \ m_2$ is *equal* to $\sum \epsilon_i w_i$. Then when we take $b_1 \sum \epsilon_i w_i \ mod \ m_1$ we obtain $\sum \epsilon_i v_i$, from which we can determine the $\epsilon_i$ using the above algorithm for a superincreasing knapsack problem.

At the present time, although there is no polynomial time algorithm which has been proved to give a solution of the iterated knapsack problem (i.e., the public key cryptosystem described in the last paragraph), Shamir's algorithm has been generalized by Brickell and others, who show that iterated knapsack cryptosystems are vulnerable to efficient cryptanalysis. In any case, after Shamir's breakthrough, most experts lost confidence in the security of a public key cryptosystem of this type.

**An as yet unbroken knapsack.** We now describe a method of message transmission based on a knapsack-type one-way function that uses polynomials over a finite field. The cryptosystem is due to Chor and Rivest; we shall describe a slightly simplified (and less efficient) version of their construction.

Again suppose that Alice wants to be able to receive messages that are $k$-tuples of bits $\epsilon_0, \ldots, \epsilon_{k-1}$. (The number $k$ is selected by Alice, as described below.) Her public key, as before, is a sequence of positive integers $v_0, \ldots, v_{k-1}$, constructed in the way described below. This time Bob must send her not only the integer $c = \sum \epsilon_j v_j$ but also the sum of the bits $c' = \sum \epsilon_j$.

Alice constructs the sequence $v_j$ as follows. All of the choices described in this paragraph can be kept secret, since it is only the final $k$-tuple $v_0, \ldots, v_{k-1}$ that Bob needs to know in order to send a message. First, Alice chooses a prime power $q = p^f$ such that $q-1$ has no large prime factors (in which case discrete logs can feasibly be computed in $\mathbf{F}_q^*$, see §3) and such that both $p$ and $f$ are of intermediate size (e.g., 2 or 3 digits). In the 1988 paper by Chor and Rivest the value $q = 197^{24}$ was suggested. Next, Alice chooses a monic irreducible polynomial $F(X) \in \mathbf{F}_p[X]$ of degree $f$, so that $\mathbf{F}_q$ may be regarded as $\mathbf{F}_p[X]/F(X)$. She also chooses a generator $g$ of $\mathbf{F}_q^*$, and an integer $z$. Alice makes these choices of $F$, $g$, and $z$ in some random way.

Let $t \in \mathbf{F}_q = \mathbf{F}_p[X]/F(X)$ denote the residue class of $X$. Alice chooses $k$ to be any integer less than both $p$ and $f$. For $j = 0, \ldots, k-1$, she computes the nonnegative integer $b_j < q-1$ such that $g^{b_j} = t + j$. (By assumption, Alice can easily find discrete logarithms in $\mathbf{F}_q^*$.) Finally, Alice chooses at random a permutation $\pi$ of $\{0, \ldots, k-1\}$, and sets $v_j$ equal to the least nonnegative residue of $b_{\pi(j)} + z$ modulo $q-1$. She publishes the $k$-tuple $(v_0, \ldots, v_{k-1})$ as her public key.

Deciphering works as follows. After receiving $c$ and $c'$ from Bob, she first computes $g^{c-zc'}$, which is represented as a unique polynomial $G(X) \in \mathbf{F}_p[X]$ of degree $< f$. But she knows that this element must also be equal to $\prod g^{\epsilon_j b_{\pi(j)}} = \prod (t + \pi(j))^{\epsilon_j}$, which is represented by the polynomial $\prod (X + \pi(j))^{\epsilon_j}$. Since both $G(X)$ and $\prod (X + \pi(j))^{\epsilon_j}$ have degree $< f$ and represent the same element modulo $F(X)$, she must have

$$G(X) = \prod (X + \pi(j))^{\epsilon_j},$$

from which she can determine the $\epsilon_j$ by factoring $G(X)$ (for which efficient algorithms are available, see Vol. 2 of Knuth).

*Exercises*

1. For each of the following sequences and "volumes," decide whether the knapsack problem is superincreasing and how many solutions (if any) it has: (a) $\{2,3,7,20,35,69\}$, $V = 45$; (b) $\{1,2,5,9,20,49\}$, $V = 73$; (c) $\{1,3,7,12,22,45\}$, $V = 67$; (d) $\{2,3,6,11,21,40\}$, $V = 39$; (e) $\{4,5,10,30,50,101\}$, $V = 186$; (f) $\{3,5,8,15,28,60\}$, $V = 43$;
2. (a) Show that the superincreasing sequence with the smallest $v_i$'s is the one with $v_i = 2^i$.
   (b) Show that a superincreasing knapsack problem with $v_i = 2^i$ always has a solution $n$, namely $n = V$, and that for no other superincreasing sequence does the corresponding knapsack problem always have a solution.
3. Show that any sequence of positive integers $\{v_i\}$ with $v_{i+1} \geq 2v_i$ for all $i$ is superincreasing.
4. Suppose that plaintext message units are single letters in the usual 26-letter alphabet with A—Z corresponding to 0—25. You receive the sequence of ciphertext message units 14, 25, 89, 3, 65, 24, 3, 49, 89, 24, 41, 25, 68, 41, 71. The public key is the sequence $\{57,14,3,24,8\}$ and the secret key is $b = 23$, $m = 61$.
   (a) Try to decipher the message without using the deciphering key; check by using the deciphering key and the algorithm for a superincreasing knapsack problem.
   (b) Use the above public key to send the message TENFOUR.
5. Suppose that plaintext message units are trigraphs in the 32-letter alphabet with A—Z corresponding to 0—25, blank=26, ?=27, !=28, .=29, '=30, \$=31. You receive the sequence of ciphertext message units 152472, 116116, 68546, 165420, 168261. The public key is the sequence {24038, 29756, 34172, 34286, 38334, 1824, 18255, 19723, 143, 17146, 35366, 11204, 32395, 12958, 6479}, and the secret key is $b = 30966$, $m = 47107$. Decipher the message.
6. Suppose that plaintext message units are digraphs in the 32-letter alphabet of Exercise 5. You receive the sequence of ciphertext message units 33219, 7067, 18127, 43099, 37953, which were enciphered using a two-iteration knapsack system with public key {23161, 6726, 4326, 16848, 21805, 11073, 120, 15708, 2608, 341}. The secret key is $b_1 = 533$, $m_1 = 2617$, $b_2 = 10175$, $m_2 = 27103$. Decipher the message.

## References for § IV.4


1. E. Brickell, "Breaking iterated knapsacks," *Advances in Cryptology — Crypto '84*, Springer-Verlag, 1985, 342–358.
2. E. Brickell and A. Odlyzko, "Cryptanalysis: A survey of recent results," *Proc. IEEE* **76** (1988), 578–593.
3. B. Chor and R. Rivest, "A knapsack-type public key cryptosystem based on arithmetic in finite fields," *Advances in Cryptology — Crypto*

*'84*, Springer-Verlag, 1985, 54–65; revised version in *IEEE Transactions on Information Theory IT-34* (1988), 901–909.
4. M. R. Garey and D. S. Johnson, *Computers and Intractability: A Guide to the Theory of NP–Completeness*, W. H. Freeman, 1979.
5. R. M. F. Goodman and A. J. McAuley, "A new trapdoor knapsack public key cryptosystem," *Advances in Cryptography, Proc. Eurocrypt 84,* Springer, 1985, 150–158.
6. M. E. Hellman, "The mathematics of public-key cryptography," *Scientific American* **241** (1979), 146–157.
7. M. E. Hellman and R. C. Merkle, "Hiding information and signatures in trapdoor knapsacks," *IEEE Transactions on Information Theory IT-24* (1978), 525–530.
8. A. Odlyzko, "The rise and fall of knapsack cryptosystems," *Cryptology and Computational Number Theory, Proc. Symp. Appl. Math.* **42** (1990), 75–88.
9. C. Schnorr, "Efficient identification and signatures for smart cards," *Advances in Cryptology — Crypto '89*, Springer-Verlag, 1990, 239–251.
10. A. Shamir, "A polynomial time algorithm for breaking the basic Merkle–Hellman cryptosystem," *Proc. 23rd Annual Symposium on the Foundations of Computer Science* (1982), 145–152.
11. P. van Oorschot, "A comparison of practical public-key cryptosystems based on integer factorization and discrete logarithms," in G. Simmons, ed., *Contemporary Cryptology: The Science of Information Integrity*, IEEE Press, 1992, 289–322.

## 5 Zero-knowledge protocols and oblivious transfer

"Zero knowledge" is the name of a cryptographic concept first developed in the early 1980's to deal with the following problem. Suppose someone wants to prove that she has figured out how to do something — find a solution to an equation, prove a theorem, solve a puzzle — while at the same time conveying no knowledge about her proof or solution. Can this ever be done? How can you convince someone that you have a solution without exhibiting it? The somewhat surprising fact is that in many situations it is possible to do this.

The "prover," whom we shall call Pícara, is the person with the solution; the "verifier" Vivales is the one who in the end must become satisfied that Pícara has a solution, while still not having the foggiest idea of what that solution is.

In this section we shall first give a simple, visual example of a zero-knowledge proof which is interactive (i.e., it requires communication back and forth between Pícara and Vivales). This example concerns map coloring and does not use number theory. Then we give a second example: how to prove that you have found a discrete logarithm without helping the verifier to know what it is. We next discuss a concept called "oblivious transfer," with which one can construct noninteractive zero-knowledge proofs. Finally, we use oblivious transfer to give a zero-knowledge proof of factorization.

**Map coloring.** Our first example is the following. It is now known that any planar map can be colored with 4 colors. Some maps can be colored with 3 colors and others cannot. Suppose Pícara is given a complicated map, which after much effort she is able to find a way of coloring with only 3 colors (red, blue, green). How can she convince Vivales that she has done this, without giving him a clue that would help him color the map?

We first translate this problem into the language of graphs.

**Definition.** A *graph* is a set $V$, whose elements are called "vertices," and a subset $E$ of the set of all (unordered) pairs of elements of $V$. The elements of $E$ are called "edges." An "edge" $e = \{u, v\}$, where $u, v \in V$, should be visualized as a line joining the vertices $u$ and $v$.

**Definition.** We say that a graph is *colorable* by the colors $r$, $b$, $g$, if there exists a function $f: V \rightarrow \{r, b, g\}$ such that no vertices joined by an edge have the same color, i.e., $\{u, v\} \in E \Longrightarrow f(u) \neq f(v)$.

The 3-colorability problem consists in determining, given a graph, whether or not it is colorable by $r$, $b$, $g$.

To translate the map-coloring problem to a graph-coloring problem, simply take $V$ to be the set of countries (visualized now as points), and "connect" two countries with an edge if and only if they have a common boundary.

The 3-colorability problem has two nice properties which make it a convenient choice for discussions of many questions: (1) it is easy to visualize; and (2) it is NP-complete (see the discussion of the knapsack in §4). The NP-completeness property implies that, if you have a zero-knowledge verification of 3-colorability, then you can get a zero-knowledge verification for any NP-problem by "reducing" it to 3-colorability.

However, this does not mean that, once a zero-knowledge verification has been constructed for a certain NP-complete problem $P_1$ (say, 3-colorability), it is then superfluous to construct a zero-knowledge proof for another NP-problem $P_2$. On the contrary, in the process of reducing $P_2$ to $P_1$, one generally increases the size of the input data substantially. Thus, a much more efficient zero-knowledge verification is likely to result by working directly with $P_2$ rather than by reducing $P_2$ to $P_1$ and then using the earlier verification of $P_1$. For example, we shall later give a direct zero-knowledge proof of possession of a discrete logarithm. It would be inefficient in the extreme to construct such a zero-knowledge proof by first reducing possession of a discrete log to 3-colorability of some graph.

**Zero-knowledge proof of 3-colorability.** Suppose that Pícara is given a graph. We shall visualize the vertices as small balls containing little colored lights and joined by bars wherever there is an edge. The light in each vertex can flash either red, blue or green. Pícara has (1) a device $A$ which sets each vertex to flash whichever of the three colors she chooses, and (2) a device $B$

which, whenever a button is pushed, chooses a random permutation of the three colors and then resets each vertex according to the permutation. For example, if the device $B$ chooses the transposition of red and blue, then it goes to all vertices with blue lights, switches them to red lights, goes to all vertices with red lights, switches them to blue lights, and leaves the vertices with green lights alone. Vivales has no control over the device $B$ and does not even know which permutations it generates.

We further suppose that the lights inside the vertex balls are hidden from view. However, whenever someone grabs onto the bar connecting two vertices, the lights in those two vertices (and no others) become visible.

Now Pícara has figured out a 3-coloring of the graph, and uses the device $A$ to set the vertices with the corresponding colors. Here is the procedure used to convince Vivales that she has been successful in doing this:

1. Vivales is allowed to grab any one of the edge-bars, revealing the colors of the two vertices at each end. He will see that those two vertices have different colors, thereby giving a little bit of evidence that Pícara has a valid coloring (recall that "valid" means that no two adjacent vertices have the same color).
2. Next, Pícara pushes the button on $B$, permuting the colors.
3. Vivales may then grab another edge-bar.
4. Pícara and Vivales repeat steps #2 and #3 in alternation, until Vivales has tested all the bars (or, if he insists, until he has tested all the bars several times — perhaps he suspects that Pícara has cheated by resetting the vertices on a bar that was tested earlier).

After a little thought, two things should be clear: (1) If Pícara has really not been able to 3-color the graph, she won't be able to fool Vivales — eventually step #3 will reveal adjacent vertices of the same color. (2) Because of the random permutations of the colors, Vivales learns nothing about the coloring, except for the fact that Pícara has been successful. That is, if he, too, now wants to 3-color the graph, it will be just as hard for him to 3-color it after going through steps #1–4 above as it would have been before.

To prove the claim that Vivales has learned nothing about the coloring, one argues as follows. Suppose that a third person, Clyde, does not know how to 3-color the graph but *does* know in advance which edge-bar Vivales will grab. Then Clyde could produce the exact same result as Pícara, i.e., the information Vivales receives from Clyde is indistinguishable from what Pícara would have given him. But Clyde could hardly be conveying anything useful about 3-coloring the graph, since he himself does not know a 3-coloring. We say that Clyde "simulates" the role of Pícara. This argument by simulation is the standard way to show that a certain protocol is really a zero-knowledge proof.

**Zero-knowledge proof of having found a discrete logarithm.** As in §3, suppose that $G$ is a finite group containing $N$ elements (whose group operation will be written multiplicatively), $b$ is a fixed element of $G$, and $y$ is an element of $G$ for which Pícara has found a discrete logarithm to the base $b$, i.e., she has solved the equation $b^x = y$ for a positive integer $x$. She wants to demonstrate to Vivales that she knows $x$ without giving him a clue as to what $x$ is. We first suppose that Vivales knows the order $N$ of the group. Here is the sequence of steps performed by the two of them:

1. Pícara generates a random positive integer $e < N$, and sends Vivales $b' = b^e$.
2. Vivales flips a coin. If it comes up heads, Pícara must reveal $e$, and Vivales checks that in fact $b'$ is $b^e$.
3. If the coin comes up tails, then Pícara must reveal the least positive residue of $x + e$ modulo $N$, at which point Vivales checks that $yb' = b^{x+e}$.
4. Steps #1–3 are repeated until Vivales is convinced that Pícara must know the value $x$ of the discrete logarithm.

Notice that if Pícara does not know the value $x$ of the discrete log, then she will not be able to respond to more than one possible result of the coin toss. If she has performed step (1) as she was supposed to, then she can respond to heads — but not to tails — without knowing $x$. On the other hand, if she anticipates tails and so in step (1) decides to send Vivales $b' = b^e/y$ (so that in step (3) she can send him simply $e$ instead of $x + e$), then she will be in a jam if the coin comes up heads (since she does not know the power of $b$ that gives $b'$).

Further notice that the zero-knowledge property of this protocol can be proved by a simulation argument. Namely, suppose that Clyde does not know the discrete log of $y$ to the base $b$ but *does* know in advance how the coin toss will go. Then Clyde can simulate the same steps as Pícara (by sending $b' = b^e$ for heads and $b' = b^e/y$ for tails), giving Vivales information that is indistinguishable from what Pícara would have given him. Clyde cannot be telling Vivales anything useful for finding the discrete log, since he himself has no idea what the discrete log is.

In the exercises we will examine the situation when Vivales does not know $N$. For example, suppose that he knows that $G = (\mathbf{Z}/M\mathbf{Z})^*$, but he does not know the factorization of $M$. (Recall that if $M$ is a product of two primes, then knowing its factorization is equivalent to knowing $N = \varphi(M)$, see §I.3.) Then ideally Pícara (or the simulator Clyde), who uses the value of $N$ in step (1), must avoid conveying to Vivales any information about $N$ (or else we don't really have a "zero knowledge" proof). This might seem to be too much to ask for, but one can insist that no more than a very small amount of information be conveyed.

**Oblivious transfer.** An "oblivious transfer channel" from Pícara to Vivales is a system for Pícara to send Vivales two encrypted packets of information subject to the following conditions:

1. Vivales can decipher and read exactly one of the two packets;
2. Pícara does not know which of the two packets he can read; and

3. both Pícara and Vivales are certain that conditions (1) and (2) hold.

At first glance, this might seem like an odd thing to want. However, such a channel turns out to be a fundamental concept in cryptography. We shall soon see how it can be used to construct a non-interactive zero-knowledge proof. But before discussing this application to zero knowledge, we describe one way to obtain an oblivious transfer channel, based on the intractability of the discrete log problem.

More precisely, we suppose that we have a large finite field $\mathbf{F}_q$ and a fixed element $b$ of the multiplicative group $\mathbf{F}_q^*$ such that, given $b^x$ and $b^y$, there is no computationally feasible way to find $b^{xy}$. This is the Diffie–Hellman assumption, which conjecturally holds if the discrete logarithm problem is intractable in $\mathbf{F}_q^*$ (see §3).

We further suppose that we have an easily computed (and easily inverted) map $\psi$ from our finite field to the $\mathbf{F}_2$-vector space $\mathbf{F}_2^n$ of $n$-tuples of bits. Suppose that the image of this map contains all of $\mathbf{F}_2^{n-1}$ (i.e., all $n$-tuples whose last bit is 0). For example, if $q$ is a prime $p$, then we can choose $n$ so that $2^{n-1} < p < 2^n$, and map any element of $\mathbf{F}_q$ — i.e., any nonnegative integer less than $p$ — to its sequence of binary digits.

We suppose that our message units are also $n$-tuples of bits, i.e., elements $m \in \mathbf{F}_2^n$. We finally suppose that an element $C \in \mathbf{F}_q^*$, fixed once and for all, has been chosen so that no one knows its discrete logarithm. (Recall that we have assumed that the discrete log problem is intractable in $\mathbf{F}_q^*$.) This element $C$ might have been supplied by a "trusted Center," or by an agreed upon random procedure, or by an interactive construction in which both Pícara and Vivales participated.

The oblivious transfer proceeds as follows. Vivales chooses a random integer $x$, $0 < x < q-1$, and also a random element $i \in \{1,2\}$. In what follows both $x$ and $i$ denote fixed integers in the range $\{1,\ldots,q-2\}$ and $\{1,2\}$, respectively. Vivales sets $\beta_i = b^x$ and $\beta_{3-i} = C/b^x$. He then publishes his "public key" $(\beta_1, \beta_2)$, while keeping $x$ and $i$ secret. Notice that Vivales is assumed not to know the discrete logarithm of $\beta_{3-i}$ — which we shall denote $x'$ — because if he did, then he would know the discrete log of $C = \beta_i\beta_{3-i}$, contrary to assumption.

Now suppose that Pícara has a message unit $m_1 \in \mathbf{F}_2^n$ from the first packet and a message unit $m_2 \in \mathbf{F}_2^n$ from the second packet. She chooses two random integers $0 < y_1, y_2 < q-1$, and sends to Vivales the following two elements of $\mathbf{F}_q^*$ and two elements of $\mathbf{F}_2^n$:

$$b^{y_1}, \qquad b^{y_2}; \qquad\qquad \alpha_1 = m_1 + \psi(\beta_1^{y_1}), \qquad \alpha_2 = m_2 + \psi(\beta_2^{y_2}).$$

(Here addition is in the $\mathbf{F}_2$-vector space $\mathbf{F}_2^n$; this addition operation is also known as "exclusive or.") Pícara keeps $y_1$ and $y_2$ secret.

Since $\beta_i^{y_i} = (b^{y_i})^x$, and Vivales knows both $b^{y_i}$ and $x$, he can easily determine $\psi(\beta_i^{y_i})$, and hence find $m_i = \alpha_i + \psi(\beta_i^{y_i})$. However, if he wanted to find $m_{3-i}$, he would have to find $\beta_{3-i}^{y_{3-i}} = b^{x'y_{3-i}}$ knowing only $b^{y_{3-i}}$ and $b^{x'}$ but not $y_{3-i}$ or $x'$. This is impossible, by the Diffie–Hellman assumption.

Notice that Pícara can easily check that $\beta_1\beta_2 = C$, and thus be sure that Vivales does not know the discrete logs of both elements of his public key $(\beta_1, \beta_2)$. Since it is in Vivales' interest to get as much information as possible, Pícara can be sure that he does know the discrete log of one of the two elements. But there is no way Pícara can distinguish between $\beta_1$ and $\beta_2$ for the purpose of determining which Vivales obtained as $b^x$ and which as $C/b^x$. Thus, both Vivales and Pícara can be confident that the above conditions (1) and (2) are fulfilled.

If a sequence of pairs $(m_1, m_2)$ are sent using the same $(\beta_1, \beta_2)$ (i.e., the same values of $x$ and $i$), then Pícara does know that the element of the pair $(m_1, m_2)$ that Vivales is deciphering (namely, $m_i$) remains the same for all pairs of message units in the sequence. If we want another sequence of message units to be sent independently, then Vivales must randomly select new values for $x$ and $i$, and send a new public key $(\beta_1, \beta_2)$.

**Use of oblivious transfer for a non-interactive proof of factorization.** The idea conveyed by the term "non-interactive" can be summarized in the form of a diagram

Center

$\swarrow \qquad \searrow$

Pícara $\longrightarrow$ Vivales

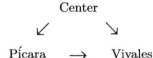

Here the "trusted Center" can be thought of as a source of random bits, which are sent simultaneously to Pícara and Vivales (it is permissible for the Center first to perform some arithmetic operations on the bits before sending them). The combination of these bits and Pícara's reaction to them — what she sends Vivales — must be enough to convince Vivales (with an exponentially decreasing chance that he's being fooled) that she did what she claims to have done.

The "non-interaction" means that in the course of the proof Vivales does not communicate to Pícara. However, it is permitted that at the very beginning Pícara has been given a long sequence of oblivious transfer public keys $(\beta_1, \beta_2)$ for Vivales, as described above. This is not counted as a communication from Vivales to Pícara. In fact, the same public keys are available, as the word "public" suggests, for anyone to use who's playing the role of Pícara. And Pícara can use the same sequence of public keys in many different zero-knowledge proofs she sends to Vivales.

We now describe the procedure that Pícara uses to convince Vivales that she can factor an integer $n = pq$ without giving him any information about what its factors might be. We will use the fact that the ability to take the square root modulo $n = pq$ of an arbitrary number that has a square root is equivalent to knowledge of $p$ and $q$ (see Exercise 5 below). The procedure is as follows:

1. The Center randomly generates an integer $x$, and sends Pícara and Vivales the least nonnegative residue of $x^2$ modulo $n$; let us denote $y = x^2 \bmod n$.

2. Pícara finds the four square roots of $y$ modulo $n$, namely, $\pm x$, $\pm x'$. She arbitrarily chooses $x_0$ to be one of these four square roots.
3. Pícara randomly picks an integer $r$ and sends Vivales the integer $s = r^2 \bmod n$. She sets $m_1 = r\ mod\ n$, $m_2 = x_0 r\ mod\ n$, and sends these two messages to Vivales by oblivious transfer.
4. Vivales is able to read exactly one of the two messages. He checks that its square modulo $n$ is $s$ (if his random $i$ is 1) or $ys$ (if $i = 2$).
5. Steps 1–4 are repeated (with different public keys $(\beta_1, \beta_2)$). If Pícara meets the test $T$ times, then Vivales is satisfied (with certainty $1-2^{-T}$) that Pícara really knows the factorization.

*Exercises*

1. In the zero-knowledge proof of possession of a discrete logarithm, if Pícara does not really know the discrete log, then what are the odds against her successfully fooling Vivales for $T$ repetitions of steps (1)-(3)?
2. In the zero-knowledge proof of possession of a discrete logarithm, suppose that Vivales does not know the value of $N$.
(a) Explain how the protocol described in the text is not really "zero knowledge."
(b) How could Pícara decrease the amount of information Vivales obtains about the magnitude of $N$?
3. Suppose that Pícara does not know $N$, and so in step (1) she chooses a random $e$ in some other range (e.g., $e < B$, where $B$ is an upper bound for the possible value of $N$), and in step (3) she sends simply $x + e$ rather than the least positive residue of $x + e$ modulo $N$. Explain why this is not a zero-knowledge proof. Why is the procedure followed by Clyde not a valid simulation?
4. Explain how the zero-knowledge proof in the text for possession of a discrete logarithm can be used for public key electronic identification. (This means that Pícara convinces Vivales that she really is Pícara.)
5. Explain why being able to extract square roots modulo $n = pq$ is essentially equivalent to knowing the factorization of $n$.
6. Can the same public key $(\beta_1, \beta_2)$ for oblivious transfer be used by several different people to give Vivales zero-knowledge proofs that they all independently know the same factorization? Assume that each person can eavesdrop on the transmissions of the others.
7. Using oblivious transfer, construct a non-interactive zero-knowledge proof for possession of a discrete logarithm. (Suppose that the order $N$ of the group is known to everyone.)
8. The following scheme was recently proposed as a zero-knowledge protocol for Pícara to use in order to demonstrate to Vivales that she knows the factors $p$ and $q$ of an integer $n$, where $n$ is known to be a product of two primes that are $\equiv 3 \pmod 4$. Find a basic flaw in the scheme.

**Step 1.** Vivales, who knows $n$, but not $p$ and $q$, chooses an integer $x$ at random. He computes the least nonnegative residue of $x^4$ modulo $n$, and sends this number — which we denote $y$ — to Pícara.
**Step 2.** When Pícara receives $y$, she computes a square root modulo $n$ (which is easy, since she knows the factorization of $n$; see Exercise 5 above). Of the four possible square roots, she chooses the unique one which is a quadratic residue modulo both $p$ and $q$. This must be the least positive residue of $x^2$ modulo $n$. She sends this integer to Vivales.
**Step 3.** Vivales checks that the number he received from Pícara is in fact the residue of $x^2$ modulo $n$. He is then convinced that she can take square roots modulo $n$, something that would have been impossible if she didn't know the factors of $n$.

9. Find the drawback of the following procedure for a zero-knowledge proof of factorization. Suppose that $n$ is the product of two primes $p$ and $q$. Suppose that a "trusted Center" supplies an unending sequence of random squares modulo $n$, as in the text: $y_1, y_2, \ldots$. For each of the successive $y_i$, Pícara finds one of its square roots $x_i$, and sends it to Vivales, who verifies that $x_i^2 \equiv y \ (mod\ n)$.

## References for §IV.5


1. M. Bellare and S. Micali, "Non-interactive oblivious transfer and applications," *Advances in Cryptology – Crypto '89*, Springer-Verlag, 547–557.
2. M. Ben-Or, O. Goldreich, S. Goldwasser, J. Håstad, J. Kilian, S. Micali, and P. Rogaway, "Everything provable is provable in zero-knowledge," *Advances in Cryptology — Crypto '88*, Springer-Verlag, 1990, 37–56.
3. M. Blum, P. Feldman, and S. Micali, "Non-interactive zero-knowledge proofs and their applications," *Proc. 20th ACM Symposium on the Theory of Computing* (1988).
4. D. Chaum, J.-H. Evertse, J. van de Graaf, and R. Peralta, "Demonstrating possession of a discrete logarithm without revealing it," *Advances in Cryptology – Crypto '86*, Springer–Verlag, 1987, 200–212.
5. M. R. Garey and D. S. Johnson, *Computers and Intractability: A Guide to the Theory of NP-Completeness,* W. H. Freeman, 1979.
6. S. Goldwasser, S. Micali, and C. Rackoff, "The knowledge complexity of interactive proof systems," *SIAM J. Computing* **18** (1989), 186–208.
7. J. Kilian, "Founding cryptography on oblivious transfer," *Proc. 20th ACM Symposium on the Theory of Computing* (1988), 20–31.
8. M. Rabin, "How to exchange secrets by oblivious transfer," *Technical Report TR-81*, Aiken Computation Laboratory, Harvard University, 1981.
9. A. Shamir, "The search for provably secure identification schemes," *Proc. Intern. Cong. Math.* (1986), 1488–1495.

# V
# Primality and Factoring

There are many situations where one wants to know if a large number $n$ is prime. For example, in the RSA public key cryptosystem and in various cryptosystems based on the discrete log problem in finite fields, we need to find a large "random" prime. One interpretation of what this means is to choose a large odd integer $n_0$ using a generator of random digits and then test $n_0$, $n_0 + 2$, ... for primality until we obtain the first prime which is $\geq n_0$. A second type of use of primality testing is to determine whether an integer of a certain very special type is a prime. For example, for some large prime $f$ we might want to know whether $2^f - 1$ is a Mersenne prime. If we're working in the field of $2^f$ elements, we saw that every element $\neq 0, 1$ is a generator of $\mathbf{F}^*_{2^f}$ if (and only if) $2^f - 1$ is prime (see Ex.13(a) of § II.1).

A *primality test* is a criterion for a number $n$ *not* to be prime. If $n$ "passes" a primality test, then it *may* be prime. If it passes a whole lot of primality tests, then it is very likely to be prime. On the other hand, if $n$ fails any single primality test, then it is definitely composite. But that leaves us with a very difficult problem: finding the prime factors of $n$. In general, it is much more time-consuming to factor a large number once it is known to be composite (because it fails a primality test) than it is to find a prime number of the same order of magnitude. (This is an empirical statement, not a theorem; no assertion of this sort has been proved.) The security of the RSA cryptosystem is based on the assumption that it is much easier for someone to find two extremely large primes $p$ and $q$ than it is for someone else, knowing $n = pq$ but not $p$ or $q$, to find the two factors in $n$. After discussing primality tests in §1, we shall describe three different factorization methods in §§2–5.

# 1 Pseudoprimes

Have you ever noticed that there's no attempt being made to find really large numbers that *aren't* prime? I mean, wouldn't you like to see a news report that says "Today the Department of Computer Sciences at the University of Washington announced that $2^{58,111,625,031}+8$ is even. This is the largest non-prime yet reported."

— bathroom graffiti, University of Washington

Un phénomène dont la probabilité est $10^{-50}$ ne se produira donc jamais, ou du moins ne sera jamais observé.

— Émile Borel, *Les Probabilités et la vie*

Let $n$ be a large odd integer, and suppose that you want to determine whether or not $n$ is prime. The simplest primality test is "trial division." This means that you take an odd integer $m$ and see whether or not it divides $n$. If $m \neq 1$, $n$ and $m|n$, then $n$ is composite; otherwise, $n$ passes the primality test "trial division by $m$." As $m$ runs through the odd numbers starting with 3, if $n$ passes all of the trial division tests, then it becomes more and more likely that $n$ is prime. We know for sure that $n$ is prime when $m$ reaches $\sqrt{n}$. Of course, this is an extremely time-consuming way to test whether or not $n$ is prime. The other tests described in this section are much quicker.

Most of the efficient primality tests that are known are similar in general form to the following one.

According to Fermat's Little Theorem, we know that, if $n$ is prime, then for any $b$ such that $g.c.d.(b,n)=1$ one has

$$b^{n-1} \equiv 1 \ mod\ n. \tag{1}$$

If $n$ is *not* prime, it is still possible (but probably not very likely) that (1) holds.

**Definition.** If $n$ is an odd composite number and $b$ is an integer such that $g.c.d.(n,b)=1$ and (1) holds, then $n$ is called a *pseudoprime to the base b*.

In other words, a "pseudoprime" is a number $n$ that "pretends" to be prime by passing the test (1).

**Example 1.** The number $n=91$ is a pseudoprime to the base $b=3$, because $3^{90}\equiv 1\ mod\ 91$. However, 91 is *not* a pseudoprime to the base 2, because $2^{90}\equiv 64\ mod\ 91$. If we hadn't already known that 91 is composite, the fact that $2^{90}\not\equiv 1\ mod\ 91$ would tell us that it is.

**Proposition V.1.1.** *Let $n$ be an odd composite integer.*

(a) *$n$ is a pseudoprime to the base $b$, where $g.c.d.(b,n)=1$, if and only if the order of $b$ in $(\mathbf{Z}/n\mathbf{Z})^*$ (i.e, the least positive power of $b$ which is $\equiv 1\ mod\ n$) divides $n-1$.*

(b) *If $n$ is a pseudoprime to the bases $b_1$ and $b_2$ (where $g.c.d.(b_1, n) = g.c.d.(b_2, n) = 1$), then $n$ is a pseudoprime to the base $b_1 b_2$ and also to the base $b_1 b_2^{-1}$ (where $b_2^{-1}$ is an integer which is inverse to $b_2$ modulo $n$).*

(c) *If $n$ fails the test (1) for a single base $b \in (\mathbf{Z}/n\mathbf{Z})^*$, then $n$ fails (1) for at least half of the possible bases $b \in (\mathbf{Z}/n\mathbf{Z})^*$.*

**Proof.** Parts (a) and (b) are very easy, and will be left to the reader. To prove (c), let $\{b_1, b_2, \ldots, b_s\}$ be the set of all bases for which $n$ is a pseudoprime, i.e., the set of all integers $0 < b_i < n$ for which the congruence (1) holds. Let $b$ be a fixed base for which $n$ is not a pseudoprime. If $n$ were a pseudoprime for any of the bases $bb_i$, then, by part (b), it would be a pseudoprime for the base $b \equiv (bb_i)b_i^{-1}$ *mod* $n$, which is not the case. Thus, for the $s$ distinct residues $\{bb_1, bb_2, \ldots, bb_s\}$ the integer $n$ fails the test (1). Hence, there are at least as many bases in $(\mathbf{Z}/n\mathbf{Z})^*$ for which $n$ fails to be a pseudoprime as there are bases for which (1) holds. This completes the proof.

Thus, unless $n$ happens to pass the test (1) for *all* possible $b$ with $g.c.d.(b, n) = 1$, we have at least a 50% chance that $n$ will fail (1) for a randomly chosen $b$. That is, suppose we want to know if a large odd integer $n$ is prime. We might choose a random $b$ in the range $0 < b < n$. We first find $d = g.c.d.(b, n)$ using the Euclidean algorithm. If $d > 1$, we know that $n$ is not prime, and in fact we have found a nontrivial factor $d|n$. If $d = 1$, then we raise $b$ to the $(n-1)$-st power (using the repeated squaring method of modular exponentiation, see § I.3). If (1) fails, we know that $n$ is composite. If (1) holds, we have some evidence that perhaps $n$ is prime. We then try another $b$ and go through the same process. If (1) fails for any $b$, then we can stop, secure in the knowledge that $n$ is composite. Suppose that we try $k$ different $b$'s and find that $n$ is a pseudoprime for all of the $k$ bases. By Proposition V.1.1, the chance that $n$ is still composite despite passing the $k$ tests is at most 1 out of $2^k$, *unless* $n$ happens to have the very special property that (1) holds for every single $b \in (\mathbf{Z}/n\mathbf{Z})^*$. If $k$ is large, we can be sure "with a high probability" that $n$ is prime (unless $n$ has the property of being a pseudoprime for all bases). This method of finding prime numbers is called a *probabilistic* method. It differs from a *deterministic* method: the word "deterministic" means that the method will either reveal $n$ to be composite or else determine with 100% certainty that $n$ is prime.

Can it ever happen for a composite $n$ that (1) holds for every $b$? In that case our probabilistic method fails to reveal the fact that $n$ is composite (unless we are lucky and hit upon a $b$ with $g.c.d.(b, n) > 1$). The answer is yes, and such a number is called a *Carmichael number.*

**Definition.** A *Carmichael number* is a composite integer $n$ such that (1) holds for every $b \in (\mathbf{Z}/n\mathbf{Z})^*$.

**Proposition V.1.2.** *Let $n$ be an odd composite integer.*

(a) *If $n$ is divisible by a perfect square $> 1$, then $n$ is not a Carmichael number.*

(b) *If $n$ is square free, then $n$ is a Carmichael number if and only if $p-1|n-1$ for every prime $p$ dividing $n$.*

**Proof.** (a) Suppose that $p^2|n$. Let $g$ be a generator modulo $p^2$, i.e., an integer such that $g^{p(p-1)}$ is the lowest power of $g$ which is $\equiv 1 \ mod \ p^2$. According to Exercise 2 of § II.1, such a $g$ always exists. Let $n'$ be the product of all primes other than $p$ which divide $n$. By the Chinese Remainder Theorem, there is an integer $b$ satisfying the two congruences: $b \equiv g \ mod \ p^2$ and $b \equiv 1 \ mod \ n'$. Then $b$ is, like $g$, a generator modulo $p^2$, and it also satisfies $g.c.d.(b,n) = 1$, since it is not divisible by $p$ or by any prime which divides $n'$. We claim that $n$ is not a pseudoprime to the base $b$. To see this, we notice that if (1) holds, then, since $p^2|n$, we automatically have $b^{n-1} \equiv 1 \ mod \ p^2$. But in that case $p(p-1)|n-1$, since $p(p-1)$ is the order of $b$ modulo $p^2$. However, $n-1 \equiv -1 \ mod \ p$, since $p|n$, and this means that $n-1$ is not divisible by $p(p-1)$. This contradiction proves that there is a base $b$ for which $n$ fails to be a pseudoprime.

(b) First suppose that $p-1|n-1$ for every $p$ dividing $n$. Let $b$ be any base, where $g.c.d.(b,n) = 1$. Then for every prime $p$ dividing $n$ we have: $b^{n-1}$ is a power of $b^{p-1}$, and so is $\equiv 1 \ mod \ p$. Thus, $b^{n-1}-1$ is divisible by all of the prime factors $p$ of $n$, and hence by their product, which is $n$. Hence, (1) holds for all bases $b$. Conversely, suppose that there is a $p$ such that $p-1$ does not divide $n-1$. Let $g$ be an integer which generates $(\mathbf{Z}/p\mathbf{Z})^*$. As in the proof of part (a), find an integer $b$ which satisfies: $b \equiv g \ mod \ p$ and $b \equiv 1 \ mod \ n/p$. Then $g.c.d.(b,n) = 1$, and $b^{n-1} \equiv g^{n-1} \ mod \ p$. But $g^{n-1}$ is not $\equiv 1 \ mod \ p$, because $n-1$ is not divisible by the order $p-1$ of $g$. Hence, $b^{n-1} \not\equiv 1 \ mod \ p$, and so (1) cannot hold. This completes the proof of the proposition.

**Example 2.** $n = 561 = 3 \cdot 11 \cdot 17$ is a Carmichael number, since 560 is divisible by $3-1$, $11-1$ and $17-1$. In the exercises we shall see that this is the smallest Carmichael number.

**Proposition V.1.3.** *A Carmichael number must be the product of at least three distinct primes.*

**Proof.** By Proposition V.1.2, we know that a Carmichael number must be a product of distinct primes. So it remains to rule out the possibility that $n = pq$ is the product of two distinct primes. Suppose that $p < q$. Then, if $n$ were a Carmichael number, we would have $n-1 \equiv 0 \ mod \ q-1$, by part (b) of Proposition V.1.2. But $n-1 = p(q-1+1)-1 \equiv p-1 \ mod \ q-1$, and this is not $\equiv 0 \ mod \ q-1$, since $0 < p-1 < q-1$. This concludes the proof.

**Remark.** It was only very recently that it was proved (by Alford, Granville, and Pomerance) that there exist infinitely many Carmichael numbers. See Granville's report in *Notices of the Amer. Math. Soc.* **39** (1992), 696–700.

**Euler pseudoprimes.** Let $n$ be an odd integer, and let $(\frac{b}{n})$ denote the Jacobi symbol (see § II.2). According to Proposition II.2.2, if $n$ is a prime number, then

$$b^{(n-1)/2} \equiv \left(\frac{b}{n}\right) \bmod n \tag{2}$$

for any integer $b$. On the other hand, if $n$ is composite, then Exercise 21 of § II.2 shows that at least 50% of all $b \in (\mathbf{Z}/n\mathbf{Z})^*$ fail to satisfy (2). From these two facts we can obtain an efficient probabilistic test for whether or not a large odd integer $n$ is prime. We start with the following definition.

**Definition.** If $n$ is an odd composite number and $b$ is an integer such that $g.c.d.(n, b) = 1$ and (2) holds, then $n$ is called an *Euler pseudoprime to the base b*.

**Proposition V.1.4.** *If n is an Euler pseudoprime to the base b, then it is a pseudoprime to the base b.*

**Proof.** We must show that, if (2) holds, then (1) holds. But this is obvious by squaring both sides of the congruence (2).

**Example 3.** The converse of Proposition V.1.4 is false. For example, in Example 1 we saw that 91 is a pseudoprime to the base 3. However, $3^{45} \equiv 27 \bmod 91$, so (2) does not hold for $n = 91$, $b = 3$. (Note that it is easy to raise $b$ to a large power modulo 91 if we know the order of $b$ in $(\mathbf{Z}/91\mathbf{Z})^*$; since $3^6 \equiv 1 \bmod 91$, we immediately see that $3^{45} \equiv 3^3 \bmod 91$.) An example of a base to which 91 is an Euler pseudoprime is 10, since $10^{45} \equiv 10^3 \equiv -1 \bmod 91$, and $\left(\frac{10}{91}\right) = -1$.

**Example 4.** It is easy to see that any odd composite $n$ is an Euler pseudoprime to the base $\pm 1$; in what follows we shall rule out these two "trivial" bases $b$.

We can now describe the **Solovay–Strassen primality test**. Suppose that $n$ is a positive odd integer, and we would like to know whether $n$ is prime or composite. Choose $k$ integers $0 < b < n$ at random. For each $b$, first compute both sides of (2). Finding the left side $b^{(n-1)/2}$ takes $O(log^3 n)$ bit operations, using the repeated squaring method (Proposition I.3.6); finding the Jacobi symbol on the right also takes $O(log^3 n)$ bit operations (see Exercise 17 of § II.2). If the two sides are not congruent modulo $n$, then you know that $n$ is composite, and the test stops. Otherwise, move on to the next $b$. If (2) holds for all $k$ random choices of $b$, then the probability that $n$ is composite despite passing all of the tests is at most $1/2^k$. Thus, the Solovay–Strassen test is a probabilistic algorithm which leads either to the conclusion that $n$ is composite or to the conclusion that it is "probably" prime.

Notice that there are no Euler pseudoprime analogs of Carmichael numbers: for *any* composite $n$, the test (2) fails for at least half of the possible bases $b$.

**Strong pseudoprimes.** We now discuss one more type of primality test, which is in one respect even better than the Solovay–Strassen test based on the definition of an Euler pseudoprime. This is the Miller–Rabin test, which is based on the notion of a "strong pseudoprime," which will be defined below. Suppose that $n$ is a large positive odd integer, and $b \in (\mathbf{Z}/n\mathbf{Z})^*$. Suppose that $n$ is a pseudoprime to the base $b$, i.e., $b^{n-1} \equiv 1 \bmod n$.

The idea behind the strong pseudoprime criterion is that, if we successively "extract square roots" of this congruence, i.e., if we raise $b$ to the $((n-1)/2)$-th, $((n-1)/4)$-th, ..., $((n-1)/2^s)$-th powers (where $t = (n-1)/2^s$ is odd), then the first residue class we get other than 1 must be $-1$ if $n$ is prime, because $\pm 1$ are the only square roots of 1 modulo a prime number. Actually, in practice one proceeds in the other direction, setting $n-1 = 2^s t$ with $t$ odd, then computing $b^t$ *mod* $n$, then (if that is not $\equiv 1$ *mod* $n$) squaring to get $b^{2t}$ *mod* $n$, then squaring again to get $b^{2^2 t}$ *mod* $n$, etc., until we first obtain the residue 1; then the step before getting 1 we must have had $-1$, or else we know that $n$ is composite.

**Definition.** Let $n$ be an odd composite number, and write $n-1 = 2^s t$ with $t$ odd. Let $b \in (\mathbf{Z}/n\mathbf{Z})^*$. If $n$ and $b$ satisfy the condition

$$\text{either } b^t \equiv 1 \; mod\; n \quad \text{or} \quad \text{there exists } r,\ 0 \le r < s, \text{ such that } b^{2^r t} \equiv -1 \; mod\; n, \tag{3}$$

then $n$ is called a *strong pseudoprime to the base b*.

**Proposition V.1.5.** *If $n \equiv 3$ mod 4, then n is a strong pseudoprime to the base b if and only if it is an Euler pseudoprime to the base b.*

**Proof.** Since in this case $s = 1$ and $t = (n-1)/2$, we see that $n$ is a strong pseudoprime to the base $b$ if and only if $b^{(n-1)/2} \equiv \pm 1$ *mod* $n$. If $n$ is an Euler pseudoprime, then this congruence holds, by definition. Conversely, suppose that $b^{(n-1)/2} \equiv \pm 1$. We must show that the $\pm 1$ on the right is $(\frac{b}{n})$. But for $n \equiv 3$ *mod* 4 we have $\pm 1 = (\frac{\pm 1}{n})$, and so

$$\left(\frac{b}{n}\right) = \left(\frac{b \cdot (b^2)^{(n-3)/4}}{n}\right) = \left(\frac{b^{(n-1)/2}}{n}\right) \equiv b^{(n-1)/2} \; mod\; n,$$

as required. The next two important propositions are somewhat harder to prove.

**Proposition V.1.6.** *If n is a strong pseudoprime to the base b, then it is an Euler pseudoprime to the base b.*

**Proposition V.1.7.** *If n is an odd composite integer, then n is a strong pseudoprime to the base b for at most 25% of all $0 < b < n$.*

**Remark.** The converse of Proposition V.1.6 is not true, in general, as we shall see in the exercises below.

Before proving these two propositions, we describe the **Miller–Rabin primality test**. Suppose we want to determine whether a large positive odd integer $n$ is prime or composite. We write $n-1 = 2^s t$ with $t$ odd, and choose a random integer $b$, $0 < b < n$. First we compute $b^t$ *mod* $n$. If we get $\pm 1$, we conclude that $n$ passes the test (3) for our particular $b$, and we go on to another random choice of $b$. Otherwise, we square $b^t$ modulo $n$, then square that modulo $n$, and so on, until we get $-1$. If we get $-1$, then $n$ passes the test. However, if we never obtain $-1$, i.e., if we reach $b^{2^{r+1}} \equiv 1$ *mod* $n$ while $b^{2^r} \not\equiv -1$ *mod* $n$, then $n$ fails the test and we know that $n$ is composite. If $n$ passes the test (3) for all our random choices of $b$ — suppose we try $k$ different bases $b$ — then we know by Proposition V.1.7 that $n$ has at most a

1 out of $4^k$ chance of being composite. This is because, if $n$ is composite, then at most $1/4$ of the bases $0 < b < n$ satisfy (3). Notice that this is somewhat better than for the Solovay–Strassen test, where the analogous estimate is a 1 out of $2^k$ chance (because there exist composite $n$ which are Euler pseudoprimes for half of all bases $0 < b < n$, as we shall see in the exercises).

We now proceed to the proofs of Propositions V.1.6 and V.1.7.

**Proof of Proposition V.1.6.** We have $n$ and $b$ satisfying (3). We must prove that they satisfy (2). Let $n - 1 = 2^s t$ with $t$ odd.

Case (i). First suppose that $b^t \equiv 1 \; mod \; n$. Then the left side of (2) is clearly 1. We must show that $(\frac{b}{n}) = 1$. But $1 = (\frac{1}{n}) = (\frac{b^t}{n}) = (\frac{b}{n})^t$. Since $t$ is odd, this means that $(\frac{b}{n}) = 1$.

Case (ii). Next suppose that $b^{(n-1)/2} \equiv -1 \; mod \; n$. Then we must show that $(\frac{b}{n}) = -1$. Let $p$ be any of the prime divisors of $n$. We write $p - 1$ in the form $p - 1 = 2^{s'} t'$ with $t'$ odd, and we prove the following claim:

**Claim.** *We have $s' \geq s$, and*

$$\left(\frac{b}{p}\right) = \begin{cases} -1, & \text{if } s' = s; \\ 1, & \text{if } s' > s. \end{cases}$$

**Proof of the claim.** Because $b^{(n-1)/2} = b^{2^{s-1}t} \equiv -1 \; mod \; n$, raising both sides to the $t'$ power gives $(b^{2^{s-1}t'})^t \equiv -1 \; mod \; n$. Since $p|n$, the same congruence holds modulo $p$. But if we had $s' < s$, this would mean that $b^{2^{s'}t'}$ could not be $\equiv 1 \; mod \; p$, as it must be by Fermat's Little Theorem. Thus, $s' \geq s$. If $s' = s$, then the congruence $(b^{2^{s-1}t'})^t \equiv -1 \; mod \; p$ implies that $(\frac{b}{p}) \equiv b^{(p-1)/2} = b^{2^{s'-1}t'} \; mod \; p$ must be $-1$ rather than 1. On the other hand, if $s' > s$, then the same congruence raised to the $(2^{s'-s})$-th power implies that $(\frac{b}{p})$ must be 1 rather than $-1$. This proves the claim.

We now return to the proof of Proposition V.1.6 in Case (ii). We write $n$ as a product of primes (*not* necessarily distinct): $n = \prod p$. Let $k$ denote the number of primes $p$ such that $s' = s$ when one writes $p - 1 = 2^{s'} t'$ with $t'$ odd. ($k$ counts such a prime $p$ with its multiplicity, i.e., $\alpha$ times if $p^\alpha || n$.) According to the claim, we always have $s' \geq s$, and $(\frac{b}{n}) = \prod (\frac{b}{p}) = (-1)^k$. On the other hand, working modulo $2^{s+1}$, we see that $p \equiv 1$ unless $p$ is one of the $k$ primes for which $s' = s$, in which case $p \equiv 1 + 2^s$. Since $n = 1 + 2^s t \equiv 1 + 2^s \; mod \; 2^{s+1}$, we have $1 + 2^s \equiv \prod p \equiv (1 + 2^s)^k \equiv 1 + k 2^s \; mod \; 2^{s+1}$ (where the last step follows by the binomial expansion). This means that $k$ must be odd, and hence $(\frac{b}{n}) = (-1)^k = -1$, as was to be proved.

Case (iii). Finally, suppose that $b^{2^{r-1}t} \equiv -1 \; mod \; n$ for some $0 < r < s$. (We are using $r - 1$ in place of the $r$ in (3).) Since then $b^{(n-1)/2} \equiv 1 \; mod \; n$, we must show that in Case (iii) we have $(\frac{b}{n}) = 1$. Again let $p$ be any prime divisor of $n$, and write $p - 1 = 2^{s'} t'$ with $t'$ odd.

**Claim.** *We have $s' \geq r$, and*

$$\left(\frac{b}{p}\right) = \begin{cases} -1, & \text{if } s' = r; \\ 1, & \text{if } s' > r. \end{cases}$$

The proof of this claim is identical to the proof of the claim in Case (ii).

To prove the proposition in Case (iii), we let $k$ denote the number of primes $p$ (not necessarily distinct) in the product $n = \prod p$ for which the first alternative holds, i.e., $s' = r$. Then, as in Case (ii), we obviously have $(\frac{b}{n}) = (-1)^k$. On the other hand, since $n = 1 + 2^s t \equiv 1 \; mod \; 2^{r+1}$ and also $n = \prod p \equiv (1+2^r)^k \; mod \; 2^{r+1}$, it follows that $k$ must be even, i.e., $(\frac{b}{n}) = 1$. This concludes the proof of Proposition V.1.6.

Before proving Proposition V.1.7, we prove a general lemma about the number of solutions to the equation $x^k = 1$ in a "cyclic group" containing $m$ elements. We already encountered this lemma once at the beginning of § II.2; the proof of the lemma should be compared to the proof of Proposition II.2.1.

**Lemma 1.** *Let $d = g.c.d.(k,m)$. Then there are exactly $d$ elements in the group $\{g, g^2, g^3, \ldots, g^m = 1\}$ which satisfy $x^k = 1$.*

**Proof.** An element $g^j$ satisfies the equation if and only if $g^{jk} = 1$, i.e., if and only if $m|jk$. This is equivalent to: $\frac{m}{d}|j\frac{k}{d}$, which, since $m/d$ and $k/d$ are relatively prime, is equivalent to: $j$ is a multiple of $m/d$. There are $d$ such values of $j$, $1 \leq j \leq m$. This proves the lemma.

We need one more lemma, which has a proof similar to that of Lemma 1.

**Lemma 2.** *Let $p$ be an odd prime, and write $p - 1 = 2^{s'}t'$ with $t'$ odd. Then the number of $x \in (\mathbf{Z}/p\mathbf{Z})^*$ which satisfy $x^{2^r t} \equiv -1 \; mod \; p$ (where $t$ is odd) is equal to 0 if $r \geq s'$ and is equal to $2^r g.c.d.(t,t')$ if $r < s'$.*

**Proof.** We let $g$ be a generator of $(\mathbf{Z}/p\mathbf{Z})^*$, and we write $x$ in the form $g^j$ with $0 \leq j < p-1$. Since $g^{(p-1)/2} \equiv -1 \; mod \; p$ and $p - 1 = 2^{s'}t'$, the congruence in the lemma is equivalent to: $2^r tj \equiv 2^{s'-1}t' \; mod \; 2^{s'}t'$ (with $j$ the unknown). Clearly there is no solution if $r > s' - 1$. Otherwise, we divide out by the g.c.d. of the modulus and the coefficient of the unknown, which is $2^r d$, where $d = g.c.d.(t,t')$. The resulting congruence has a unique solution modulo $2^{s'-r}\frac{t'}{d}$, and it has $2^r d$ solutions modulo $2^{s'}t'$, as claimed. This proves Lemma 2.

**Proof of Proposition V.1.7.** Case (i). We first suppose that $n$ is divisible by the square of some prime $p$. Say $p^\alpha || n$, $\alpha \geq 2$. We show that in this case $n$ cannot even be a pseudoprime (let alone a strong pseudoprime) for more than $(n-1)/4$ bases $b$, $0 < b < n$. To do this, we suppose that $b^{n-1} \equiv 1 \; mod \; n$, which implies that $b^{n-1} \equiv 1 \; mod \; p^2$, and we find a condition modulo $p^2$ that $b$ must satisfy. Recall that $(\mathbf{Z}/p^2\mathbf{Z})^*$ is a cyclic group of order $p(p-1)$ (see Exercise 2 of § II.1), i.e., there exists an integer $g$ such that $(\mathbf{Z}/p^2\mathbf{Z})^* = \{g, g^2, g^3, \ldots, g^{p(p-1)}\}$. According to Lemma 1, the number of possibilities for $b$ modulo $p^2$ for which $b^{n-1} \equiv 1 \; mod \; p^2$ is $d = g.c.d.(p(p-1), n-1)$. Since $p|n$, it follows that $p \not| n-1$, and hence $p \not| d$. Thus, the largest $d$ can be is $p-1$. Hence, the proportion of all $b$ not divisible by $p^2$ in the range from 0 to $n$ which satisfy $b^{n-1} \equiv 1 \; mod \; p^2$ is less than or equal to

$$\frac{p-1}{p^2-1} = \frac{1}{p+1} \le \frac{1}{4}.$$

Since the proportion of $b$ in the range from 0 to $n$ which satisfy $b^{n-1} \equiv 1 \ mod \ n$ is less than or equal to this, we conclude that $n$ is a pseudoprime to the base $b$ for at most $1/4$ of the $b$, $0 < b < n$. This proves the proposition in Case (i). (**Remark:** This upper bound of 25% is actually reached in Case (i) in the case when $n = 9$, i.e., 9 is a (strong) pseudoprime for 2 out of the 8 possible values of $b$, namely, $b = \pm 1$.)

Case (ii). We next suppose that $n$ is the product of 2 distinct primes $p$ and $q$: $n = pq$. We write $p - 1 = 2^{s'}t'$ with $t'$ odd and $q - 1 = 2^{s''}t''$ with $t''$ odd. Without loss of generality we may suppose that $s' \le s''$. In order for an element $b \in (\mathbf{Z}/n\mathbf{Z})^*$ to be a base to which $n$ is a strong pseudoprime, one of the following must occur: (1) $b^t \equiv 1 \ mod \ p$ and $b^t \equiv 1 \ mod \ q$, or (2) $b^{2^r t} \equiv -1 \ mod \ p$ and $b^{2^r t} \equiv -1 \ mod \ q$ for some $r$, $0 \le r < s$. According to Lemma 1, the number of $b$ for which the first possibility holds is the product of $g.c.d.(t, t')$ (the number of residue classes modulo $p$) times $g.c.d.(t, t'')$ (the number of residue classes modulo $q$), which is certainly no greater than $t't''$. According to Lemma 2, for each $r < \min(s', s'') = s'$ the number of $b$ for which $b^{2^r t} \equiv -1 \ mod \ n$ is $2^r g.c.d.(t, t') \cdot 2^r g.c.d.(t, t'') < 4^r t't''$. Since we have $n - 1 > \varphi(n) = 2^{s'+s''}t't''$, it follows that the fraction of integers $b$, $0 < b < n$, for which $n$ is a strong pseudoprime is at most

$$\frac{t't'' + t't'' + 4t't'' + 4^2t't'' + \cdots + 4^{s'-1}t't''}{2^{s'+s''}t't''} = 2^{-s'-s''}\left(1 + \frac{4^{s'} - 1}{4 - 1}\right).$$

If $s'' > s'$, then this is at most $2^{-2s'-1}(\frac{2}{3} + \frac{4^{s'}}{3}) \le 2^{-3}\frac{2}{3} + \frac{1}{6} = \frac{1}{4}$, as desired. On the other hand, if $s' = s''$, then we note that one of the two inequalities $g.c.d.(t, t') \le t'$, $g.c.d.(t, t'') \le t''$ must be a strict inequality, since if we had $t'|t$ and $t''|t$, we could conclude from the congruence $n - 1 = 2^s t = pq - 1 \equiv q - 1 \ mod \ t'$ that $t'|q - 1 = 2^{s''}t''$, i.e., $t'|t''$, and similarly $t''|t'$; but this would mean that $t' = t''$ and $p = q$, a contradiction. Hence one of the two g.c.d.'s is strictly less than $t'$ or $t''$, and so must be less at least by a factor of 3 (since we're working with odd numbers). Thus, in this case we may replace $t't''$ by $\frac{1}{3}t't''$ in the above estimates for the number of $b$ satisfying each condition for $n$ to be a strong pseudoprime to the base $b$. This leads to the following upper bound for the fraction of integers $b$, $0 < b < n$, for which $n$ is a strong pseudoprime:

$$\frac{1}{3}2^{-2s'}\left(\frac{2}{3} + \frac{4^{s'}}{3}\right) \le \frac{1}{18} + \frac{1}{9} = \frac{1}{6} < \frac{1}{4},$$

as desired. This completes the proof of the theorem in Case (ii).

Case (iii). Finally, we suppose that $n$ is a product of more than 2 distinct primes: $n = p_1 p_2 \cdots p_k$, $k \ge 3$. We write $p_j - 1 = 2^{s_j}t_j$ with $t_j$ odd, and we proceed exactly as in Case (ii). Without loss of generality, we may suppose that $s_1 \le s_j$ is the smallest of the $s_j$. We obtain the following upper bound for the fraction of possible $b$'s for which $n$ is a strong pseudoprime:

$$2^{-s_1-s_2-\cdots-s_k}\left(1+\frac{2^{ks_1}-1}{2^k-1}\right) \le 2^{-ks_1}\left(\frac{2^k-2}{2^k-1}+\frac{2^{ks_1}}{2^k-1}\right) =$$

$$= 2^{-ks_1}\frac{2^k-2}{2^k-1}+\frac{1}{2^k-1} \le 2^{-k}\frac{2^k-2}{2^k-1}+\frac{1}{2^k-1} = 2^{1-k} \le \frac{1}{4},$$

because $k \ge 3$ in Case (iii). This concludes the proof of Proposition V.1.7.

**Remarks.** **1.** In fact, in practice one does not have to choose a very large number of bases $b$ to be almost sure that $n$ is prime if it is a strong pseudoprime to each base $b$. For example, it has been computed that there is only one composite number less than $2.5\cdot 10^{10}$ — namely, $n = 3215031751$ — which is a strong pseudoprime to all four bases 2, 3, 5, 7.

**2.** It is not entirely satisfactory to rely upon a probabilistic test. Despite Émile Borel's assurance, quoted at the beginning of the section, it would be nice to have rapid methods to *prove* that a given $n$ really is prime (especially, if it is of some special practical or theoretical importance to know that the particular $n$ is prime). For example, suppose we knew that there is some fairly small $B$ (depending on the size of $n$) such that, if $n$ is composite, then there is some base $b < B$ for which $n$ is not a strong pseudoprime. If we knew that, then in order to be absolutely sure that $n$ is prime it would suffice to test (3) only for the first $B$ bases.

There is such a fact, but it depends upon an unproved conjecture called the "Generalized Riemann Hypothesis." The usual Riemann Hypothesis is the assertion that all complex zeros of the so-called "Riemann zeta-function" $\zeta(s)$ (which is defined to be the sum of the reciprocal $s$-th powers when $s > 1$) which lie in the "critical strip" (where the real part of $s$ is between 0 and 1) must lie on the "critical line" (where the real part of $s$ is 1/2). The Generalized Riemann Hypothesis is the same assertion for certain generalizations of $\zeta(s)$ called "Dirichlet $L$-series." The following fact, whose proof is beyond the scope of this book, shows that the Miller–Rabin test (3) gives a *deterministic* primality test which takes polynomial time (in $\log n$), provided that one is willing to assume the validity of the Generalized Riemann Hypothesis (GRH).

*If the GRH is true, and if $n$ is a composite odd integer, then $n$ fails the test (3) for at least one base $b$ less than $2\log^2 n$.*

**3.** In the 1980's an efficient deterministic primality test was developed which, while strictly speaking not polynomial in $\log n$, in practice can routinely prove primality of numbers of over a hundred decimal digits in a matter of seconds (on current large computers). This method of Adleman–Pomerance–Rumely and Cohen–Lenstra is based on the same ideas as the primality tests considered above, except that it uses analogs of Fermat's Little Theorem in extension fields of the rational numbers. A basic role is played by Gauss sums (certain types of which were introduced in § II.2 in order to prove quadratic reciprocity) and the closely related "Jacobi sums." A detailed discussion of their method would take us too far afield. A thorough and readable account is given in the Cohen–Lenstra article in *Mathematics of Computation.*

## *Exercises*

1. (a) Find all bases $b$ for which 15 is a pseudoprime. (Do not include the trivial bases $\pm 1$.)
(b) Find all bases for which 21 is a pseudoprime.
(c) Prove that there are 36 bases $b \in (\mathbf{Z}/91\mathbf{Z})^*$ (i.e., 50% of the possible bases) for which 91 is a pseudoprime.
(d) Generalizing part (c), show that if $p$ and $2p-1$ are both prime, and $n = p(2p-1)$, then $n$ is a pseudoprime for 50% of the possible bases $b$, namely for all $b$ which are quadratic residues modulo $2p-1$.
2. Let $n$ be a positive odd composite integer, and let $g.c.d.(b,n) = 1$.
(a) Show that if $p$ is a prime divisor of $n$ and we set set $n' = n/p$, then $n$ is a pseudoprime to the base $b$ only if $b^{n'-1} \equiv 1 \; mod \; p$.
(b) Prove that no integer of the form $n = 3p$ (with $p > 3$ prime) can be a pseudoprime to the base 2, 5 or 7.
(c) Prove that no integer of the form $n = 5p$ (with $p > 5$ prime) can be a pseudoprime to the base 2, 3 or 7.
(d) Prove that 91 is the smallest pseudoprime to the base 3.
3. Show that $p^2$ (with $p$ prime) is a pseudoprime to the base $b$ if and only if $b^{p-1} \equiv 1 \; mod \; p^2$.
4. (a) Find the smallest pseudoprime to the base 5.
(b) Find the smallest pseudoprime to the base 2.
5. Let $n = pq$ be a product of two distinct primes.
(a) Set $d = g.c.d.(p-1, q-1)$. Prove that $n$ is a pseudoprime to the base $b$ if and only if $b^d \equiv 1 \; mod \; n$. In terms of $d$, how many bases are there to which $n$ is a pseudoprime?
(b) How many bases are there to which $n$ is a pseudoprime if $q = 2p+1$? List all of them (in terms of $p$).
(c) For $n = 341$, what is the probability that a randomly chosen $b$ prime to $n$ will be a base to which $n$ is a pseudoprime?
6. Show that, if $n$ is a pseudoprime to the base $b \in (\mathbf{Z}/n\mathbf{Z})^*$, then $n$ is also a pseudoprime to the base $-b$ and to the base $b^{-1}$.
7. (a) Prove that if $n$ is a pseudoprime to the base 2, then so is $N = 2^n - 1$.
(b) Prove that if $n$ is a pseudoprime to the base $b$, and if $g.c.d.(b-1, n) = 1$, then the integer $N = (b^n - 1)/(b-1)$ is a pseudoprime to the base $b$.
(c) Prove that there are infinitely many pseudoprimes to the base $b$ for $b = 2, 3, 5$.
(d) Give an example showing that part (b) may be false if we omit the condition $g.c.d.(b-1, n) = 1$.

8. Let $b$ be any integer greater than 1, let $p$ be an odd prime not dividing $b$, $b-1$ or $b+1$. Set $n = (b^{2p}-1)/(b^2-1)$.
(a) Show that $n$ is composite.
(b) Show that $2p|n-1$.
(c) Show that $n$ is a pseudoprime to the base $b$; conclude that for any base $b$ there are infinitely many pseudoprimes to the base $b$.
9. (a) Use the test (1) to show that $2047 = 2^{11}-1$ is composite.
(b) Explain why you should never test whether the Fermat number $2^{2^k}+1$ or the Mersenne number $2^p-1$ is prime by checking (1) with $b=2$. What about using the test (2) with $b=2$? What about using (3) with $b=2$?
10. Suppose that $m$ is a positive integer such that $6m+1$, $12m+1$ and $18m+1$ are all primes. Let $n = (6m+1)(12m+1)(18m+1)$. Prove that $n$ is a Carmichael number. **Note.** It is not known whether there are infinitely many Carmichael numbers of the form $n = (6m+1)(12m+1)(18m+1)$, but heuristic arguments suggest that there are.
11. Show that the following are Carmichael numbers: $1105 = 5\cdot 13\cdot 17$; $1729 = 7\cdot 13\cdot 19$; $2465 = 5\cdot 17\cdot 29$; $2821 = 7\cdot 13\cdot 31$; $6601 = 7\cdot 23\cdot 41$; $29341 = 13\cdot 37\cdot 61$; $172081 = 7\cdot 13\cdot 31\cdot 61$; $278545 = 5\cdot 17\cdot 29\cdot 113$.
12. (a) Find all Carmichael numbers of the form $3pq$ (with $p$ and $q$ prime).
(b) Find all Carmichael numbers of the form $5pq$ (with $p$ and $q$ prime).
(c) Prove that for any fixed prime number $r$, there are only finitely many Carmichael numbers of the form $rpq$ (with $p$ and $q$ prime).
13. Prove that 561 is the smallest Carmichael number.
14. Give an example of a composite number $n$ and a base $b$ such that $b^{(n-1)/2} \equiv \pm 1 \ mod\ n$ but $n$ is not an Euler pseudoprime to the base $b$.
15. (a) Prove that if $n$ is an Euler pseudoprime to the base $b \in (\mathbf{Z}/n\mathbf{Z})^*$, then it is also an Euler pseudoprime to the base $-b$ and to the base $b^{-1}$.
(b) Prove that if $n$ is an Euler pseudoprime to the base $b_1$ and to the base $b_2$, then it is also an Euler pseudoprime to the base $b = b_1 b_2$.
16. Let $n$ be of the form $p(2p-1)$, as in Exercise 1(d).
(a) Prove that $n$ is an Euler pseudoprime for 25% of all possible bases $b \in (\mathbf{Z}/n\mathbf{Z})^*$.
(b) Find a class of numbers $n$ of this type such that $n$ is a strong pseudoprime for 25% of all possible bases.
17. Let $n$ be of the form $(6m+1)(12m+1)(18m+1)$, as in Exercise 10. Prove that (a) if $m$ is odd, then $n$ is an Euler pseudoprime for 50% of all possible bases $b \in (\mathbf{Z}/n\mathbf{Z})^*$; and (b) if $m$ is even, then $n$ is an Euler pseudoprime for 25% of all possible bases.
18. (a) Using the big-$O$ notation, estimate the number of bit operations required to perform the Miller–Rabin test on a number $n$ enough times so that, if $n$ passes all the tests, it has less than a $1/m$ chance of being composite (here $n$ and $m$ are very large).

(b) Assuming the Generalized Riemann Hypothesis, estimate the number of bit operations required to perform the Miller–Rabin test on $n$ enough times to be sure that, if $n$ passes all the tests, then it is prime.

19. (a) Prove that, if $n$ is a pseudoprime to the base 2, then $N = 2^n - 1$ is a strong pseudoprime and an Euler pseudoprime to the base 2.
(b) Prove that there are infinitely many strong pseudoprimes and Euler pseudoprimes to the base 2.
20. Prove that, if $n$ is a strong pseudoprime to the base $b$, then it is a strong pseudoprime to the base $b^k$ for any integer $k$.
21. Let $n$ be the Carmichael number 561.
(a) Find the number of bases $b \in (\mathbf{Z}/561\mathbf{Z})^*$ for which 561 is an Euler pseudoprime.
(b) Find the number of bases for which 561 is a strong pseudoprime, and make a list of them.
22. Prove that if $n$ is a prime power $p^\alpha$, where $\alpha > 1$, then $n$ is a strong pseudoprime to the base $b$ if and only if it is a pseudoprime to the base $b$.
23. (a) Show that 65 is a strong pseudoprime to the base 8 and to the base 18, but not to the base 14, which is the product of 8 and 18 modulo 65.
(b) For any odd composite integer $n$, let $(*)$ denote the assertion, "Whenever $n$ is a strong pseudoprime to the base $b_1$ and to the base $b_2$ it is a strong pseudoprime to the base $b = b_1 b_2$" (in other words, the strong pseudoprime property is preserved under multiplication of bases). Prove that $(*)$ holds if and only if $n$ is a prime power or is divisible by a prime which is $\equiv 3 \ mod\ 4$.
24. (a) Prove that, if you find a $b$ such that $n$ is a pseudoprime but *not* a strong pseudoprime to the base $b$, then you can quickly find a nontrivial factor of $n$.
(b) Explain how to guard against this when choosing your $n = pq$ in the RSA cryptosystem.

**Remark.** In many primality tests, if a composite $n$ happens to pass some initial test and then fails a subsequent test, one not only learns that $n$ is composite, but at the same time one can quickly find a nontrivial factor. Exercise 24 is an example of this: if $n$ passes the pseudoprime test to the base $b$ and then fails the strong pseudoprime test to the base $b$, then you can factor $n$. One can easily be misled into thinking that in this way the primality tests can also be used for factorization. This is not the case. Given a large composite number $n$ (e.g., a product of two randomly selected large primes), it is extremely unlikely that we would stumble upon a base $b$ for which $n$ is a pseudoprime (see Exercise 5(a) above to get an idea of the probability of stumbling upon such a $b$). Thus, the various refined pseudoprime tests are useful only in convincing ourselves of the primality of a number that really is prime; in practice, if we have a composite number that we want to factor, it will fail every single primality test we apply to it, and the primality tests will not help us find a factor.

## References for § V.1

1. L. M. Adleman, C. Pomerance, and R. S. Rumely, "On distinguishing prime numbers from composite numbers," *Annals of Math.* **117** (1983), 173–206.
2. H. Cohen and H. W. Lenstra, Jr., "Primality testing and Jacobi sums," *Math. Comp.* **42** (1984), 297–330.
3. J. D. Dixon, "Factorization and primality tests," *American Math. Monthly* **91** (1984), 333–352.
4. E. Kranakis, *Primality and Cryptography*, John Wiley & Sons, 1986.
5. A. Lenstra, "Primality testing," *Cryptology and Computational Number Theory, Proc. Symp. Appl. Math.* **42** (1990), 13–25.
6. G. L. Miller, "Riemann's hypothesis and tests for primality," *Proc. 7th Annual ACM Symposium on the Theory of Computing*, 234–239.
7. C. Pomerance, "Recent developments in primality testing," *The Math . Intelligencer* **3** (1981), 97–105.
8. C. Pomerance, "The search for prime numbers," *Scientific American* **247** (1982), 136–147.
9. M. O. Rabin, "Probabilistic algorithms for testing primality," *J. Number Theory* **12** (1980), 128–138.
10. R. Solovay and V. Strassen, "A fast Monte Carlo test for primality," *SIAM J. Computing* **6** (1977), 84–85 and *erratum*, **7** (1978), 118.
11. S. Wagon, "Primality testing," *The Math. Intelligencer* **8**, No. 3 (1986), 58–61.

## 2 The rho method

Suppose we know that a certain large odd integer $n$ is composite; for example, we found that it fails one of the primality tests in §1. As mentioned before, this does not mean that we have any idea of what a factor of $n$ might be. Of the methods we have encountered for testing primality, only the very slowest — trying to divide by the successive primes less than $\sqrt{n}$ — actually gives us a prime factor at the same time as it tells us that $n$ is composite. All of the faster primality test algorithms are more indirect: they tell us that $n$ must have proper factors, but not what they are.

The method of trial division by primes $< \sqrt{n}$ can take more than $O(\sqrt{n})$ bit operations. The simplest algorithm which is substantially faster than this is J. M. Pollard's "rho method" (also called the "Monte Carlo" method) of factorization.

The first step in the rho method is to choose an easily evaluated map from $\mathbf{Z}/n\mathbf{Z}$ to itself, namely, a fairly simple polynomial with integer coefficients, such as $f(x) = x^2 + 1$. Next, one chooses some particular value $x = x_0$ (perhaps $x_0 = 1$ or 2, or perhaps it is a randomly generated integer) and computes the successive iterates of $f$: $x_1 = f(x_0)$, $x_2 = f(f(x_0))$, $x_3 = f(f(f(x_0)))$, etc. That is, we define

$$x_{j+1} = f(x_j), \qquad j = 0, 1, 2, \ldots.$$

Then we make comparisons between different $x_j$'s, hoping to find two which are in different residue classes modulo $n$ but in the same residue class modulo some divisor of $n$. Once we find such $x_j$, $x_k$, we have $g.c.d.(x_j - x_k, n)$ equal to a proper divisor of $n$, and we are done.

**Example 1.** Let us factor 91 by choosing $f(x) = x^2 + 1$, $x_0 = 1$. Then we have $x_1 = 2$, $x_2 = 5$, $x_3 = 26$, etc. We find that $g.c.d.(x_3 - x_2, n) = g.c.d.(21, 91) = 7$, so 7 is a factor. Of course, this is a trivial example: we could have found the factor 7 faster by trial division.

In the rho method it is important to choose a polynomial $f(x)$ which maps $\mathbf{Z}/n\mathbf{Z}$ to itself in a rather disjointed, "random" way. For example, we shall later see that $f(x)$ must not be a linear polynomial, and in fact, should not give a 1-to-1 map.

Let us suppose that $f(x)$ is a "random" map from $\mathbf{Z}/n\mathbf{Z}$ to itself, and compute how long we expect to have to wait before we have two iterations $x_j$ and $x_k$ such that $x_j - x_k$ has a nontrivial common factor with $n$. We do this by finding for a fixed divisor $r$ of $n$ (which, in practice, is not yet known to us) the *average* (taken over all maps from $\mathbf{Z}/n\mathbf{Z}$ to itself and over all values $x_0$) of the first index $k$ such that there exists $j < k$ with $x_j \equiv x_k \ mod\ r$. In other words, we regard $f(x)$ as a map from $\mathbf{Z}/r\mathbf{Z}$ to itself and ask how many iterations are required before we encounter the first repetition of values $x_k = x_j$ in $\mathbf{Z}/r\mathbf{Z}$.

**Proposition V.2.1.** *Let $S$ be a set of $r$ elements. Given a map $f$ from $S$ to $S$ and an element $x_0 \in S$, let $x_{j+1} = f(x_j)$ for $j = 0, 1, 2, \ldots$. Let $\lambda$ be a positive real number, and let $\ell = 1 + \left[\sqrt{2\lambda r}\,\right]$. Then the proportion of pairs $(f,\ x_0)$ for which $x_0, x_1, \ldots, x_\ell$ are distinct, where $f$ runs over all maps from $S$ to $S$ and $x_0$ runs over all elements of $S$, is less than $e^{-\lambda}$.*

**Proof.** The total number of pairs is $r^{r+1}$, because there are $r$ choices of $x_0$, and for each of the $r$ different $x \in S$ there are $r$ choices of $f(x)$. How many pairs $(f,\ x_0)$ are there for which $x_0, x_1, \ldots, x_\ell$ are distinct? There are $r$ choices for $x_0$, there are $r - 1$ choices for $f(x_0) = x_1$ (since this cannot equal $x_0$), there are $r - 2$ choices for $f(x_1) = x_2$, and so on, until $f(x)$ has been defined for $x = x_0, x_1, \ldots, x_{\ell-1}$. Then the value of $f(x)$ for each of the $r - \ell$ remaining $x$ is arbitrary, i.e., there are $r^{r-\ell}$ possibilities for those values. Hence, the total number of possible ways of choosing $x_0$ and assigning the values $f(x)$ so that $x_0, \ldots, x_\ell$ are distinct is:

$$r^{r-\ell}\prod_{j=0}^{\ell}(r-j),$$

and the proportion of pairs having the stated property (i.e., the above number divided by $r^{r+1}$) is

$$r^{-\ell-1}\prod_{j=0}^{\ell}(r-j)=\prod_{j=1}^{\ell}\left(1-\frac{j}{r}\right).$$

The proposition states that the log of this is less than $-\lambda$ (where $\ell = 1 + [\sqrt{2\lambda r}\,]$). To prove the proposition, then, we take the log of the product on the right, and use the fact that $log(1-x) < -x$ for $0 < x < 1$ (geometrically, this is simply the fact that the logarithm curve remains under the line which is tangent to it at the point $(1,0)$). Using the formula for the sum of the first $\ell$ integers, we have:

$$log\Big(\prod_{j=1}^{\ell}(1-\frac{j}{r})\Big)<\sum_{j=1}^{\ell}-\frac{j}{r}=\frac{-\ell(\ell+1)}{2r}<\frac{-\ell^2}{2r}<\frac{-(\sqrt{2\lambda r})^2}{2r}=-\lambda,$$

as required. This completes the proof of the proposition.

The significance of Proposition V.2.1 is that it gives an estimate for the probable length of time of the rho method, *provided that* we assume that our polynomial behaves like an average map from $\mathbf{Z}/r\mathbf{Z}$ to itself. Before explaining this estimate, we make a slight refinement of the rho method in the interest of efficiency.

Recall that the rho method works by successively computing $x_k = f(x_{k-1})$ and comparing $x_k$ with the earlier $x_j$ until we find a pair satisfying $g.c.d.(x_k - x_j, n) = r > 1$. But as $k$ becomes large, it becomes very time-consuming to have to compute $g.c.d.(x_k - x_j, n)$ for each $j < k$. We now describe a way to carry out the algorithm so as to make only one $g.c.d.$ computation for each $k$. First, observe that, once there is a $k_0$ and $j_0$ such that $x_{k_0} \equiv x_{j_0}\ mod\ r$ for some divisor $r|n$, we then have the same relation $x_k \equiv x_j\ mod\ r$ for any pair of indices $j$, $k$ having the same difference $k - j = k_0 - j_0$. To see this, simply set $k = k_0 + m$, $j = j_0 + m$, and apply the polynomial $f$ to both sides of the congruence $x_{k_0} \equiv x_{j_0}\ mod\ r$ repeatedly, i.e., $m$ times.

We now describe how the rho algorithm works. We successively compute the $x_k$, and for each $k$ we proceed as follows. Suppose $k$ is an $(h+1)$-bit integer, i.e., $2^h \le k < 2^{h+1}$. Let $j$ be the largest $h$-bit integer: $j = 2^h - 1$. We compare $x_k$ with this particular $x_j$, i.e., we compute $g.c.d.(x_k - x_j, n)$. If this $g.c.d.$ gives a nontrivial factor of $n$, we stop; otherwise we move on to $k + 1$.

This modified approach has the advantage that we compute only one $g.c.d.$ for each $k$. It has the disadvantage that we probably will not detect the first time there is a $k_0$ such that $g.c.d.(x_{k_0} - x_{j_0}, n) = r > 1$ for some $j_0 < k_0$.

However, before long we will detect such a pair $x_k$, $x_j$ whose difference has a common factor with $n$. Namely, suppose that $k_0$ has $h+1$ bits. Set $j = 2^{h+1} - 1$ and $k = j + (k_0 - j_0)$, in which case $j$ is the largest $(h+1)$-bit integer and $k$ is an $(h+2)$-bit integer such that $g.c.d.(x_k - x_j, n) > 1$. Notice that we have $k < 2^{h+2} = 4 \cdot 2^h \leq 4k_0$.

**Example 2.** Let us return to Example 1 but compare each $x_k$ only with the particular $x_j$ for which $j$ is the largest integer $< k$ of the form $2^h - 1$. For $n = 91$, $f(x) = x^2 + 1$, $x_0 = 1$ we have $x_1 = 2$, $x_2 = 5$, $x_3 = 26$ as before, and $x_4 = 40$ (since $26^2 + 1 \equiv 40 \ mod\ 91$). Following the algorithm described above, we first find a factor of $n$ when we compute $g.c.d.(x_4 - x_3, n) = g.c.d.(14, 91) = 7$.

**Example 3.** Factor 4087 using $f(x) = x^2 + x + 1$ and $x_0 = 2$.

**Solution.** Our computations proceed in the following order:

$$
\begin{aligned}
x_1 &= f(2) = 7;\ g.c.d.(x_1 - x_0, n) = g.c.d.(7-2, 4087) = 1;\\
x_2 &= f(7) = 57;\ g.c.d.(x_2 - x_1, n) = g.c.d.(57-7, 4087) = 1;\\
x_3 &= f(57) = 3307;\ g.c.d.(x_3 - x_1, n) = g.c.d.(3307-7, 4087) = 1;\\
x_4 &\equiv f(3307) \equiv 2745 \ mod\ 4087;\ g.c.d.(x_4 - x_3, n)\\
&= g.c.d.(2745 - 3307, 4087) = 1;\\
x_5 &\equiv f(2745) \equiv 1343 \ mod\ 4087;\ g.c.d.(x_5 - x_3, n)\\
&= g.c.d.(1343 - 3307, 4087) = 1;\\
x_6 &\equiv f(1343) \equiv 2626 \ mod\ 4087;\ g.c.d.(x_6 - x_3, n)\\
&= g.c.d.(2626 - 3307, 4087) = 1:\\
x_7 &\equiv f(2626) \equiv 3734 \ mod\ 4087;\ g.c.d.(x_7 - x_3, n)\\
&= g.c.d.(3734 - 3307, 4087) = 61.
\end{aligned}
$$

Thus, we obtain $4087 = 61 \cdot 67$, and we are done.

**Proposition V.2.2.** *Let $n$ be an odd composite integer, and let $r$ be a nontrivial divisor of $n$ which is less than $\sqrt{n}$ (i.e., $r|n$, $1 < r < \sqrt{n}$; we suppose that we are trying to determine what $r$ is). If a pair $(f,\ x_0)$ consisting of a polynomial $f$ with integer coefficients and an initial value $x_0$ is chosen which behaves like an average pair $(f,\ x_0)$ in the sense of Proposition V.2.1 (with $f$ a map from $\mathbf{Z}/r\mathbf{Z}$ to itself and $x_0$ an integer), then the rho method will reveal the factor $r$ in $O(\sqrt[4]{n}\, log^3 n)$ bit operations with a high probability. More precisely, there exists a constant $C$ such that for any positive real number $\lambda$ the probability that the rho method fails to find a nontrivial factor of $n$ in $C\sqrt{\lambda}\ \sqrt[4]{n}\, log^3 n$ bit operations is less than $e^{-\lambda}$.*

**Proof.** Let $C_1$ be a constant such that $g.c.d.(y - z, n)$ can be computed in $C_1 log^3 n$ bit operations whenever $y, z \leq n$ (see § I.3). Let $C_2$ be a constant such that the least nonnegative residue of $f(x)$ modulo $n$ can be computed in $C_2 log^2 n$ bit operations whenever $x < n$ (see § I.1). If $k_0$ is the first index for which there exists $j_0 < k_0$ with $x_{k_0} \equiv x_{j_0} \ mod\ r$, then the rho algorithm as described above finds $r$ in the $k$-th step, where $k < 4k_0$. (Strictly speaking, it could happen that $x_k - x_j$ has a larger g.c.d. with $n$, i.e., $g.c.d.((x_k - x_j)/r,\ n/r) > 1$; but the chance of a random integer having nontrivial g.c.d. with $n/r$ is small, especially if $n$ is a product of a small number of large primes. So we shall neglect this possibility, which at worse would have the effect of requiring a slightly larger constant $C$ in the proposition.)

Thus, the number of bit operations needed to find $r$ is bounded by $4k_0(C_1 log^3 n + C_2 log^2 n)$. According to Proposition V.2.1, the probability that $k_0$ is greater than $1 + \sqrt{2\lambda r}$ is less than $e^{-\lambda}$. If $k_0$ is not greater than $1 + \sqrt{2\lambda r}$, then the number of bit operations needed to find $r$ is bounded by (here we use the fact that $r < \sqrt{n}$):

$$4(1+\sqrt{2\lambda r}\,)(C_1 log^3 n + C_2 log^2 n) < 4(1+\sqrt{2}\,\sqrt{\lambda}\,\sqrt[4]{n}\,)(C_1 log^3 n + C_2 log^2 n).$$

If we choose $C$ slightly greater than $4\sqrt{2}(C_1 + C_2)$ (so as to take care of the added 1), we conclude, as claimed, that the factor $r$ will be found in $C\sqrt{\lambda}\,\sqrt[4]{n}\, log^3 n$ bit operations, unless we made an unfortunate choice of $(f,\ x_0)$, of which the likelihood is less than $e^{-\lambda}$.

**Remarks. 1.** The basic assumption underlying the rho method is that polynomials can be found which behave like random maps in the sense of Proposition V.2.1. This has not been proved. However, practical experience factoring numbers by the rho method suggests that the "average" polynomial behaves like the "average" map, and that some very simple polynomials (the most popular one being $f(x) = x^2 + 1$) have this "average" property.

**2.** According to Proposition V.2.2, if we choose $\lambda$ large enough to have confidence in success — for example, $e^{-\lambda}$ is only about 0.0001 for $\lambda = 9$ — then we know that for an average pair $(f,\ x_0)$ we are almost certain to factor $n$ in $3C\sqrt[4]{n}\, log^3 n$ bit operations.

## *Exercises*

In Exercises 1–4, use the rho method with the indicated $f(x)$ and $x_0$ to factor the given $n$. In each case compare $x_k$ only with the $x_j$ for which $j = 2^h - 1$ (where $k$ is an $(h+1)$-bit integer).

1. $x^2 - 1$, $x_0 = 2$, $n = 91$.
2. $x^2 + 1$, $x_0 = 1$, $n = 8051$.
3. $x^2 - 1$, $x_0 = 5$, $n = 7031$.
4. $x^3 + x + 1$, $x_0 = 1$, $n = 2701$.
5. Let $S$ be a set containing $r$ elements, and let the maps $f$ in the pairs $(f,\ x_0)$ range over all *bijections* of the set $S$ to itself (i.e., $f$ is a 1-to-1 correspondence between $S$ and itself — no two $x$'s have the same $f(x)$). As before, let $x_{j+1} = f(x_j)$ for $j = 0, 1, 2, \ldots$. For each pair

$(f,\ x_0)$, let $k$ denote the first index such that there exists $j < k$ for which $f(x_k) = f(x_j)$. Prove that
(a) $k$ is at most $r$, and for each value from 1 to $r$ there is a $1/r$ probability that $k$ is that value;
(b) the average value of $k$ is $(r+1)/2$ (where the average is taken over all pairs $(f,\ x_0)$ with $f$ a bijection).

6. Using Exercise 5, explain why a linear polynomial $ax+b$ should *never* be chosen for $f(x)$ in the rho method.
7. Suppose that you are using the rho method to factor a number which has a prime divisor $r$. You decide to choose $f(x) = x^2$ as your function to be iterated. (This is a bad choice of $f(x)$, as will become clear below.) We are interested in determining the first value of $k$ such that $x_k \equiv x_\ell \mod r$ for some $\ell < k$, i.e., the first value of $k$ such that $x_0, x_1, \ldots, x_k$ are *not* all distinct modulo $r$. Suppose that you happen to choose $x_0$ which is a generator of $(\mathbf{Z}/r\mathbf{Z})^*$. Set $r - 1 = 2^s t$, where $t$ is odd.
(a) Write a congruence modulo $r-1$ which is equivalent to $x_k = x_\ell$ (equality means congruence modulo $r$).
(b) Find the first values of $k$ and $\ell$ for which the condition in (a) holds, expressing them in terms of $s$ and the binary expansion of the fraction $1/t$.
(c) Roughly how large is $k$ compared to $r$? Why is $f(x)$ a bad choice of function for the rho method?

## References for § V.2


1. W. D. Blair, C. B. Lacampagne and J. L. Selfridge, "Factoring large numbers on a pocket calculator," *American Math. Monthly* **93** (1986), 802–808.
2. R. P. Brent, "An improved Monte Carlo factorization algorithm," *BIT* **20** (1980), 176–184.
3. R. P. Brent and J. M. Pollard, "Factorization of the eighth Fermat number," *Math. Comp.* **36** (1981), 627–630.
4. R. K. Guy, "How to factor a number," *Proc. 5th Manitoba Conference on Numerical Mathematics* (1975), 49–89.
5. J. M. Pollard, "A Monte Carlo method for factorization," *BIT* **15** (1975), 331–334.


# 3 Fermat factorization and factor bases

**Fermat factorization.** As we saw earlier (see Exercise 3 of § I.2 and Exercise 4 of § IV.2), there's a way to factor a composite number $n$ that is efficient if

$n$ is a product of two integers which are close to one another. This method, called "Fermat factorization," is based on the fact that $n$ is then equal to a difference of two squares, one of which is very small.

**Proposition V.3.1.** *Let $n$ be a positive odd integer. There is a 1-to-1 correspondence between factorizations of $n$ in the form $n = ab$, where $a \geq b > 0$, and representations of $n$ in the form $t^2 - s^2$, where $s$ and $t$ are nonnegative integers. The correspondence is given by the equations*

$$t = \frac{a+b}{2}, \qquad s = \frac{a-b}{2}; \qquad\qquad a = t + s, \qquad b = t - s.$$

**Proof.** Given such a factorization, we can write $n = ab = ((a+b)/2)^2 - ((a-b)/2)^2$, so we obtain the representation as a difference of two squares. Conversely, given $n = t^2 - s^2$ we can factor the right side as $(t+s)(t-s)$. The equations in the proposition explicitly give the 1-to-1 correspondence between the two ways of writing $n$.

If $n = ab$ with $a$ and $b$ close together, then $s = (a-b)/2$ is small, and so $t$ is only slightly larger than $\sqrt{n}$. In that case, we can find $a$ and $b$ by trying all values for $t$ starting with $[\sqrt{n}] + 1$, until we find one for which $t^2 - n = s^2$ is a perfect square.

In what follows, we shall assume that $n$ is never a perfect square, so as not to have to worry about trivial exceptions to the procedures and assertions.

**Example 1.** Factor 200819.

**Solution.** We have $[\sqrt{200819}] + 1 = 449$. Now $449^2 - 200819 = 782$, which is not a perfect square. Next, we try $t = 450$: $450^2 - 200819 = 1681 = 41^2$. Thus, $200819 = 450^2 - 41^2 = (450+41)(450-41) = 491 \cdot 409$.

Notice that if the $a$ and $b$ are not close together for any factorization $n = ab$, then the Fermat factorization method will eventually find $a$ and $b$, but only after trying a large number of $t = [\sqrt{n}] + 1, [\sqrt{n}] + 2, \ldots$. There is a generalization of Fermat factorization that often works better in such a situation. We choose a small $k$, successively set $t = [\sqrt{kn}] + 1, [\sqrt{kn}] + 2$, etc., until we obtain a $t$ for which $t^2 - kn = s^2$ is a perfect square. Then $(t+s)(t-s) = kn$, and so $t+s$ has a nontrivial common factor with $n$ which can be found by computing $g.c.d.(t+s, n)$.

**Example 2.** Factor 141467.

**Solution.** If we try to use Fermat factorization, setting $t = 377, 378, \ldots$, after a while we tire of trying different $t$'s. However, if we try $t = [\sqrt{3n}] + 1 = 652, \ldots$ we soon find that $655^2 - 3 \cdot 141467 = 68^2$, at which point we compute $g.c.d.(655 + 68, 141467) = 241$. We conclude that $141467 = 241 \cdot 587$. The reason why generalized Fermat factorization worked with $k = 3$ is that there is a factorization $n = ab$ with $b$ close to $3a$. With $k = 3$ we need to try only four $t$'s, whereas with simple Fermat factorization (i.e., $k = 1$) it would have taken thirty-eight $t$'s.

**Factor bases.** There is a generalization of the idea behind Fermat factorization which leads to a much more efficient factoring method. Namely, we use the fact that any time we are able to obtain a congruence of the form $t^2 \equiv s^2 \ mod\ n$ with $t \not\equiv \pm s \ mod\ n$, we immediately find a factor of $n$ by computing $g.c.d.(t+s, n)$ (or $g.c.d.(t-s, n)$). This is because we have $n|t^2 - s^2 = (t+s)(t-s)$, while $n$ does not divide $t+s$ or $t-s$; thus $g.c.d.(t+s, n)$ must be a proper factor $a$ of $n$, and then $b = n/a$ divides $g.c.d.(t-s, n)$.

**Example 4.** Suppose we want to factor 4633, and happen to notice that $118^2$ leaves a remainder of $25 = 5^2$ modulo 4633. Then we find that $g.c.d.(118+5, 4633) = 41$, $g.c.d.(118-5, 4633) = 113$, and $4633 = 41 \cdot 113$. A skeptic might wonder how in Example 4 we ever came upon a number such as 118 whose square has least positive residue also a perfect square. Would a random selection of various $b$ soon yield one for which the least positive residue of $b^2 \ mod\ n$ is a perfect square? That is very unlikely if $n$ is large, so it is necessary to generalize this method in a way that allows much greater flexibility in choosing the $b$'s for which we consider $b^2 \ mod\ n$. The idea is to choose several $b_i$'s which have the property that $b_i^2 \ mod\ n$ is a product of small prime powers, and such that some subset of them, when multiplied together, give a $b$ whose square is congruent to a perfect square modulo $n$. We now give the details.

By the "least absolute residue" of a number $a$ modulo $n$ we mean the integer in the interval from $-n/2$ to $n/2$ to which $a$ is congruent. We shall denote this $a \ mod\ n$.

**Definition.** A *factor base* is a set $B = \{p_1, p_2, \ldots, p_h\}$ of distinct primes, except that $p_1$ may be the integer $-1$. We say that the square of an integer $b$ is a *B-number* (for a given $n$) if the least absolute residue $b^2 \ mod\ n$ can be written as a product of numbers from $B$.

**Example 5.** For $n = 4633$ and $B = \{-1, 2, 3\}$, the squares of the three integers 67, 68 and 69 are $B$-numbers, because $67^2 \equiv -144 \ mod\ 4633$, $68^2 \equiv -9 \ mod\ 4633$, and $69^2 \equiv 128 \ mod\ 4633$.

Let $\mathbf{F}_2^h$ denote the vector space over the field of two elements which consists of $h$-tuples of zeros and ones. Given $n$ and a factor base $B$ containing $h$ numbers, we show how to correspond a vector $\vec{\epsilon} \in \mathbf{F}_2^h$ to every $B$-number. Namely, we write $b^2 \ mod\ n$ in the form $\prod_{j=1}^h p_j^{\alpha_j}$ and set the $j$-th component $\epsilon_j$ equal to $\alpha_j \ mod\ 2$, i.e., $\epsilon_j = 0$ if $\alpha_j$ is even, and $\epsilon_j = 1$ if $\alpha_j$ is odd.

**Example 6.** In the situation of Example 5, the vector corresponding to 67 is $\{1, 0, 0\}$, the vector corresponding to 68 is $\{1, 0, 0\}$, and the vector corresponding to 69 is $\{0, 1, 0\}$.

Suppose that we have some set of $B$-numbers $b_i^2 \ mod\ n$ such that the corresponding vectors $\vec{\epsilon}_i = \{\epsilon_{i1}, \ldots, \epsilon_{ih}\}$ add up to the zero vector in $\mathbf{F}_2^h$. Then the product of the least absolute residues of $b_i^2$ is equal to a product of *even* powers of all of the $p_j$ in $B$. That is, if for each $i$ we let $a_i$ denote the least absolute residue of $b_i^2 \ mod\ n$ and we write $a_i = \prod_{j=1}^h p_j^{\alpha_{ij}}$, we obtain

$$\prod a_i = \prod_{j=1}^{h} p_j^{\sum_i \alpha_{ij}},$$

with the exponent of each $p_j$ an even number on the right. Then the right hand side is the square of $\prod_j p_j^{\gamma_j}$ with $\gamma_j = \frac{1}{2}\sum_i \alpha_{ij}$. Thus, if we set $b = \prod_i b_i \ mod\ n$ (least positive residue) and $c = \prod_j p_j^{\gamma_j} \ mod\ n$ (least positive residue), we obtain two numbers $b$ and $c$, constructed in quite different ways (one as a product of $b_i$'s and the other as a product of $p_j$'s) whose squares are congruent modulo $n$.

It may happen that $b \equiv \pm c \ mod\ n$, in which case we are out of luck, and we must start again with another collection of $B$-numbers whose corresponding vectors sum to zero. This will happen, for example, if we foolishly choose $b_i$ less than $\sqrt{n/2}$, in which case all of the vectors are zero-vectors, and we end up with a trivial congruence.

But for more randomly chosen $b_i$, because $n$ is composite we would expect that $b$ and $c$ would happen to be congruent (up to $\pm 1$) modulo $n$ at most 50% of the time. This is because any square modulo $n$ has $2^r \geq 4$ square roots if $n$ has $r$ different prime factors (see Exercise 7 of § I.3); thus a random square root of $b^2$ has only a $2/2^r \leq \frac{1}{2}$ chance of being either $b$ or $-b$. And as soon as we have $b$ and $c$ with $b^2 \equiv c^2 \ mod\ n$ but $b \not\equiv \pm c \ mod\ n$ we can immediately find a nontrivial factor $g.c.d.(b+c, n)$, as we saw before. Thus, if we go through the above procedure for finding $b$ and $c$ until we find a pair that gives us a nontrivial factor of $n$, we see that there is at most a $2^{-k}$ probability that this will take more than $k$ tries.

In practice, how do we choose our factor base $B$ and our $b_i$? One method is to start with $B$ consisting of the first $h$ primes (or the first $h-1$ primes together with $p_1 = -1$) and choose random $b_i$'s until we find several whose squares are $B$-numbers. Another method is to start by choosing some $b_i$'s for which $b_i^2 \ mod\ n$ (least absolute residue) is small in absolute value (for example, take $b_i$ close to $\sqrt{kn}$ for small multiples $kn$; another way will be explained in §4). Then choose $B$ to consist of a small set of small primes (and usually $p_1 = -1$) so that several of the $b_i^2 \ mod\ n$ can be expressed in terms of the numbers in $B$.

**Example 7.** In the situation of Examples 5–6, we actually chose 67 and 68 because they are close to $\sqrt{4633}$. After finding that $67^2 \equiv -144 \ mod\ 4633$ and $68^2 \equiv -9 \ mod\ 4633$, we saw that we can choose $B = \{-1, 2, 3\}$. As we saw before, the vectors corresponding to $b_1 = 67$ and $b_2 = 68$ are $\{1, 0, 0\}$ and $\{1, 0, 0\}$, which add up to the zero vector. We compute $b = 67 \cdot 68 \ mod\ 4633 = -77$ and $c = 2^{\gamma_2} \cdot 3^{\gamma_3}$ (we can ignore the power of $-1$ in $c$), i.e., $c = 36$. Fortunately, $-77 \not\equiv \pm 36 \ mod\ 4633$, and so we find a factor by computing $g.c.d.(-77 + 36, 4633) = 41$.

When can we be sure that we have enough $b_i$ to find a sum of $\overrightarrow{\epsilon}_i$ which is the zero vector? In other words, given a collection of vectors in $\mathbf{F}_2^h$, when can we be sure of being able to find a subset of them which sums to zero? To ask for this is to ask for the collection of vectors to be *linearly dependent over the field* $\mathbf{F}_2$. According to basic linear algebra (which applies just as well over the field $\mathbf{F}_2$ as over the real numbers), this is guaranteed to occur as soon as we have $h+1$ vectors. Thus, at worst we'll have to generate $h+1$ different $B$-numbers in order to find our first example of $(\prod_i b_i)^2 \equiv (\prod_j p_j^{\gamma_j})^2 \ mod\ n$. (Example 7 shows that we may very well obtain linearly dependent vectors sooner; in that case $h=3$, and we were able to stop after finding two $B$-numbers.) If $h$ is large, we might not be able to notice by inspection a subset of vectors which sums to zero; in that case, we must write the vectors as rows in a matrix and use the row-reduction technique of linear algebra to find a linearly dependent set of rows.

**Example 8.** Let $n = 4633$. Find the smallest factor-base $B$ such that the squares of 68, 69 and 96 are $B$-numbers, and then factor 4633.

**Solution.** As we saw before, $68^2 \ mod\ n$ and $69^2 \ mod\ n$ are products of $-1$, 2, and 3; since $96^2 \ mod\ n = -50$, the least absolute residues of all three squares can be written in terms of the factor-base $B = \{-1, 2, 3, 5\}$. We already computed the vectors $\epsilon_1 = \{1, 0, 0, 0\}$ and $\epsilon_2 = \{0, 1, 0, 0\}$ corresponding to 68 and 69, respectively. Since $96^2 \equiv -50 \ mod\ 4633$, we have $\epsilon_3 = \{1, 1, 0, 0\}$. Since the sum of these vectors is zero, we can take $b = 68 \cdot 69 \cdot 96 \equiv 1031 \ mod\ 4633$ and $c = 2^4 \cdot 3 \cdot 5 = 240$. Then we obtain $g.c.d.(240 + 1031, 4633) = 41$.

Examples 7 and 8 indicate how one might proceed systematically to find several $b_i$ such that the least absolute residue $b_i^2 \ mod\ n$ is a product of small primes. The likelihood that $b_i^2 \ mod\ n$ is a product of small primes is greater if this residue is small in absolute value. Thus, we might successively try integers $b_i$ close to $\sqrt{kn}$ for small integers $k$. For example, we might choose $[\sqrt{kn}]$ and $[\sqrt{kn}] + 1$ for $k = 1, 2, \ldots$.

**Example 9.** Let us factor $n = 1829$ by taking for $b_i$ all integers of the form $[\sqrt{1829k}]$ and $[\sqrt{1829k}] + 1$, $k = 1, 2, \ldots$, such that $b_i^2 \ mod\ n$ is a product of primes less than 20. For such $b_i$ we write $b_i^2 \ mod\ n = \prod_j p_j^{\alpha_{ij}}$ and tabulate the $\alpha_{ij}$. After taking $k = 1, 2, 3, 4$, we have the following table, in which the number at the top of the $j$-th column is $p_j$ and the entry in the $i$-th row beneath $p_j$ is the power of $p_j$ which occurs in $b_i^2 \ mod\ n$:

| $b_i$ | $-1$ | 2 | 3 | 5 | 7 | 11 | 13 |
|---|---|---|---|---|---|---|---|
| 42 | 1 | – | – | 1 | – | – | 1 |
| 43 | – | 2 | – | 1 | – | – | – |
| 61 | – | – | 2 | – | 1 | – | – |
| 74 | 1 | – | – | – | – | 1 | – |
| 85 | 1 | – | – | – | 1 | – | 1 |
| 86 | – | 4 | – | 1 | – | – | – |

We now look for a subset of rows whose entries sum to an even number in each column. We see at a glance that the 2nd and 6th rows sum to the even row – 6 – 2 – – – . This leads to the congruence $(b_2 \cdot b_6)^2 \equiv (2^{6/2} \cdot 5^{2/2})^2 \ mod\ n$, i.e., $(43 \cdot 86)^2 \equiv 40^2 \ mod\ 1829$. But since

$43 \cdot 86 \equiv 40 \; mod \; 1829$, we have found only a trivial relationship. Thus, we have to look for another subset of rows which sum to a row of even numbers. We notice that the sum of the first three rows and the fifth row is 2 2 2 2 2 – 2 , and this gives the congruence $(42 \cdot 43 \cdot 61 \cdot 85)^2 \equiv (2 \cdot 3 \cdot 5 \cdot 7 \cdot 13)^2 \; mod \; n$, i.e., $1459^2 \equiv 901^2 \; mod \; 1829$. We conclude that a factor of 1829 is $g.c.d.(1459 + 901, 1829) = 59$.

**Factor base algorithm.** We now summarize a systematic method to factor a very large $n$ using a *random* choice of the $b_i$. Choose an integer $y$ of intermediate size, for example, if $n$ is a 50-decimal-digit integer, we might choose $y$ to be a number with 5 or 6 decimal digits. Let $B$ consist of $-1$ and all primes $\leq y$. Choose a large number of random $b_i$, and try to express $b_i^2 \; mod \; n$ (least absolute residue) as a product of the primes in $B$. Once you obtain a large quantity of $B$-numbers $b_i^2 \; mod \; n$ ($\pi(y) + 2$ is enough, where $\pi(y)$ denotes the number of primes $\leq y$), take the corresponding vectors in $\mathbf{F}_2^h$ (where $h = \pi(y) + 1$) and by row-reduction determine a subset of the $b_i$ whose corresponding $\vec{\epsilon}_i$ sum to zero. Then form $b = \prod b_i \; mod \; n$ and $c = \prod p_j^{\gamma_j} \; mod \; n$, as described above. Then $b^2 \equiv c^2 \; mod \; n$. If $b \equiv \pm c \; mod \; n$, start again with a new random collection of $B$-numbers (or, to be more efficient, choose a different subset of rows in the matrix of $\vec{\epsilon}$'s which sum to zero, if necessary finding a few more $B$-numbers and their corresponding rows). When you finally obtain $b^2 \equiv c^2 \; mod \; n$ and $b \not\equiv \pm c \; mod \; n$, compute $g.c.d.(b + c, n)$, which will be a nontrivial factor of $n$.

**Heuristic time estimate.** We now give a very rough derivation of an estimate for the number of bit operations it takes to find a factor of a *very* large $n$ using the algorithm described above. We shall use several simplifying assumptions and approximations, and in any case the result will only be a probabilistic estimate. If we are very unlucky in our random choice of $b_i$, then the algorithm will take longer.

We shall need the following preliminary facts:

**Fact 1** (*Stirling's formula*). $log(n!)$ is approximately $n \, log \, n - n$.

By "approximately," we mean that the difference grows much more slowly than $n$ as $n \longrightarrow \infty$. This can be proved by observing that $log(n!)$ is the right-endpoint Riemann sum (with endpoints at $1, 2, 3, \ldots$) for the definite integral $\int_1^n log \, x \, dx = n \, log \, n - n + 1$.

**Fact 2.** Given a positive integer $N$ and a positive number $u$, the total number of nonnegative integer $N$-tuples $\alpha_j$ such that $\sum_{j=1}^N \alpha_j \leq u$ is the binomial coefficient $\binom{[u]+N}{N}$.

Here [ ] denotes the greatest integer function. Fact 2 can be proved by letting each $N$-tuple solution $\alpha_j$ correspond to the following choice of $N$ integers $\beta_j$ from among $1, 2, \ldots, [u] + N$. Let $\beta_1 = \alpha_1 + 1$, and for $j \geq 1$ let $\beta_{j+1} = \beta_j + \alpha_{j+1} + 1$, i.e., we choose the $\beta_j$'s so that there are $\alpha_j$ numbers between $\beta_{j-1}$ and $\beta_j$. This gives a 1-to-1 correspondence between the number of solutions and the number of ways of choosing $N$ numbers from a set of $[u] + N$ numbers.

Now, in order to estimate the time our algorithm takes, a crucial step is to estimate the probability that a random number less than $x$ will be a product of primes less than $y$ (where $y$ is a number much less than $x$). To do this, we first let $u$ denote the ratio $\frac{\log x}{\log y}$. That is, if $x$ is an $r$-bit integer and $y$ is an $s$-bit integer, then $u$ is approximately the ratio of digits $r/s$.

In the course of the computations, we shall want to make some simplifications by ignoring smaller terms. We shall do this under the assumption that $u$ is *much* smaller than $y$. We let $\pi(y)$, as usual, denote the number of prime numbers which are $\leq y$. Since $\pi(y)$ is approximately equal to $y/\log y$, by the Prime Number Theorem, we are also assuming that we are working with values of $u$ which are much smaller than $\pi(y)$. In a typical practical application of the algorithm, we might take $y$, $u$, $x$ of approximately the following sizes:

$$\begin{aligned} &y \approx 10^6 \qquad \text{(so that } \pi(y) \approx 7 \cdot 10^4 \text{ and } \log y \approx 14); \\ &u \approx 8; \\ &x \approx 10^{48}. \end{aligned}$$

It is customary to let $\Psi(x,y)$ denote the number of integers $\leq x$ which are not divisible by any prime greater than $y$, i.e., the number of integers which can be written as a product $\prod p_j^{\alpha_j} \leq x$, where the product is over all primes $\leq y$ and the $\alpha_j$ are nonnegative integers. There is obviously a 1-to-1 correspondence between $\pi(y)$-tuples of nonnegative integers $\alpha_j$ for which $\prod_j p_j^{\alpha_j} \leq x$ and integers $\leq x$ which are not divisible by any prime greater than $y$. Thus, $\Psi(x,y)$ is equal to the number of integer solutions $\alpha_j$ to the inequality $\sum_{j=1}^{\pi(y)} \alpha_j \log p_j \leq \log x$, as we see by taking logarithms. We now observe that most of the $p_j$'s have logarithms not too much less than $\log y$. This is because most of the primes less than $y$ have almost the same number of digits as $y$; only relatively few have many fewer digits and hence a much smaller logarithm. Thus, we shall allow ourselves to replace $\log p_j$ by $\log y$ in the previous inequality. Dividing both sides of the resulting inequality by $\log y$ and replacing $\log x/\log y$ by $u$, we can say that $\Psi(x,y)$ is approximately equal to the number of solutions of the inequality $\sum_{j=1}^{\pi(y)} \alpha_j \leq u$.

We now make another important simplification, replacing the number of variables $\pi(y)$ by $y$. This might appear at first to be a rather reckless modification of our problem. And in fact, replacing $\pi(y)$ by $y$ does introduce nontrivial terms; however, it turns out that those terms cancel, and the net result is the same as one would get by a much more careful approximation of $\Psi(x,y)$. Thus, we shall suppose that $\Psi(x,y)$ is roughly equal to the number of $y$-tuple nonnegative integer solutions to the inequality $\sum_{j=1}^{y} \alpha_j \leq u$.

But, by Fact 2 (with $N = y$), this means that $\Psi(x,y)$ is approximately $\binom{[u]+y}{y}$. We now estimate $\log\left(\frac{\Psi(x,y)}{x}\right)$, which is the logarithm of the probability that a random integer between 1 and $x$ is a product of primes $\leq y$.

Notice that $log\, x = u\, log\, y$, by the definition of $u$. We use the approximation for $\Psi(x, y)$ and Fact 1:

$$\begin{aligned} log\Big(\frac{\Psi(x,y)}{x}\Big) &\approx log\Big(\frac{([u]+y)!}{[u]!y!}\Big) - u\, log\, y \\ &\approx ([u]+y)log([u]+y) - ([u]+y) - \\ &\quad - \big([u]\, log\, [u] - [u]\big) - \big(y\, log\, y - y\big) - u\, log\, y. \end{aligned}$$

We now make some further approximations. First, we replace $[u]$ by $u$. Next, we note that, because $u$ is assumed to be much smaller than $y$, we can replace $log(u + y)$ by $log\, y$. After cancellation we obtain

$$log\Big(\frac{\Psi(x,y)}{x}\Big) \approx -u\, log\, u,$$

i.e.,

$$\frac{\Psi(x,y)}{x} \approx u^{-u}.$$

For example, this says that if $x \approx 10^{48}$ and $y \approx 10^6$ as above, then the probability that a random number between 1 and $x$ is a product of primes $\leq y$ is about 1 out of $8^8$.

We are now ready to estimate the number of bit operations required to carry out the factor base algorithm described above, where for simplicity we shall suppose that our factor base $B$ consists of the first $h = \pi(y)$ primes, i.e., all primes $\leq y$. To make our analysis easier, we shall suppose that $B$ does not include $-1$, and that we consider the least positive residue (rather than the least absolute residue) of $b_i^2$ *mod* $n$.

Thus, we estimate the number of bit operations required to carry out the following steps: (1) choose random numbers $b_i$ between 1 and $n$ and express the least positive residue of $b_i^2$ modulo $n$ as a product of primes $\leq y$ if it can be so expressed, continuing until you have $\pi(y) + 1$ different $b_i$'s for which $b_i^2$ *mod* $n$ is written as such a product; (2) find a set of linearly dependent rows in the corresponding $((\pi(y) + 1) \times \pi(y))$-matrix of zeros and ones to obtain a congruence of the form $b^2 \equiv c^2$ *mod* $n$; (3) if $b \equiv \pm c$ *mod* $n$, repeat (1) and (2) with new $b_i$ until you obtain $b^2 \equiv c^2$ *mod* $n$ with $b \not\equiv \pm c$ *mod* $n$, at which point find a nontrivial factor of $n$ by computing $g.c.d.(b + c, n)$.

Assuming that the $b_i^2$ *mod* $n$ (meaning least positive residue of $b_i^2$ modulo $n$) are randomly distributed between 1 and $n$, by the argument above we expect that it will take approximately $u^u$ tries before we find a $b_i$ such that $b_i^2$ *mod* $n$ is a product of primes $\leq y$, where $u = log\, n/log\, y$. We will later decide how to choose $y$ so as to minimize the length of time. The point is that choosing $y$ large would make $u^u$ small, and so we would frequently encounter $b_i$ such that $b_i^2$ *mod* $n$ is a product of primes $\leq y$. However, in that case the factorization of $b_i^2$ *mod* $n$ into a product involving all of those primes — which we would have to do $\pi(y) + 1$ times — and then the row reduction of the matrix would all be very time consuming. Conversely, if we choose $y$ fairly small, then the latter tasks would be easy, but it would take us a very long time to find any $b_i$'s for which $b_i^2 \bmod n$ is divisible only by primes $\leq y$, because in that case $u^u$ would be very large. So $y$ should be chosen in some intermediate range, as a compromise between these two extremes.

In order to decide how $y$ should be chosen, we first make a very rough estimate in terms of $y$ (and $n$, of course) of the number of bit operations. We then minimize this with respect to $y$ (using first year calculus and some simplifying approximations), and find our time estimate with $y$ chosen so that the time is minimized.

Suppose that $n$ is an $r$-bit integer and $y$ is an $s$-bit integer; then $u$ is very close to $r/s$. First of all, how many bit operations are needed for each test of a randomly chosen $b_i$? We claim that the number of operations is polynomial in $r$ and $y$, i.e., it is $O(r^l e^{ks})$ for some (fairly small) integers $k$ and $l$. It takes a fixed amount of time to generate a random bit, and so $O(r)$ bit operations to generate a random integer $b_i$ between 1 and $n$. Next, computing $b_i^2 \bmod n$ takes $O(r^2)$ bit operations. We must then divide $b_i^2 \bmod n$ successively by all primes $\leq y$ which divide it evenly (and by any power of the prime that divides it evenly), hoping that when we're done we'll be left with 1. A simple way to do this (though not the most efficient) would be to divide successively by 2 and by all odd integers $p$ from 3 to $y$, recording as we go along what power of $p$ divides $b_i^2 \bmod n$ evenly. Notice that if $p$ is not prime, then it will not divide evenly, since we will have already removed from $b_i^2 \bmod n$ all of the factors of $p$. Since a division of an integer of $\leq r$ bits by an integer of $\leq s$ bits takes time $O(rs)$, we see that each test of a randomly chosen $b_i$ takes $O(rsy)$ bit operations.

To complete step (1) requires testing approximately $u^u(\pi(y)+1)$ values of $b_i$, in order to find $\pi(y)+1$ values for which $b_i^2 \bmod n$ is a product of primes $\leq y$. Since $\pi(y) \approx \frac{y}{\log y} = O(y/s)$, this means that step (1) takes $O(u^u r y^2)$ bit operations.

Step (2) then involves operations which are polynomial in $y$ and $r$ (such as matrix reduction and finding $b$ and $c$ modulo $n$). Thus, step (2) takes $O(y^j r^h)$ bit operations for some integers $j$ and $h$. Each time we perform steps (1)–(2) there is at least a 50% chance of success, i.e., of finding that $b \not\equiv \pm c \bmod n$. More precisely, the chance of success is 50% if $n$ is divisible by only two distinct primes, and is greater if $n$ is divisible by more primes. Thus, if we are satisfied with, say, a $1-2^{-50}$ probability of finding a nontrivial factor of $n$, it suffices to go through the steps 50 times. Taking this as good enough for all practical purposes, we end up with the estimate

$$O(50(u^u r^2 y^2 + y^j r^h)) = O(r^h u^u y^j) = O(r^h u^u e^{ks}) = O(r^h (r/s)^{r/s} e^{ks}),$$

for suitable integers $h$ and $k$.

We now find $y$ — equivalently, $s$ — for which this time estimate is minimal. Since $r$, the number of bits in $n$, is fixed, this means minimizing

$(r/s)^{r/s}e^{ks}$ with respect to $s$, or equivalently, minimizing its log, which is $\frac{r}{s}\,log\,\frac{r}{s}\ +\ ks$. Thus, we set

$$0 = \frac{d}{ds}\Big(\frac{r}{s}\,log\,\frac{r}{s}\ +\ ks\Big) = -\frac{r}{s^2}\Big(log\,\frac{r}{s}\ +\ 1\Big) + k \approx -\frac{r}{s^2}\,log\,\frac{r}{s}\ +\ k,$$

i.e., we choose $s$ in such a way that $ks$ is approximately equal to $\frac{r}{s}\,log\,\frac{r}{s}$, in other words, in such a way that the two factors in $(r/s)^{r/s}e^{ks}$ are approximately equal. Because $k$ is a constant, it follows from the above approximate equality that $s^2$ has the same order of magnitude as $r\,log(r/s) = r(log\,r - log\,s)$, which means that $s$ has order of magnitude between $\sqrt{r}$ and $\sqrt{r\,log\,r}$. But this means that $log\,s$ is approximately $\frac{1}{2}log\,r$, and so, making the substitution $log\,s \approx \frac{1}{2}log\,r$, we transform the above relation to:

$$0 \approx -\frac{r}{2s^2}\,log\,r\ +\ k, \qquad \text{i.e.,} \qquad s \approx \sqrt{\frac{r}{2k}\,log\,r}.$$

With this value of $s$, we now estimate the time. Since the two factors $(r/s)^{r/s}$ and $e^{ks}$ are approximately equal for our optimally chosen $s$, the time estimate simplifies to $O(e^{2ks}) = O(e^{\sqrt{2k}\,\sqrt{r\,log\,r}})$. Replacing the constant $\sqrt{2k}$ by $C$, we finally obtain the following estimate for the number of bit operations required to factor an $r$-bit integer $n$:

$$O\Big(e^{C\,\sqrt{r\,log\,r}}\Big).$$

The above argument was very rough. We made no attempt to justify our simplifications or bound the error in our approximate equalities. In addition, both our algorithm and our estimate of its running time are probabilistic.

Until the advent of the number field sieve very recently (see the remark at the end of §5), all analyses of the running time of the best general-purpose factoring algorithms known led to estimates of the form $O\left(e^{C\sqrt{r\log r}}\right)$. In some cases, the estimates were proved rigorously, and in other cases they relied upon plausible but unproved conjectures. The main difference between the time estimates for the various competing algorithms was the constant $C$ in the exponent. In this respect the factoring problem has had a history quite different from the primality problem considered in §1, where improvements in running time (especially of deterministic primality tests) have been dramatic. For a detailed survey and comparison of the factoring algorithms that were known in the early 1980's, see Pomerance's 1982 article cited in the references below.

**Remark.** Since $r = O(log\,n)$, the above time estimate can also be expressed in the form

$$\text{Time(Factor } n) = O\Big(e^{C\,\sqrt{log\,n\,log\,log\,n}}\Big).$$

Except for the number field sieve, all of the asymptotically fast general factoring algorithms have conjectured running times of the above form with $C = 1 + \epsilon$ for $\epsilon$ arbitrarily small.

**Implications for RSA.** Recall that the security of the RSA public key cryptosystem (see § IV.2) depends upon the circumstance that factoring a very large integer of the form $n = pq$ is much more time consuming than the various tasks which legitimate users of the system must perform, tasks which are polynomial time or near-polynomial time (primality testing) as functions of the number $r$ of bits in $n$. We have just seen why time estimates of the form $O(e^{C\sqrt{r \log r}})$ tend to arise when analyzing factoring algorithms. Since a polynomial function of $r$ can be written in the form $O(e^{C \log r})$, we see that for large $r$ the time required for factorization is indeed much larger than for polynomial time or near-polynomial time algorithms. (However, the factoring algorithms with time estimate of the form $O(e^{C\sqrt{r \log r}})$ are better for large $r$ than the rho method, which has time estimate approximately $O(\sqrt[4]{n}) = O(e^{Cr})$, where $C = \frac{1}{4} \log 2$.)

Finally, we note that the question of replacing $\sqrt{r \log r}$ in the exponent by a smaller function of $r$ is not the only matter of practical importance in evaluating the security of the RSA system. After all, a polynomial function of the number of bits $r$ becomes much smaller than $C_1 e^{C_2\sqrt{r \log r}}$ only when $r$ is large, and how large $r$ must be taken depends strongly on the values of the constants $C_1$ and $C_2$. So even the discovery of a factoring algorithm with the same time estimate except with smaller constants would have practical implications for the usability of the RSA public key cryptosystem.

## *Exercises*

1. Use Fermat factorization to factor: (a) 8633, (b) 809009, (c) 92296873, (d) 88169891, (e) 4601.
2. Prove that, if $n$ has a factor that is within $\sqrt[4]{n}$ of $\sqrt{n}$, then Fermat factorization works on the first try (i.e., for $t = [\sqrt{n}] + 1$).
3. (a) Prove that if $k = 2$, or if $k$ is any integer divisible by 2 but not by 4, then we cannot factor a large odd integer $n$ using generalized Fermat factorization with this choice of $k$.
   (b) Prove that if $k = 4$, and if generalized Fermat factorization works for a certain $t$, then simple Fermat factorization (with $k = 1$) would have worked equally well.
4. Use generalized Fermat factorization to factor: (a) 68987, (b) 29895581, (c) 19578079, (d) 17018759.
5. Let $n = 2701$. Use the $B$-numbers $52^2$, $53^2 \bmod n$ for a suitable factor-base $B$ to factor 2701. What are the $\vec{\epsilon}$ 's corresponding to 52 and 53?
6. Let $n = 4633$. Use 68, 152 and 153 with a suitable factor-base $B$ to factor 4633. What are the corresponding vectors?

7. (a) Prove that: $\log n! \; - \; (n \log n \; - n) = O(\log n)$.
(b) Derive the more precise estimate: $\log n! - ((n+\frac{1}{2})\log n - n) = O(1)$.
(c) What is the expected value of $\log j$ for a randomly chosen integer $j$ between 1 and $y$?
8. (a) What is the probability that a randomly chosen set of $k$ vectors in $\mathbf{F}_2^n$ is linearly independent (where $k \le n$)?
(b) What is the probability that 5 randomly chosen vectors in $\mathbf{F}_2^5$ are a basis?
9. Let $n$ be an $r$-bit integer. By what factor does each of the expressions $\sqrt[4]{n}$ (that appears in the time estimate for the rho method) and $e^{\sqrt{r \log r}}$ (that appears in the estimate for the factor base method) increase if $n$ increases from a 50-decimal-digit to a 100-decimal-digit integer?
10. (a) Suppose that $f(s)$ is a positive monotonically decreasing function and $g(s)$ is a positive monotonically increasing function on an interval, and suppose that $f(s_0) = g(s_0)$. Prove that the function $h(s) = f(s) + g(s)$ "essentially" reaches its minimum at $s_0$, in the sense that the minimum value of $h(s)$ is between $h(s_0)$ and $\frac{1}{2}\, h(s_0)$.
(b) Suppose that $f(s) > 1$ is a monotonically decreasing function and $g(s) > 1$ is a monotonically increasing function on an interval, and suppose that $f(s_0) = g(s_0)$. Prove that the function $h(s) = f(s)g(s)$ "essentially" reaches its minimum at $s_0$, in the sense that the minimum value of $h(s)$ is between $h(s_0)$ and $\sqrt{h(s_0)}$.
(c) Using part (b), show that the function $h(s) = (r/s)^{r/s}\, e^{ks}$ on the interval $(0, r)$ (here $k$ and $r$ are positive constants) "essentially" reaches its minimum when $(r/s)^{r/s} = e^{ks}$.

## References for § V.3

1. L. E. Dickson, *History of the Theory of Numbers*, Vol. 1, Chelsea, 1952, p. 357.
2. M. Kraitchik, *Théorie des Nombres*, Vol. 2, Gauthier–Villars, 1926.
3. R. S. Lehman, "Factoring large integers," *Math. Comp.* **28** (1974), 637-646.
4. C. Pomerance, "Analysis and comparison of some integer factoring algorithms," *Computational Methods in Number Theory, Part I*, Mathematisch Centrum (Amsterdam), 1982.

# 4 The continued fraction method

In the last section, we saw that the factor-base method of finding a nontrivial factor of a large composite integer $n$ works best if one has a good method of finding integers $b$ between 1 and $n$ such that the least absolute residue $b^2$ *mod* $n$ is a product of small primes. This is most likely to occur if the absolute value of $b^2$ *mod* $n$ is small. In this section we describe a method (originally due to Legendre) for finding many $b$ such that $|b^2 \bmod n| < 2\sqrt{n}$. This method uses "continued fractions," so we shall start with a brief introduction to the continued fraction representation of a real number. Our account will describe only those features which will be needed here; the reader interested in a more thorough treatment of continued fractions should consult, for example, Davenport's classic and readable book (see the references at the end of the section).

**Continued fractions.** Given a real number $x$, we construct its continued fraction expansion as follows. Let $a_0 = [x]$ be the greatest integer not greater than $x$, and set $x_0 = x - a_0$; let $a_1 = [1/x_0]$, and set $x_1 = 1/x_0 - a_1$; and for $i > 1$, let $a_i = [1/x_{i-1}]$, and set $x_i = 1/x_{i-1} - a_i$. If/when you find that $1/x_{i-1}$ is an integer, you have $x_i = 0$, and the process stops. It is not hard to see that the process terminates if and only if $x$ is rational (because in that case the $x_i$ are rational numbers with decreasing denominators). Because of the construction of $a_0, a_1, \ldots, a_i$, for each $i$ you can write

$$x = a_0 + \cfrac{1}{a_1 + \cfrac{1}{a_2 + \cdots \cfrac{1}{a_i + x_i}}},$$

which is usually written in a more compact notation as follows:

$$x = a_0 + \frac{1}{a_1+}\,\frac{1}{a_2+}\,\frac{1}{a_3+}\cdots\frac{1}{a_i + x_i}.$$

Suppose that $x$ is an *irrational* real number. If we carry out the above expansion to the $i$-th term and then delete $x_i$, we obtain a rational number $b_i/c_i$, called the $i$-th *convergent* of the continued fraction for $x$:

$$\frac{b_i}{c_i} = a_0 + \frac{1}{a_1+}\,\frac{1}{a_2+}\,\frac{1}{a_3+}\cdots\frac{1}{a_{i-1}+}\,\frac{1}{a_i}.$$

**Proposition V.4.1.** *In the above notation, one has:*

(a) $\frac{b_0}{c_0} = \frac{a_0}{1}$; $\frac{b_1}{c_1} = \frac{a_0a_1+1}{a_1}$; $\frac{b_i}{c_i} = \frac{a_ib_{i-1}+b_{i-2}}{a_ic_{i-1}+c_{i-2}}$ *for* $i \geq 2$;

(b) *the fractions on the right in part (a) are in lowest terms, i.e., if* $b_i = a_ib_{i-1} + b_{i-2}$ *and* $c_i = a_ic_{i-1} + c_{i-2}$, *then g.c.d.*$(b_i, c_i) = 1$;

(c) $b_ic_{i-1} - b_{i-1}c_i = (-1)^{i-1}$ *for* $i \geq 1$.

**Proof.** We define the sequences $\{b_i\}$ and $\{c_i\}$ by the relations in (a), and prove by induction that then $b_i/c_i$ is the $i$-th convergent. We will prove this without assuming that the $a_i$ are integers, i.e., we will prove that for any real numbers $a_i$ the ratio $b_i/c_i$ with $b_i$ and $c_i$ defined by the formulas in (a) is equal to $a_0 + \frac{1}{a_1+}\cdots\frac{1}{a_i}$. It is trivial to check the beginning of the induction ($i = 0, 1, 2$). We now suppose that the claim is true through the

$i$-th convergent, and we prove the claim for the $(i+1)$-th convergent. Note that we obtain the $(i+1)$-th convergent by replacing $a_i$ by $a_i + 1/a_{i+1}$ in the formula that expresses the numerator and denominator of the $i$-th convergent in terms of the $(i-1)$-th and $(i-2)$-th. That is, the $(i+1)$-th convergent is

$$\frac{(a_i + \frac{1}{a_{i+1}})b_{i-1} + b_{i-2}}{(a_i + \frac{1}{a_{i+1}})c_{i-1} + c_{i-2}} = \frac{a_{i+1}(a_i b_{i-1} + b_{i-2}) + b_{i-1}}{a_{i+1}(a_i c_{i-1} + c_{i-2}) + c_{i-1}} = \frac{a_{i+1}b_i + b_{i-1}}{a_{i+1}c_i + c_{i-1}},$$

by the induction assumption. This completes the induction, and proves part (a).

Part (c) is also easy to prove by induction. The induction step goes as follows:

$$\begin{aligned} b_{i+1}c_i - b_i c_{i+1} &= (a_{i+1}b_i + b_{i-1})c_i - b_i(a_{i+1}c_i + c_{i-1}) = b_{i-1}c_i - b_i c_{i-1} \\ &= -(-1)^{i-1} = (-1)^i, \end{aligned}$$

so part (c) for $i$ implies part (c) for $i+1$. Finally, part (b) follows from part (c), because any common divisor of $b_i$ and $c_i$ must divide $(-1)^{i-1}$, which is $\pm 1$. This proves the proposition.

If we divide the equation in Proposition V.4.1(c) by $c_i c_{i-1}$, we find that

$$\frac{b_i}{c_i} - \frac{b_{i-1}}{c_{i-1}} = \frac{(-1)^{i-1}}{c_i c_{i-1}}.$$

Since the $c_i$ clearly form a strictly increasing sequence of positive integers, this equality shows that the sequence of convergents behaves like an alternating series, i.e., it oscillates back and forth with shrinking amplitude; thus, the sequence of convergents converges to a limit.

Finally, it is not hard to see that the limit of the convergents is the number $x$ which was expanded in the first place. To see that, notice that $x$ can be obtained by forming the $(i+1)$-th convergent with $a_{i+1}$ replaced by $1/x_i$. Thus, by Proposition V.4.1(a) (with $i$ replaced by $i+1$ and $a_{i+1}$ replaced by $1/x_i$), we have

$$x = \frac{b_i/x_i + b_{i-1}}{c_i/x_i + c_{i-1}} = \frac{b_i + x_i b_{i-1}}{c_i + x_i c_{i-1}},$$

and this is strictly between $b_{i-1}/c_{i-1}$ and $b_i/c_i$. (To see this, consider the two vectors $\mathbf{u} = (b_i, c_i)$ and $\mathbf{v} = (b_{i-1}, c_{i-1})$ in the plane, both in the same quadrant; note that the slope of the vector $\mathbf{u} + x_i\mathbf{v}$ is intermediate between the slopes of $\mathbf{u}$ and $\mathbf{v}$.) Thus, the sequence $b_i/c_i$ oscillates around $x$ and converges to $x$.

Continued fractions have many special properties that cause them to come up in several different branches of mathematics. For example, they provide a way of generating "best possible" rational approximations to real numbers (in the sense that any rational number that is closer to $x$ than $b_i/c_i$

must have a denominator larger than $c_i$). Another property is analogous to the fact that the decimal (or base-$b$) digits of a real number $x$ repeat if and only if $x$ is rational. In the continued fraction expansion of $x$, we saw that the sequence of integers $a_i$ terminates if and only if $x$ is rational. It can be shown that the $a_i$ become a repeating sequence if and only if $x$ is a quadratic irrationality, i.e., of the form $x_1 + x_2\sqrt{n}$ with $x_1$ and $x_2$ rational and $n$ not a perfect square. This is known as Lagrange's theorem.

**Example 1.** If we start expanding $\sqrt{3}$ as a continued fraction, we obtain

$$\sqrt{3} = 1 + \frac{1}{1+}\,\frac{1}{2+}\,\frac{1}{1+}\,\frac{1}{2+}\,\frac{1}{1+}\,\frac{1}{2+}\cdots.$$

At this point we might conjecture that the $a_i$'s alternate between 1 and 2. To prove this, let $x$ equal the infinite continued fraction on the right with alternating 1's and 2's. Then clearly $x = 1 + \frac{1}{1+(1/(1+x))}$, as we see by replacing $x$ on the right by its definition as a continued fraction. Simplifying the rational expression on the right and multiplying both sides of the equation by $2 + x$ gives: $2x + x^2 = 3 + 2x$, i.e., $x = \sqrt{3}$.

**Proposition V.4.2.** *Let $x > 1$ be a real number whose continued fraction expansion has convergents $b_i/c_i$. Then for all $i$: $|b_i^2 - x^2c_i^2| < 2x$.*

**Proof.** Since $x$ is between $b_i/c_i$ and $b_{i+1}/c_{i+1}$, and since the absolute value of the difference between these successive convergents is $1/c_ic_{i+1}$ (by Proposition V.4.1(c)), we have

$$|b_i^2 - x^2c_i^2| = c_i^2|x - \frac{b_i}{c_i}||x + \frac{b_i}{c_i}| < c_i^2\frac{1}{c_ic_{i+1}}(x + (x + \frac{1}{c_ic_{i+1}})).$$

Hence,

$$\begin{aligned}|b_i^2 - x^2c_i^2| - 2x &< 2x\Big(-1 + \frac{c_i}{c_{i+1}} + \frac{1}{2xc_{i+1}^2}\Big) < 2x\Big(-1 + \frac{c_i}{c_{i+1}} + \frac{1}{c_{i+1}}\Big)\\ &< 2x\Big(-1 + \frac{c_{i+1}}{c_{i+1}}\Big) = 0.\end{aligned}$$

This proves the proposition.

**Proposition V.4.3.** *Let $n$ be a positive integer which is not a perfect square. Let $b_i/c_i$ be the convergents in the continued fraction expansion of $\sqrt{n}$. Then the residue of $b_i^2$ modulo $n$ which is smallest in absolute value (i.e., between $-n/2$ and $n/2$) is less than $2\sqrt{n}$.*

**Proof.** Apply Proposition V.4.2 with $x = \sqrt{n}$. Then $b_i^2 \equiv b_i^2 - nc_i^2 \bmod n$, and the latter integer is less than $2\sqrt{n}$ in absolute value.

Proposition V.4.3 is the key to the continued fraction algorithm. It says that we can find a sequence of $b_i$'s whose squares have small residues by taking the numerators of the convergents in the continued fraction expansion of $\sqrt{n}$. Note that we do not have to find the actual convergent: only the numerator $b_i$ is needed, and that is needed only modulo $n$. Thus, the fact that the numerator and denominator of the convergents soon become very large does not worry us. We never need to work with integers larger than $n^2$ (when we multiply integers modulo $n$).

We now describe in sequence how the continued fraction algorithm works. All we do is use the factor-base method in §3, except with Proposition V.4.3 replacing random choice of the $b_i$'s.

**Continued fraction factoring algorithm.** Let $n$ be the integer to be factored. All computations below will be done modulo $n$, i.e., products and sums of integers will be reduced modulo $n$ to their least nonnegative residue (or least absolute residue in step (3)). First set $b_{-1} = 1$, $b_0 = a_0 = [\sqrt{n}\,]$, and $x_0 = \sqrt{n} - a_0$. Compute $b_0^2$ *mod* $n$ (which will be $b_0^2 - n$). Next, for $i = 1, 2, \ldots$ successively:

1. Set $a_i = [1/x_{i-1}]$ and then $x_i = 1/x_{i-1} - a_i$.
2. Set $b_i = a_i b_{i-1} + b_{i-2}$ (reduced modulo $n$).
3. Compute $b_i^2$ *mod* $n$. After doing this for several $i$, look at the numbers in step 3 which factor into $\pm$ a product of small primes. Take your factor base $B$ to consist of $-1$, the primes which occur in more than one of the $b_i^2$ *mod* $n$ (or which occur to an even power in just one $b_i^2$ *mod* $n$). Then list all of the numbers $b_i^2$ *mod* $n$ which are $B$-numbers, along with the corresponding vectors $\vec{\epsilon}_i$ of zeros and ones. If possible, find a subset whose vectors sum to zero. Set $b = \prod b_i$ (working modulo $n$ and taking the product over the subset for which $\sum \vec{\epsilon}_i = 0$). Set $c = \prod p_j^{\gamma_j}$, where $p_j$ are the elements of $B$ (except for $-1$) and $\gamma_j = \frac{1}{2}\sum \alpha_{ij}$ (with the sum taken over the same subset of $i$; see §3). If $b \not\equiv \pm c$ *mod* $n$, then *g.c.d.*$(b+c, n)$ is a nontrivial factor of $n$. If $b \equiv \pm c$ *mod* $n$, then look for another subset of $i$ such that $\sum \vec{\epsilon}_i = 0$. If it is not possible to find any subset of $i$ such that $\sum \vec{\epsilon}_i = 0$, then you must continue computing more $a_i$, $b_i$, and $b_i^2$ *mod* $n$, enlarging your factor base $B$ if necessary.

**Remark.** In order to be able to compute $c = \prod p_j^{\gamma_j}$, it is efficient if for each $B$-number $b_i^2$ *mod* $n$ we record the vector $\vec{\alpha}_i = \{\ldots, \alpha_{ij}, \ldots\}_j$ rather than $\vec{\epsilon}_i$, which is simply $\vec{\alpha}_i$ reduced modulo 2.

**Example 2.** Use the above algorithm to factor 9073.

**Solution.** We first make a list of successive $a_i$'s and $b_i$'s (where $b_i$ is the least nonnegative residue modulo $n$ of $a_i b_{i-1} + b_{i-2}$), along with the corresponding least absolute residue modulo $n$ of $b_i^2$:

| $i$ | 0 | 1 | 2 | 3 | 4 |
|---|---|---|---|---|---|
| $a_i$ | 95 | 3 | 1 | 26 | 2 |
| $b_i$ | 95 | 286 | 381 | 1119 | 2619 |
| $b_i^2$ *mod* $n$ | $-48$ | 139 | $-7$ | 87 | $-27$ |

Looking at the last line of the table, we see that it is reasonable to set $B = \{-1, 2, 3, 7\}$. Then $b_i^2$ *mod* $n$ is a $B$-number for $i = 0, 2, 4$. The corresponding vectors $\vec{\alpha}_i$ are, respectively, $\{1, 4, 1, 0\}$, $\{1, 0, 0, 1\}$, and $\{1, 0, 3, 0\}$. The sum of the first and third is zero modulo 2. So let us choose $b = 95 \cdot 2619 \equiv 3834$ *mod* 9073, and $c = 2^2 \cdot 3^2 = 36$. Thus, $3834^2 \equiv 36^2$ *mod* 9073.

Since $3834 \not\equiv \pm 36 \; mod \; 9073$, we obtain the nontrivial factor $g.c.d.(3834 + 36, 9073) = 43$. Thus, $9073 = 43 \cdot 211$.

**Example 3.** Factor 17873.

**Solution.** As in Example 2, we start out with a table

| $i$ | 0 | 1 | 2 | 3 | 4 | 5 |
|---|---|---|---|---|---|---|
| $a_i$ | 133 | 1 | 2 | 4 | 2 | 3 |
| $b_i$ | 133 | 134 | 401 | 1738 | 3877 | 13369 |
| $b_i^2 \; mod \; n$ | $-184$ | 83 | $-56$ | 107 | $-64$ | 161 |

If we set $B = \{-1, 2, 7, 23\}$, we have $B$-numbers when $i = 0, 2, 4, 5$; the corresponding vectors $\overrightarrow{\alpha}_i$ are, respectively, $\{1,3,0,1\}$, $\{1,3,1,0\}$, $\{1,6,0,0\}$ and $\{0,0,1,1\}$. The sum of the first, second and fourth of these four vectors is zero modulo 2. However, if we compute $b = 133 \cdot 401 \cdot 13369 \equiv 1288 \; mod \; 17873$ and $c = 2^3 \cdot 7 \cdot 23 = 1288$, we find that $b \equiv c \; mod \; 17873$. Thus, we must continue to look for more $B$-numbers with vectors that sum to zero modulo 2. Continuing the table, we have

| $i$ | 6 | 7 | 8 |
|---|---|---|---|
| $a_i$ | 1 | 2 | 1 |
| $b_i$ | 17246 | 12115 | 11488 |
| $b_i^2 \; mod \; n$ | $-77$ | 149 | $-88$ |

If we now enlarge $B$ to include the prime 11, i.e., $B = \{-1, 2, 7, 11, 23\}$, then for $i = 0, 2, 4, 5, 6, 8$ we obtain $B$-numbers with vectors $\overrightarrow{\alpha}_i$ as follows: $\{1,3,0,0,1\}$, $\{1,3,1,0,0\}$, $\{1,6,0,0,0\}$, $\{0,0,1,0,1\}$, $\{1,0,1,1,0\}$, $\{1,3,0,1,0\}$. We now note that the sum of the second, third, fifth and sixth of these six vectors is zero modulo 2. This leads to $b = 7272$, $c = 4928$, and we finally find a nontrivial factor $g.c.d.(7272 + 4928, 17873) = 61$. We obtain: $17873 = 61 \cdot 293$.

## *Exercises*

1. Find the continued fraction representation of the following rational numbers: (a) 45/89; (b) 55/89; (c) 1.13.
2. (a) Suppose that $x$ is a real number whose continued fraction expansion consists of the positive integer $a$ repeated infinitely:

$$x = a + \frac{1}{a+}\,\frac{1}{a+}\,\frac{1}{a+}\,\frac{1}{a+}\cdots.$$

   What real number is $x$ (written in a simple closed form)?
   (b) Prove that if $a = 1$ in part (a), then $x$ is the golden ratio and the numerators and denominators of the convergents are Fibonacci numbers.
3. Expand $e$ in a continued fraction, and try to guess a pattern in the integers $a_i$.

4. In the continued fraction algorithm explain why there is no need to include in the factor base $B$ any primes $p$ such that $(\frac{n}{p}) = -1$.
5. Following Examples 2 and 3, use the continued fraction algorithm to factor the following numbers: (a) 9509; (b) 13561; (c) 8777; (d) 14429; (e) 12403; (f) 14527; (g) 10123; (h) 12449; (i) 9353; (j) 25511; (k) 17873.

## References for § V.4


1. H. Davenport, *The Higher Arithmetic*, 5th ed., Cambridge Univ. Press, 1982.
2. D. Knuth, *The Art of Computer Programming*, Vol. 2, Addison-Wesley, 1973.
3. D. H. Lehmer and R. E. Powers, "On factoring large numbers," *Bull. Amer. Math. Soc.* **37** (1931), 770–776.
4. M. A. Morrison and J. Brillhart, "A method of factoring and the factorization of $F_7$," *Math. Comp.* **29** (1975), 183–205.
5. C. Pomerance and S. S. Wagstaff, Jr., "Implementation of the continued fraction integer factoring algorithm," *Proc. 12th Winnipeg Conference on Numerical Methods and Computing*, 1983.
6. M. C. Wunderlich, "A running time analysis of Brillhart's continued fraction factoring method," *Number Theory, Carbondale 1979*, Springer Lecture Notes Vol. 751 (1979), 328–342.
7. M. C. Wunderlich, "Implementing the continued fraction factoring algorithm on parallel machines," *Math. Comp.* **44** (1985), 251–260.


# 5 The quadratic sieve method

The quadratic sieve method for factoring large integers, developed by Pomerance in the early 1980's, for a long time was more successful than any other method in factoring integers $n$ of general type which have no prime factor of order of magnitude significantly less than $\sqrt{n}$. (For integers $n$ having a special form there may be special purpose methods which are faster, and for $n$ divisible by a prime much smaller than $\sqrt{n}$ the elliptic curve factorization method in §VI.4 is faster. Also see the discussion of the number field sieve at the end of the section.)

The quadratic sieve is a variant of the factor base approach discussed in §3. As our factor base $B$ we take the set of all primes $p \leq P$ (where $P$ is some bound to be chosen in some optimal way) such that $n$ is a quadratic residue mod $p$, i.e., $\left(\frac{n}{p}\right) = 1$ for $p$ odd, and $p = 2$ is always included in $B$. The set of integers $S$ in which we look for $B$-numbers (recall that a $B$-number is an integer divisible only by primes in $B$) will be the same set that we used in Fermat factorization (see §3), namely:

$$S = \left\{ t^2 - n \;\middle|\; [\sqrt{n}] + 1 \le t \le [\sqrt{n}] + A \right\}$$

for some suitably chosen bound $A$.

The main idea of the method is that, instead of taking each $s \in S$ one by one and dividing it by the primes $p \in B$ to see if it is a $B$-number, we take each $p \in B$ one by one and examine divisibility by $p$ (and powers of $p$) simultaneously for all of the $s \in S$. The word "sieve" refers to this idea. Here we should recall the "sieve of Eratosthenes," which one can use to make a list of all primes $p \le A$. For example, to list the primes $\le 1000$ one takes the list of all integers $\le 1000$ and then for each $p = 2, 3, 5, 7, 11, 13, 17, 19, 23, 29, 31$ one discards all multiples of $p$ greater than $p$ — one "lets them fall through a sieve which has holes spaced a distance $p$ apart" — after which the numbers that remain are the primes.

We shall give an outline of a procedure to carry out the method, and then give an example. The particular version described below is only one possible variant, and it is not necessarily the most efficient one. Moreover, our example of a number $n$ to be factored (and also the numbers to be factored in the exercises at the end of the section) will be chosen in the range $\approx 10^6$, so as to avoid having to work with large matrices. However, such $n$ are far too small to illustrate the time advantage of the sieve in finding a large set of $B$-numbers.

Thus, suppose we have an odd composite integer $n$.

1. Choose bounds $P$ and $A$, both of order of magnitude roughly

$$e^{\sqrt{log\, n\; log\, log\, n}}.$$

Generally, $A$ should be larger than $P$, but not larger than a fairly small power of $P$, e.g., $P < A < P^2$.

This function $\exp(\sqrt{log\, n\; log\, log\, n})$, which we encountered before in this chapter and which is traditionally denoted $L(n)$, has an order of magnitude intermediate between polynomial in $log\, n$ and polynomial in $n$. If $n \approx 10^6$, then $L(n) \approx 400$. In the examples below, we shall choose $P = 50$, $A = 500$.

2. For $t = [\sqrt{n}] + 1, [\sqrt{n}] + 2, \ldots, [\sqrt{n}] + A$, make a column listing the integers $t^2 - n$.

3. For each odd prime $p \le P$, first check that $\left(\frac{n}{p}\right) = 1$ (see §II.2); if not, then throw that $p$ out of the factor base.

4. Assuming that $p$ is an odd prime such that $n$ is a quadratic residue mod $p$ (we'll treat the case $p = 2$ separately), solve the equation $t^2 \equiv n \ (mod\ p^\beta)$ for $\beta = 1, 2, \ldots$, using the method in Exercise 20 of §II.2. Take increasing values of $\beta$ until you find that there is no solution $t$ which is congruent modulo $p^\beta$ to any integer in the range $[\sqrt{n}] + 1 \le t \le [\sqrt{n}] + A$. Let $\beta$ be the largest integer such that there is some $t$ in this range for which $t^2 \equiv n \ (mod\ p^\beta)$. Let $t_1$ and $t_2$ be two solutions of $t^2 \equiv n \ (mod\ p^\beta)$ with

$t_2 \equiv -t_1 \ (mod\ p^\beta)$ ($t_1$ and $t_2$ are not necessarily in the range from $[\sqrt{n}]+1$ to $[\sqrt{n}]+A$).

5. Still with the same value of $p$, run down the list of $t^2-n$ from part 2. In a column under $p$ put a 1 next to all values of $t^2-n$ for which $t$ differs from $t_1$ by a multiple of $p$, change the 1 to a 2 next to all values of $t^2-n$ for which $t$ differs from $t_1$ by a multiple of $p^2$, change the 2 to a 3 next to all values of $t^2-n$ for which $t$ differs from $t_1$ by a multiple of $p^3$, and so on until $p^\beta$. Then do the same with $t_1$ replaced by $t_2$. The largest integer that appears in this column will be $\beta$.

6. As you go through the procedure in 5), each time you put down a 1 or change a 1 to a 2, a 2 to a 3, etc., divide the corresponding $t^2-n$ by $p$ and keep a record of what's left.

7. In the column $p=2$, if $n \not\equiv 1\ mod\ 8$, then simply put a 1 next to the $t^2-n$ for $t$ odd and divide the corresponding $t^2-n$ by 2. If $n \equiv 1\ mod\ 8$, then solve the equation $t^2 \equiv n \ (mod\ 2^\beta)$ and proceed exactly as in the case of odd $p$ (except that there will be 4 different solutions $t_1$, $t_2$, $t_3$, $t_4$ modulo $2^\beta$ if $\beta \geq 3$).

8. When you finish with all primes $\leq P$, throw out all of the $t^2-n$ except for those which have become 1 after division by all the powers of $p \leq P$. You will have a table of the form in Example 9 in §3, in which the column labeled $b_i$ will have the values of $t$, $[\sqrt{n}]+1 \leq t \leq [\sqrt{n}]+A$, for which $t^2-n$ is a $B$-number, and the other columns will correspond to all values of $p \leq P$ for which $n$ is a quadratic residue.

9. The rest of the procedure is exactly as in §3.

**Example.** Let us try to factor $n=1042387$, taking the bounds $P=50$ and $A=500$. Here $[\sqrt{n}]=1020$. Our factor base consists of the 8 primes $\{2,3,11,17,19,23,43,47\}$ for which 1042387 is a quadratic residue. Since $n \not\equiv 1 \ (mod\ 8)$, the column corresponding to $p=2$ alternates between 1 and 0, with a 1 beside all odd $t$, $1021 \leq t \leq 1520$.

We describe in detail how to form the column under $p=3$. We want a solution $t_1 = t_{1,0} + t_{1,1}\cdot 3 + t_{1,2}\cdot 3^2 + \cdots + t_{1,\beta-1}\cdot 3^{\beta-1}$ to $t_1^2 \equiv 1042387 \ (mod\ 3^\beta)$, where $t_{1,j} \in \{0,1,2\}$ (for the other solution $t_2$ we can take $t_2 = 3^\beta - t_1$). We can obviously take $t_{1,0}=1$. (For each of our 8 primes the first step — solving $t_1^2 \equiv 1042387 \ (mod\ p)$ — can be done quickly by trial and error; if we were working with larger primes, we could use the procedure described at the end of §II.2.) Next, we work modulo 9: $(1+3t_{1,1})^2 \equiv 1042387 \equiv 7 \ (mod\ 9)$, i.e., $6t_{1,1} \equiv 6 \ (mod\ 9)$, i.e., $2t_{1,1} \equiv 2 \ (mod\ 3)$, so $t_{1,1}=1$. Next, modulo 27: $(1+3+9t_{1,2})^2 \equiv 1042387 \equiv 25 \ (mod\ 27)$, i.e., $16+18t_{1,2} \equiv 25 \ (mod\ 27)$, i.e., $2t_{1,2} \equiv 1 \ (mod\ 3)$, so $t_{1,2}=2$. Then modulo 81: $(1+3+18+27t_{1,3})^2 \equiv 1042387 \equiv 79 \ (mod\ 81)$, which leads to $t_{1,3}=0$. Continuing until $3^7$, we find the solution (in the notation of §I.1 for numbers written to the base 3): $t_1 \equiv (210211)_3 \ (mod\ 3^7)$, and $t_2 \equiv (2012012)_3 \ (mod\ 3^7)$. However, there is no $t$ between 1021 and 1520 which is $\equiv t_1$ or $t_2$ modulo $3^7$. Thus, we have $\beta=6$, and we can take $t_1 = (210211)_3 = 589 \equiv 1318 \ (mod\ 3^6)$ and $t_2 = 3^6 - t_1 = 140 \equiv$

1112 $(mod\ 3^5)$ (note that there is no number in the range from 1021 to 1520 which is $\equiv t_2\ (mod\ 3^6)$).

We now construct our "sieve" for the prime 3 as follows. Starting from 1318, we take jumps of 3 down until we reach 1021 and up until we reach 1519, each time putting a 1 in the column, dividing the corresponding $t^2 - n$ by 3, and recording the result of the division. (Actually, for $t$ odd, the number we divide by 3 is half of $t^2 - n$, since we already divided $t^2 - n$ by 2 when we formed the column of alternating 0's and 1's under 2.) Then we do the same with jumps of 9, each time changing the 1 to 2 in the column under 3, dividing the quotient of $t^2 - n$ by another 3, and recording the result. We go through the analogous procedure with jumps of 27, 81, 243, and 729 (there is no jump possible for 729 — we merely change the 5 to 6 next to 1318 and divide the quotient of $1318^2$—1042387 by another 3). Finally, we go through the same steps with $t_2 = 1112$ instead of $t_1 = 1318$, this time stopping with jumps of 243.

After going through this procedure for the remaining 6 primes in our factor base, we have a $500 \times 8$ array of exponents, each row corresponding to a value of $t$ between 1021 and 1520. Now we throw out all rows for which $t^2 - n$ has not been reduced to 1 by repeated division by powers of $p$ as we formed our table, i.e., we take only the rows for which $t^2 - n$ is a $B$-number. In the present example $n = 1042387$ we are left with the following table (here blank spaces denote zero exponents):

| $t$ | $t^2 - n$ | 2 | 3 | 11 | 17 | 19 | 23 | 43 | 47 |
|---|---|---|---|---|---|---|---|---|---|
| 1021 | 54 | 1 | 3 | – | – | – | – | – | – |
| 1027 | 12342 | 1 | 1 | 2 | 1 | – | – | – | – |
| 1030 | 18513 | – | 2 | 2 | 1 | – | – | – | – |
| 1061 | 83334 | 1 | 1 | – | 1 | 1 | – | 1 | – |
| 1112 | 194157 | – | 5 | – | 1 | – | – | – | 1 |
| 1129 | 232254 | 1 | 3 | 1 | 1 | – | 1 | – | – |
| 1148 | 275517 | – | 2 | 3 | – | – | 1 | – | – |
| 1175 | 338238 | 1 | 2 | – | – | 1 | 1 | 1 | – |
| 1217 | 438702 | 1 | 1 | 1 | 2 | – | 1 | – | – |
| 1390 | 889713 | – | 2 | 2 | – | 1 | – | 1 | – |
| 1520 | 1268013 | – | 1 | – | 1 | – | 2 | – | 1 |

Proceeding as we did in Example 9 in §3, we now look for relations modulo 2 between the rows of this matrix. That is, moving down from the first row, we look for a subset of the rows which sums to an even number in each column. The first such subset we find here is the first three rows, the sum of which is twice the row 1 3 2 1 – – – – . Thus, we obtain the congruence

$$(1021 \cdot 1027 \cdot 1030)^2 \equiv (2 \cdot 3^3 \cdot 11^2 \cdot 17)^2\ (mod\ 1042387).$$

But despite our good fortune in finding a set of mod 2 linearly dependent rows so quickly, it turns out that we are not so lucky after all: the two numbers being squared in the above congruence are both $\equiv 111078 \ (mod\ 1042387)$, so we get only the trivial factorization. As we continue down the matrix, we find some other sets of dependent rows, which also fail to give us a nontrivial factorization. Finally, when we are about to give up — and start over again with a larger $A$ — we notice that the last row — corresponding to our very last value of $t$ — is dependent on the earlier rows. More precisely, it is equal modulo 2 to the fifth row. This gives us $(1112 \cdot 1520)^2 \equiv (3^3 \cdot 17 \cdot 23 \cdot 47)^2 \ (mod\ 1042387)$, i.e., $647853^2 \equiv 496179^2 \ (mod\ 1042387)$, and we obtain the nontrivial factor $g.c.d.(647853 - 496179, 1042387) = 1487$.

Based on some plausible conjectures, one can show that the expected running time of the quadratic sieve factoring method is asymptotically

$$O\left(e^{(1+\epsilon)\sqrt{log\, n\, log\, log\, n}}\right)$$

for any $\epsilon > 0$. There is a fairly large space requirement, also of the form $\exp(C\sqrt{log\, n\, log\, log\, n})$. For a detailed discussion of time and space requirements for the quadratic sieve (and several other) factoring algorithms, see Pomerance's article in the volume *Computation Methods in Number Theory*.

**The number field sieve.** Until recently, all of the contenders for the best general purpose factoring algorithm had running time of the form

$$\exp\big(O(\sqrt{log\, n\, log\, log\, n})\big).$$

Some people even thought that this function of $n$ might be a natural lower bound on the running time. However, during the last few years a new method — called the *number field sieve* — has been developed that has a heuristic running time that is much better (asymptotically), namely:

$$\exp\big(O((log\, n)^{1/3}(log\, log\, n)^{2/3})\big).$$

In practice, it appears to be the fastest method for factoring numbers that are at or beyond the current (1994) upper limits of what can be factored, i.e., $> 150$ digits.

In some respects, the number field sieve factoring algorithm is similar to the earlier algorithms that attempt to combine congruences so as to obtain a relation of the form $x^2 \equiv y^2 \ (mod\ n)$. However, one uses a "factor base" in the ring of integers of a suitably chosen algebraic number field. Thus, along with the basic machinery of the quadratic sieve, this factoring method uses algebraic number theory. It is perhaps the most complicated factoring algorithm known. We shall give only an overview.

The basic requirements of the algorithm can be briefly described as follows. Given an integer $n$ to be factored, choose a degree $d$ and find $n$ as the value at some integer $m$ of an irreducible monic integer polynomial of degree $d$:

$$n = f(m) = m^d + a_{d-1}m^{d-1} + a_{d-2}m^{d-2} + \cdots + a_1 m + a_0,$$

where $m$ and the $a_k$ are integers that are $O(n^{1/d})$. One way to find such a polynomial is to let $m$ be the integer part of the $d$-th root of $n$ and then expand $n$ to the base $m$. For 125-digit numbers an analysis of the algorithm suggests that $d$ should be 5, so that $m$ and the coefficients will have about 25 digits.

The number field sieve then searches (by a sieving process similar to the quadratic sieve) for as many pairs $(a, b)$ as possible such that both $a + bm$ and also

$$b^d f(-a/b) = (-a)^d + a_{d-1}(-a)^{d-1}b + a_{d-2}(-a)^{d-2}b^2 + \cdots - a_1 ab^{d-1} + a_0 b^d$$

are smooth over a given factor base (i.e., are divisible only by primes in the factor base). The details of how this is done and how this leads to a factorization of $n$ can be found in the book *The Development of the Number Field Sieve* cited in the references below. In order for this procedure to succeed, the proportion of smooth numbers among values of the polynomial $f$ should be approximately the same as the proportion of smooth numbers among all numbers of the same size. Although this is likely to be true, and is true in all examples that have been computed, it seems to be a very hard assertion to prove. Since the estimate of running time depends on this unproved conjecture, it is a heuristic estimate. While perhaps of little consequence in practice for factoring actual numbers, this circumstance points to some important open problems in the analysis of the theoretical asymptotic complexity of factoring.

The author would like to thank Joe Buhler for providing the above brief summary of the number field sieve for this book.

## *Exercises*

1. In the example, find all linear dependence relations mod 2 between the rows of the matrix, and show that if $P = 50$ and $A \leq 499$ one cannot get a nontrivial factorization of 1042387 by this method.
2. Let $n \longrightarrow \infty$, and suppose that $P$ and $A$ are always chosen to have the same order of magnitude (for example, suppose that there are positive constants $c_1$ and $c_2$ such that $c_1 \leq \log A / \log P \leq c_2$). Asymptotically, what is the most time-consuming part of steps 1)-7) in the above version of the quadratic sieve? Give a big-$O$ estimate for the number of bit operations required by that step.
3. Use the method in this section with $P = 50$ and $A = 500$ to factor: (a) 1046603, (b) 1059691, and (c) 998771.

# References for §V.5

1. T. Caron and R. Silverman, "Parallel implementation of the quadratic sieve," *J. Supercomputing* **1** (1988), 273–290.
2. J. L. Gerver, "Factoring large numbers with a quadratic sieve," *Math. Comp.*, **41** (1983), 287–294.
3. A. Lenstra and H. W. Lenstra, Jr., eds. *The Development of the Number Field Sieve*, Springer-Verlag, 1993.
4. H. W. Lenstra, Jr. and C. Pomerance, "A rigorous time bound for factoring integers," *J. Amer. Math. Soc.* **5** (1992), 483–516.
5. C. Pomerance, "Analysis and comparison of some integer factoring algorithms," in *Computational Methods in Number Theory*, ed. by H. W. Lenstra, Jr. and R. Tijdeman, Mathematisch Centrum, Amsterdam, 1982, 89–139.
6. C. Pomerance, "Factoring," *Cryptology and Computational Number Theory, Proc. Symp. Appl. Math.* **42** (1990), 27–47.

# VI
# Elliptic Curves

In recent years a topic in number theory and algebraic geometry — elliptic curves (more precisely, the theory of elliptic curves defined over finite fields) — has found application in cryptography. The basic reason for this is that elliptic curves over finite fields provide an inexhaustible supply of finite abelian groups which, even when large, are amenable to computation because of their rich structure. Before (§ IV.3) we worked with the multiplicative groups of fields. In many ways elliptic curves are natural analogs of these groups; but they have the advantage that one has more flexibility in choosing an elliptic curve than in choosing a finite field.

We shall start by presenting the basic definitions and facts about elliptic curves. We shall include only the minimal amount of background necessary to understand the applications to cryptography in §§2–4, emphasizing examples and concrete descriptions at the expense of proofs and generality. For systematic treatments of the subject, see the references at the end of §1.

## 1 Basic facts

In this section let $K$ be a field. For us, $K$ will be either the field $\mathbf{R}$ of real numbers, the field $\mathbf{Q}$ of rational numbers, the field $\mathbf{C}$ of complex numbers, or the finite field $\mathbf{F}_q$ of $q = p^r$ elements.

**Definition.** Let $K$ be a field of characteristic $\neq 2$, 3, and let $x^3 + ax + b$ (where $a, b \in K$) be a cubic polynomial with no multiple roots. An *elliptic* *curve over K* is the set of points $(x, y)$ with $x, y \in K$ which satisfy the equation

$$y^2 = x^3 + ax + b, \tag{1}$$

together with a single element denoted $O$ and called the "point at infinity" (about which more will be said below).

If $K$ is a field of characteristic 2, then an *elliptic curve over K* is the set of points satisfying an equation of type either

$$y^2 + cy = x^3 + ax + b \tag{2a}$$

or else

$$y^2 + xy = x^3 + ax^2 + b \tag{2b}$$

(here we do not care whether or not the cubic on the right has multiple roots) together with a "point at infinity" $O$.

If $K$ is a field of characteristic 3, then an *elliptic curve over K* is the set of points satisfying the equation

$$y^2 = x^3 + ax^2 + bx + c \tag{3}$$

(where the cubic on the right has no multiple roots) together with a "point at infinity" $O$.

**Remarks.** **1.** There's a general form of the equation of an ellipse which applies to any field: $y^2 + a_1xy + a_3y = x^3 + a_2x^2 + a_4x + a_6$, which when $char\ K \neq 2$ can be transformed to $y^2 = x^3 + ax^2 + bx + c$ (and to the form $y^2 = x^3 + bx + c$ if $char\ K > 3$). In the case when the field $K$ has characteristic 2, this equation can be transformed either to (2a) or (2b).

**2.** If we let $F(x, y) = 0$ be the implicit equation for $y$ as a function of $x$ in (1) (or (2), (3)), i.e., $F(x, y) = y^2 - x^3 - ax - b$ (or $F(x, y) = y^2 + cy + x^3 + ax + b$, $y^2 + xy + x^3 + ax + b$, $y^2 - x^3 - ax^2 - bx - c$), then a point $(x, y)$ on the curve is said to be *non-singular* (or a *smooth* point) if at least one of the partial derivatives $\partial F/\partial x$, $\partial F/\partial y$ is nonzero at the point. (Derivatives of polynomials can be defined by the usual formulas over any field; see paragraph 5 at the beginning of Chapter II.) It is not hard to show that the condition that the cubic on the right in (1) and (3) not have multiple roots is equivalent to requiring that all points on the curve be nonsingular.

**Elliptic curves over the reals.** Before discussing some specific examples of elliptic curves over various fields, we shall introduce a centrally important fact about the set of points on an elliptic curve: they form an abelian group. In order to explain how this works visually, for the moment we shall assume that $K = \mathbf{R}$, i.e., the elliptic curve is an ordinary curve in the plane (plus one other point $O$ "at infinity").

**Definition.** Let $E$ be an elliptic curve over the real numbers, and let $P$ and $Q$ be two points on $E$. We define the negative of $P$ and the sum $P + Q$ according to the following rules:

1. If $P$ is the point at infinity $O$, then we define $-P$ to be $O$ and $P+Q$ to be $Q$; that is, $O$ serves as the additive identity ("zero element") of the group of points. In what follows, we shall suppose that neither $P$ nor $Q$ is the point at infinity.
2. The negative $-P$ is the point with the same $x$-coordinate but negative the $y$-coordinate of $P$, i.e., $-(x, y) = (x, -y)$. It is obvious from (1) that $(x, -y)$ is on the curve whenever $(x, y)$ is.
3. If $P$ and $Q$ have different $x$-coordinates, then it is not hard to see that the line $\ell = \overline{PQ}$ intersects the curve in exactly one more point $R$ (unless that line is tangent to the curve at $P$, in which case we take $R = P$, or at $Q$, in which case we take $R = Q$). Then define $P+Q$ to be $-R$, i.e., the mirror image (with respect to the $x$-axis) of the third point of intersection. The geometrical construction that gives $P+Q$ is illustrated in Example 1 below.
4. If $Q{=}-P$ (i.e., $Q$ has the same $x$-coordinate but minus the $y$-coordinate), then we define $P+Q = O$ (the point at infinity). (This is forced on us by (2).)
5. The final possibility is $P = Q$. Then let $\ell$ be the tangent line to the curve at $P$, let $R$ be the only other point of intersection of $\ell$ with the curve, and define $P+Q = -R$. ($R$ is taken to be $P$ if the tangent line has a "double tangency" at $P$, i.e., if $P$ is a point of inflection.)

**Example 1.** The elliptic curve $y^2 = x^3 - x$ in the $xy$-plane is sketched to the right. The diagram also shows a typical case of adding points $P$ and $Q$. To find $P+Q$ one draws a chord through $P$ and $Q$, and takes $P+Q$ to be the point symmetric (with respect to the $x$-axis) to the third point where the line through $P$ and $Q$ intersects the curve. If $P$ and $Q$ were the same point, i.e., if we wanted to find $2P$, we would use the tangent line to the curve at $P$; then $2P$ is the point symmetric to the third point where that tangent line intersects the curve.

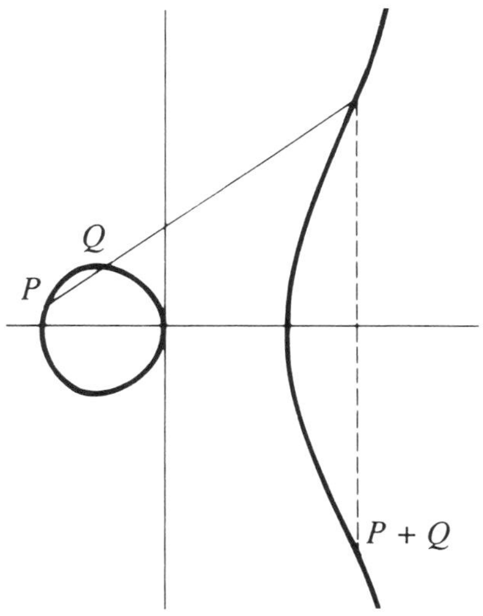


We now show why there is exactly one more point where the line $\ell$ through $P$ and $Q$ intersects the curve; at the same time we will derive a formula for the coordinates of this third point, and hence for the coordinates of $P+Q$.

Let $(x_1, y_1)$, $(x_2, y_2)$ and $(x_3, y_3)$ denote the coordinates of $P$, $Q$, and $P+Q$, respectively. We want to express $x_3$ and $y_3$ in terms of $x_1, y_1, x_2, y_2$.

Suppose that we are in case (3) in the definition of $P+Q$, and let $y = \alpha x+\beta$ be the equation of the line through $P$ and $Q$ (which is not a vertical line in case (3)). Then $\alpha = (y_2 - y_1)/(x_2 - x_1)$, and $\beta = y_1 - \alpha x_1$. A point on $\ell$, i.e., a point $(x, \alpha x + \beta)$, lies on the elliptic curve if and only if $(\alpha x + \beta)^2 = x^3 + ax + b$. Thus, there is one intersection point for each root of the cubic equation $x^3 - (\alpha x + \beta)^2 + ax + b$. We already know that there are the two roots $x_1$ and $x_2$, because $(x_1, \alpha x_1 + \beta)$, $(x_2, \alpha x_2 + \beta)$ are the points $P$, $Q$ on the curve. Since the sum of the roots of a monic polynomial is equal to minus the coefficient of the second-to-highest power, we conclude that the third root in this case is $x_3 = \alpha^2 - x_1 - x_2$. This leads to an expression for $x_3$, and hence $P + Q = (x_3, -(\alpha x_3 + \beta))$, in terms of $x_1, x_2, y_1, y_2$:

$$\begin{aligned} x_3 &= \left(\frac{y_2 - y_1}{x_2 - x_1}\right)^2 - x_1 - x_2; \\ y_3 &= -y_1 + \left(\frac{y_2 - y_1}{x_2 - x_1}\right)(x_1 - x_3). \end{aligned} \tag{4}$$

The case (5) when $P = Q$ is similar, except that $\alpha$ is now the derivative $dy/dx$ at $P$. Implicit differentiation of Equation (1) leads to the formula $\alpha = (3x_1^2 + a)/2y_1$, and so we obtain the following formulas for the coordinates of twice $P$:

$$\begin{aligned} x_3 &= \left(\frac{3x_1^2 + a}{2y_1}\right)^2 - 2x_1; \\ y_3 &= -y_1 + \left(\frac{3x_1^2 + a}{2y_1}\right)(x_1 - x_3). \end{aligned} \tag{5}$$

**Example 2.** On the elliptic curve $y^2 = x^3 - 36x$ let $P = (-3, 9)$ and $Q = (-2, 8)$. Find $P + Q$ and $2P$.

**Solution.** Substituting $x_1 = -3$, $y_1 = 9$, $x_2 = -2$, $y_2 = 8$ in the first equation in (4) gives $x_3 = 6$; then the second equation in (4) gives $y_3 = 0$. Next, substituting $x_1 = -3$, $y_1 = 9$, $a = -36$ in the first equation in (5) gives $25/4$ for the $x$-coordinate of $2P$; then the second equation in (5) gives $-35/8$ for its $y$-coordinate.

There are several ways of proving that the above definition of $P + Q$ makes the points on an elliptic curve into an abelian group. One can use an argument from projective geometry, a complex analytic argument with doubly periodic functions, or an algebraic argument involving divisors on curves. See the references at the end of the section for proofs of each type.

As in any abelian group, we use the notation $nP$ to denote $P$ added to itself $n$ times if $n$ is positive, and otherwise $-P$ added to itself $|n|$ times.

We have not yet said much about the "point of infinity" $O$. By definition, it is the identity of the group law. In the diagram above, it should be visualized as sitting infinitely far up the $y$-axis, in the limiting direction of the ever-steeper tangents to the curve. It is the "third point of intersection" of any vertical line with the curve; that is, such a line has points of intersection of the form $(x_1, y_1)$, $(x_1, -y_1)$ and $O$. A more natural way to introduce the point $O$ is as follows.

By the *projective plane* we mean the set of equivalence classes of triples $(X, Y, Z)$ (not all components zero) where two triples are said to be equivalent if they are a scalar multiple of one another, i.e., $(\lambda X, \lambda Y, \lambda Z) \sim (X, Y, Z)$. Such an equivalence class is called a *projective point.* If a projective point has nonzero $Z$, then there is one and only one triple in its equivalence class of the form $(x, y, 1)$: simply set $x = X/Z$, $y = Y/Z$. Thus, the projective plane can be identified with all points $(x, y)$ of the ordinary ("affine") plane plus the points for which $Z = 0$. The latter points make up what is called the *line at infinity*; roughly speaking, it can be visualized as the "horizon" on the plane. Any equation $F(x, y) = 0$ of a curve in the affine plane corresponds to an equation $\tilde{F}(X, Y, Z) = 0$ satisfied by the corresponding projective points: simply replace $x$ by $X/Z$ and $y$ by $Y/Z$ and multiply by a power of $Z$ to clear the denominators. For example, if we apply this procedure to the affine equation (1) of an elliptic curve, we obtain its "projective equation" $Y^2Z = X^3 + aXZ^2 + bZ^3$. This latter equation is satisfied by all projective points $(X, Y, Z)$ with $Z \neq 0$ for which the corresponding affine points $(x, y)$, where $x = X/Z$, $y = Y/Z$, satisfy (1). In addition, what projective points $(X, Y, Z)$ on the line at infinity satisfy the equation $\tilde{F} = 0$? Setting $Z = 0$ in the equation leads to $0 = X^3$, i.e., $X = 0$. But the only equivalence class of triples $(X, Y, Z)$ with both $X$ and $Z$ zero is the class of $(0, 1, 0)$. This is the point we call $O$. It is the point on the intersection of the $y$-axis with the line at infinity.

**Elliptic curves over the complexes.** The algebraic formulas (4)–(5) for adding points on an elliptic curve over the reals actually make sense over any field. (If the field has characteristic 2 or 3, one derives similar equations starting from Equation (2) or (3).) It can be shown that these formulas give an abelian group law on an elliptic curve over any field.

In particular, let $E$ be an elliptic curve defined over the field $\mathbf{C}$ of complex numbers. Thus, $E$ is the set of pairs $(x, y)$ of complex numbers satisfying Equation (1), together with the point at infinity $O$. Although $E$ is a "curve," if we think in terms of familiar geometrical pictures, it is 2-dimensional, i.e., it is a surface in the 4-real-dimensional space whose coordinates are the real and imaginary parts of $x$ and $y$. We now describe how $E$ can be visualized as a surface.

Let $L$ be a *lattice* in the complex plane. This means that $L$ is the abelian group of all integer combinations of two complex numbers $\omega_1$ and $\omega_2$ (where $\omega_1$ and $\omega_2$ span the plane, i.e., do not lie on the same line through the origin): $L = \mathbf{Z}\omega_1 + \mathbf{Z}\omega_2$. For example, if $\omega_1 = 1$ and $\omega_2 = i$, then $L$ is the Gaussian integers, the square grid consisting of all complex numbers with integer real and imaginary parts.

Given an elliptic curve (1) over the complex numbers, it turns out that there exist a lattice $L$ and a complex function, called the "Weierstrass $\wp$-function" and denoted $\wp_L(z)$, which has the following properties.

1. $\wp(z)$ is analytic except for a double pole at each point of $L$;
2. $\wp(z)$ satisfies the differential equation $\wp'^2 = \wp^3 + a\wp + b$, and hence for

any $z \notin L$ the point $(\wp(z), \wp'(z))$ lies on the elliptic curve $E$;

3. two complex numbers $z_1$ and $z_2$ give the same point $(\wp(z), \wp'(z))$ on $E$ if and only if $z_1 - z_2 \in L$;
4. the map that associates any $z \notin L$ to the corresponding point $(\wp(z), \wp'(z))$ on $E$ and associates any $z \in L$ to the point at infinity $O \in E$ gives a 1-to-1 correspondence between $E$ and the quotient of the complex plane by the subgroup $L$ (denoted $\mathbf{C}/L$);
5. this 1-to-1 correspondence is an isomorphism of abelian groups. In other words, if $z_1$ corresponds to the point $P \in E$ and $z_2$ corresponds to $Q \in E$, then the complex number $z_1 + z_2$ corresponds to the point $P + Q$.

Thus, we can think of the abelian group $E$ as equivalent to the complex plane modulo a suitable lattice. To visualize the latter group, note that every equivalence class $z + L$ has one and only one representative in the "fundamental parallelogram" consisting of complex numbers of the form $a\omega_1 + b\omega_2$, $0 \leq a, b < 1$ (for example, if $L$ is the Gaussian integers, the fundamental parallelogram is the unit square). Since opposite points on the parallel sides of the boundary of the parallelogram differ by a lattice point, they are equal in $\mathbf{C}/L$. That is, we think of them as "glued together." If we visualize this — folding over one side of the parallelogram to meet the opposite side (obtaining a segment of a cylinder) and then folding over again and gluing the opposite circles – we see that we obtain a "torus" (donut), pictured below.

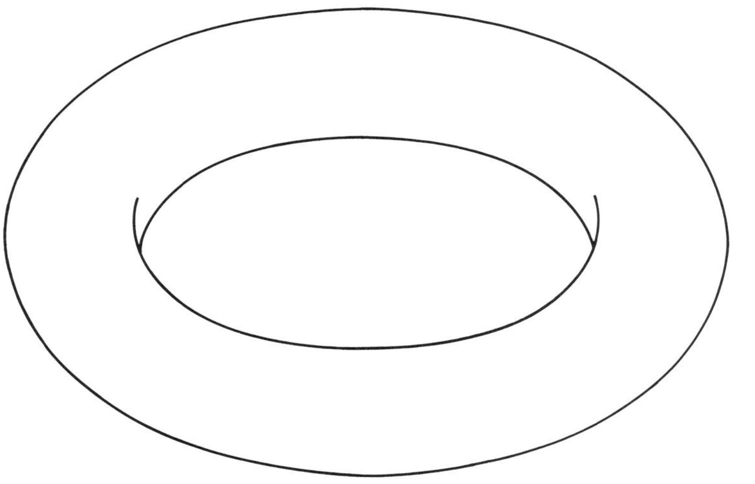

As a group, the torus is the product of two copies of a circle, i.e., its points can be parametrized by ordered pairs of angles $(\alpha, \beta)$. (More precisely, if the torus was obtained from the lattice $L = \mathbf{Z}\omega_1 + \mathbf{Z}\omega_2$, then we write an element in $\mathbf{C}/L$ in the form $a\omega_1 + b\omega_2$ and take $\alpha = 2\pi a$, $\beta = 2\pi b$.) Thus, we can think of an elliptic curve over the complex numbers as a generalization to two real dimensions of the circle in the real plane. In fact, this analogy goes much farther than one might think. The "elliptic functions" (which tell us how to go back from a point $(x, y) \in E$ to the complex number $z$ for which $(x, y) = (\wp(z), \wp'(z))$) turn out to have some properties analogous to the familiar function *Arcsin* (which tells us how to go back from a point on the unit circle to the real number that corresponds to that point when we "wrap" the real number line around the circle). In the algebraic number theory of elliptic curves, one finds a deep analogy between the coordinates of the "$n$-division points" on an elliptic curves (the points $P$ such that $nP$ is the identity $O$) and the $n$-division points on the unit circle (which are the $n$-th roots of unity in the complex plane). See the references at the end of the section for more information on this, and for the definition of the Weierstrass $\wp$-function and proofs of its properties.

**Elliptic curves over the rationals.** In Equation (1), if $a$ and $b$ are rational numbers, it is natural to look for rational solutions $(x, y)$, i.e., to consider the elliptic curve over the field $\mathbf{Q}$ of rational numbers. There is a vast theory of elliptic curves over the rationals. It turns out that the abelian group is finitely generated (the Mordell theorem). This means that it consists of a finite "torsion subgroup" (the points of finite order) plus the subgroup generated by a finite number of points of infinite order. The number of generators needed for the infinite part is called the *rank* $r$; it is zero if and only if the entire group is finite. The study of the rank $r$ and other features of the group of an elliptic curve over $\mathbf{Q}$ is related to many interesting questions in number theory and algebraic geometry. For example, a question asked since ancient times — "Given a positive integer $n$, when does there exist a right triangle with rational sides whose area is $n$?" — turns out to be equivalent to the question "Is the rank of the elliptic curve $y^2 = x^3 - n^2x$ greater than zero?" The case $n = 6$ and the $3 - 4 - 5$ right triangle lead to the point $P$ in Example 2, which is a point of infinite order on the curve $y^2 = x^3 - 36x$. For more information on this subject, we again refer the reader to the references at the end of the section.

**Points of finite order.** The *order* $N$ of a point $P$ on an elliptic curve is the smallest positive integer such that $NP = O$; of course, such a finite $N$ need not exist. It is often of interest to find points $P$ of finite order on an elliptic curve, especially for elliptic curves defined over $\mathbf{Q}$.

**Example 3.** Find the order of $P = (2, 3)$ on $y^2 = x^3 + 1$.

**Solution.** Using (5), we find that $2P = (0, 1)$, and using (5) again gives $4P = 2(2P) = (0, -1)$. Thus, $4P = -2P$, and so $6P = O$. Thus, the order of $P$ is 2, 3 or 6. But $2P = (0, 1) \neq O$, and if $P$ had order 3, then $4P = P$, which is not true. Thus, $P$ has order 6.

**Elliptic curves over a finite field.** For the rest of this section we shall let $K$ be the finite field $\mathbf{F}_q$ of $q = p^r$ elements. Let $E$ be an elliptic curve defined over $\mathbf{F}_q$. If $p = 2$ or 3, then $E$ is given by an equation of the form (2) or (3), respectively.

It is easy to see that an elliptic curve can have at most $2q+1$ $\mathbf{F}_q$-points, i.e., the point at infinity along with $2q$ pairs $(x, y)$ with $x, y \in \mathbf{F}_q$ which satisfy (1) (or (2) or (3) if $p = 2$ or 3). Namely, for each of the $q$ possible $x$'s there are at most 2 $y$'s which satisfy (1).

But since only half of the elements of $\mathbf{F}_q^*$ have square roots, one would expect (if $x^3 + ax + b$ were random elements of the field) that there would be only about half that number of $\mathbf{F}_q$-points. More precisely, let $\chi$ be the quadratic character of $\mathbf{F}_q$. This is the map which takes $x \in \mathbf{F}_q^*$ to $\pm 1$ depending on whether or not $x$ has a square root in $\mathbf{F}_q$ (and we take $\chi(0) = 0$). For example, if $q = p$ is a prime, then $\chi(x) = \left(\frac{x}{p}\right)$ is the Legendre symbol (see § II.2). Thus, in all cases the number of solutions $y \in \mathbf{F}_q$ to the equation $y^2 = u$ is equal to $1 + \chi(u)$, and so the number of solutions to (1) (counting the point at infinity) is

$$1 + \sum_{x \in \mathbf{F}_q} (1 + \chi(x^3 + ax + b)) = q + 1 + \sum_{x \in \mathbf{F}_q} \chi(x^3 + ax + b). \qquad (6)$$

We would expect that $\chi(x^3+ax+b)$ would be equally likely to be $+1$ and $-1$. Taking the sum is much like a "random walk": toss a coin $q$ times, moving one step forward for heads, one step backward for tails. In probability theory one computes that the net distance traveled after $q$ tosses is of the order of $\sqrt{q}$. The sum $\sum \chi(x^3 + ax + b)$ behaves a little like a random walk. More precisely, one finds that this sum is bounded by $2\sqrt{q}$. This result is Hasse's Theorem; for a proof, see § V.1 of Silverman's book on elliptic curves cited in the references.

**Hasse's Theorem.** *Let $N$ be the number of $\mathbf{F}_q$-points on an elliptic curve defined over $\mathbf{F}_q$. Then*

$$|N - (q + 1)| \le 2\sqrt{q}.$$

In addition to the number $N$ of elements on an elliptic curve defined over $\mathbf{F}_q$, we might want to know the actual structure of the abelian group. This abelian group is not necessarily cyclic, but it can be shown that it is always a product of two cyclic groups. This means that it is isomorphic to a product of $p$-primary groups of the form $\mathbf{Z}/p^\alpha\mathbf{Z} \times \mathbf{Z}/p^\beta\mathbf{Z}$, where the product is taken over primes dividing $N$ (here $\alpha \ge 1$, $\beta \ge 0$). By the *type* of the abelian group of $\mathbf{F}_q$-points on $E$, we mean a listing $(\ldots, p^\alpha, p^\beta, \ldots)_{p|N}$ of the orders of the cyclic $p$-primary factors (we omit $p^\beta$ when $\beta = 0$). It is not always easy to find the type.

**Example 4.** Find the type of $y^2 = x^3 - x$ over $\mathbf{F}_{71}$.

**Solution.** We first find the number of points $N$. In (6) we notice that in the sum the term for $x$ and the term for $-x$ cancel, because

$\chi((-x)^3-(-x)) = \chi(-1)\chi(x^3-x)$, and $\chi(-1) = -1$ because $71 \equiv 3 \mod 4$. Thus, $N = q+1 = 72$. Notice that there are exactly four points of order 2 (including the identity $O$), because they correspond to the roots of $x^3 - x = x(x-1)(x+1)$ (see Exercise 4(a) below). This means that the 2-primary part of the group has type $(4,2)$, and so the type of the group is either $(4,2,3,3)$ or else $(4,2,9)$, depending on whether there are 9 or 3 points of order 3, respectively. So it remains to determine whether or not there can be 9 points of order 3. Note that for any $P \neq O$ the equation $3P = O$ is equivalent to $2P = \pm P$, i.e., to the condition that the $x$-coordinates of $P$ and $2P$ be the same. By (5), this means that $((3x^2-1)/2y)^2 - 2x = x$, i.e., $(3x^2-1)^2 = 12xy^2 = 12x^4 - 12x^2$. Simplifying, we obtain $3x^4 - 6x^2 - 1 = 0$. There are at most 4 roots to this equation in $\mathbf{F}_{71}$. If there are four roots, then each root can give at most 2 points (by taking $y = \pm\sqrt{x^3-x}$ if $x^3 - x$ has a square root modulo 71), and so we may in this way obtain 9 points of order 3 (including the identity $O$ at infinity). Otherwise, there must be fewer than 9 points of order 3 (and hence exactly 3 points of order 3). But if the root $x$ of the quartic polynomial has $x^3 - x$ a square modulo 71, then the root $-x$ of the quartic has $(-x)^3 - (-x) = -(x^3 - x)$ a nonsquare modulo 71. Thus, we cannot get 9 points of order 3, and so the type of the group is $(4,2,9)$.

**Extensions of finite fields, and the Weil conjectures.** If an elliptic curve $E$ is defined over $\mathbf{F}_q$, then it is also defined over $\mathbf{F}_{q^r}$ for $r = 1, 2, \ldots$, and so it is meaningful to consider the $\mathbf{F}_{q^r}$-points, i.e., to look at solutions of (1) over extension fields. If we start out with $\mathbf{F}_q$ as the field over which $E$ is defined, we let $N_r$ denote the number of $\mathbf{F}_{q^r}$-points on $E$. (Thus, $N_1 = N$ is the number of points with coordinates in our "ground field" $\mathbf{F}_q$.)

From the numbers $N_r$ one forms the "generating series" $Z(T; E/\mathbf{F}_q)$, which is the formal power series in $\mathbf{Q}[[T]]$ defined by setting

$$Z(T; E/\mathbf{F}_q) = e^{\sum N_r T^r / r}, \tag{7}$$

in which $T$ is an indeterminate, the notation $E/\mathbf{F}_q$ designates the elliptic curve and the field we're taking as our ground field, and the sum on the right is over all $r = 1, 2, \ldots$ . It can be shown that the series on the right (obtained by taking the infinite product of the exponential power series $e^{N_r T^r/r}$) actually has positive integer coefficients. This power series is called the *zeta-function* of the elliptic curve (over $\mathbf{F}_q$), and is a very important object associated with $E$.

The "Weil conjectures" (now a theorem of P. Deligne) say in a much more general context (algebraic varieties of any dimension) that the zeta-function has a very special form. In the case of an elliptic curve $E/\mathbf{F}_q$ Weil proved the following.

**Weil conjectures [theorem] for an elliptic curve.** *The zeta-function is a rational function of $T$ having the form*

$$Z(T; E/\mathbf{F}_q) = \frac{1 - aT + qT^2}{(1 - T)(1 - qT)}, \tag{8}$$

*where only the integer $a$ depends on the particular elliptic curve $E$. The value $a$ is related to $N = N_1$ as follows: $N = q + 1 - a$. In addition, the discriminant of the quadratic polynomial in the numerator is negative (i.e., $a^2 < 4q$, which is Hasse's Theorem) and so the quadratic has two complex conjugate roots $\alpha$, $\beta$ both of absolute value $\sqrt{q}$. (More precisely, $1/\alpha$ and $1/\beta$ are the roots, and $\alpha$, $\beta$ are the "reciprocal roots.")*

For a proof, see § V.2 of Silverman's book.

**Remark.** If we write the numerator of (8) in the form $(1-\alpha T)(1-\beta T)$ and then take the derivative of the logarithm of both sides (replacing the left side by its definition (7)), we soon see that the formula (8) is equivalent to writing the sequence of relations

$$N_r = q^r + 1 - \alpha^r - \beta^r, \qquad r = 1, 2, \ldots.$$

Since $\alpha$ and $\beta$, along with $a$, are determined once you know $N = N_1$, this means that the number of points over $\mathbf{F}_q$ uniquely determines the number of points over any extension field. Thus, among other things, Weil's conjectures for elliptic curves are useful for determining the number of points over extension fields of large degree.

**Example 5.** The zeta-function of the elliptic curve $y^2 + y = x^3$ over $\mathbf{F}_2$ is easily computed from the fact that there are three $\mathbf{F}_2$-points. It is $(1 + 2T^2)/(1 - T)(1 - 2T)$, i.e., the reciprocal roots of the numerator are $\pm i\sqrt{2}$. This leads to the formula

$$N_r = \begin{cases} 2^r + 1, & \text{if } r \text{ is odd;} \\ 2^r + 1 - 2(-2)^{r/2}, & \text{if } r \text{ is even.} \end{cases} \tag{9}$$

To conclude this section, we remark that there are many analogies between the group of $\mathbf{F}_q$-points on an elliptic curve and the multiplicative group $(\mathbf{F}_q)^*$. For example, they have approximately the same number of elements, by Hasse's Theorem. But the former construction of an abelian group has a major advantage that explains its usefulness in cryptography: for a single (large) $q$ there are many different elliptic curves and many different $N$ that one can choose from. Elliptic curves offer a rich source of "naturally occurring" finite abelian groups. We shall take advantage of this in the next three sections.

## *Exercises*

1. If $E$ is an elliptic curve defined over $\mathbf{C}$ whose equation (1) actually has coefficients $a, b \in \mathbf{R}$, then the points of $E$ with real coordinates form a subgroup. What are the possible subgroups of the complex curve $E$ (which as a group is isomorphic to the product of the circle group with

itself) which can occur as the group of real points? Give an example of each.

2. How many points $P$ of order $n$ (i.e., $nP = O$) are there on an elliptic curve defined over $\mathbf{C}$? How about on an elliptic curve over $\mathbf{R}$?
3. Give an example of an elliptic curve over $\mathbf{R}$ which has exactly 2 points of order 2, and another example which has exactly 4 points of order 2.
4. Let $P$ be a point on an elliptic curve over $\mathbf{R}$. Suppose that $P$ is not the point at infinity. Give a geometric condition that is equivalent to $P$ being a point of order (a) 2; (b) 3; (c) 4.
5. Each of the following points has finite order on the given elliptic curve over $\mathbf{Q}$. In each case, find the order of $P$.
(a) $P = (0, 16)$ on $y^2 = x^3 + 256$.
(b) $P = (\frac{1}{2}, \frac{1}{2})$ on $y^2 = x^3 + \frac{1}{4}x$.
(c) $P = (3, 8)$ on $y^2 = x^3 - 43x + 166$.
(d) $P = (0, 0)$ on $y^2 + y = x^3 - x^2$ (which can be written in the form (1) by making the change of variables $y \longrightarrow y - \frac{1}{2}$, $x \longrightarrow x + \frac{1}{3}$).
6. Derive addition formulas similar to (4)–(5) for elliptic curves in characteristic 2, 3 (see Equations (2)–(3)).
7. Prove that there are $q + 1$ $\mathbf{F}_q$-points on the elliptic curve
(a) $y^2 = x^3 - x$ when $q \equiv 3 \ mod\ 4$;
(b) $y^2 = x^3 - 1$ when $q \equiv 2 \ mod\ 3$ (where $q$ is odd);
(c) $y^2 + y = x^3$ when $q \equiv 2 \ mod\ 3$ ($q$ may be even here).
8. For all odd prime powers $q = p^r$ up to 27 find the order and type of the group of $\mathbf{F}_q$-points on the elliptic curves $y^2 = x^3 - x$ and $y^2 = x^3 - 1$ (in the latter case when $p \neq 3$). In some cases you will have to check how many points have order 3 or 4.
9. Let $q = 2^r$, and let the elliptic curve $E$ over $\mathbf{F}_q$ have equation $y^2 + y = x^3$.
(a) Express the coordinates of $-P$ and $2P$ in terms of the coordinates of $P$.
(b) If $q = 16$, show that every $P \in E$ is a point of order 3.
(c) Show that any point of $E$ with coordinates in $\mathbf{F}_{16}$ actually has coordinates in $\mathbf{F}_4$. Then use Hasse's Theorem with $q = 4$ and 16 to determine the number of points on the curve.
10. Compute the zeta-functions of the two curves in Exercise 8 over $\mathbf{F}_p$ for $p = 5, 7, 11, 13$.
11. Compute the zeta function of the curve $y^2 + y = x^3 - x + 1$ over $\mathbf{F}_p$ for $p = 2$ and 3. (First show that $N_1 = 1$ in both cases.) Letting $\mathbf{N}(x) = x \cdot \bar{x}$ denote the norm of a complex number, find a simple formula for $N_r$.

## References for § VI.1

1. W. Fulton, *Algebraic Curves*, Benjamin, 1969.

2. D. Husemöller, *Elliptic Curves*, Springer–Verlag, 1987.
3. N. Koblitz, *Introduction to Elliptic Curves and Modular Forms*, 2nd ed., Springer–Verlag, 1993.
4. N. Koblitz, "Why study equations over finite fields?," *Math. Magazine* **55** (1982), 144–149.
5. S. Lang, *Elliptic Curves: Diophantine Analysis*, Springer–Verlag, 1978.
6. J. Silverman, *The Arithmetic of Elliptic Curves*, Springer–Verlag, 1986.

# 2 Elliptic curve cryptosystems

In § IV.3 we saw how the finite abelian group $\mathbf{F}_q^*$ — the multiplicative group of a finite field — can be used to create public key cryptosystems. More precisely, it was the difficulty of solving the discrete logarithm problem in finite fields that led to the cryptosystems discussed in § IV.3. The purpose of this section is to make analogous public key systems based on the finite abelian group of an elliptic curve $E$ defined over $\mathbf{F}_q$.

Before introducing the cryptosystems themselves, there are some preliminary matters that must be discussed.

**Multiples of points.** The elliptic curve analogy of multiplying two elements of $\mathbf{F}_q^*$ is *adding* two points on $E$, where $E$ is an elliptic curve defined over $\mathbf{F}_q$. Thus, the analog of raising to the $k$-th power in $\mathbf{F}_q^*$ is multiplication of a point $P \in E$ by an integer $k$. Raising to the $k$-th power in a finite field can be accomplished by the repeated squaring method in $O(log\, k\, log^3 q)$ bit operations (see Proposition II.1.9). Similarly, we shall show that the multiple $kP \in E$ can be found in $O(log\, k\, log^3 q)$ bit operations by the method of repeated doubling.

**Example 1.** To find $100P$ we write $100P = 2(2(P+2(2(2(P+2P)))))$, and end up performing 6 doublings and 2 additions of points on the curve.

**Proposition VI.2.1.** *Suppose that an elliptic curve $E$ is defined by a Weierstrass equation (equation (1), (2) or (3) in the last section) over a finite field $\mathbf{F}_q$. Given $P \in E$, the coordinates of $kP$ can be computed in $O(log\, k\; log^3 q)$ bit operations.*

**Proof.** Note that there are fewer than 20 computations in $\mathbf{F}_q$ (multiplications, divisions, additions, or subtractions) involved in computing the coordinates of a sum of two points by means of equations (4)–(5) (or the analogous equations in Exercise 6 of §1). Thus, by Proposition II.1.9, each such addition (or doubling) of points takes time $O(log^3 q)$. Since there are $O(log\, k)$ steps in the repeated doubling method (see the proof of Proposition I.3.6), we conclude that the coordinates of $kP$ can be calculated in $O(log\, k\; log^3 q)$ bit operations.

**Remarks. 1.** The time estimate in Proposition VI.2.1 is not the best possible, especially in the case when our finite field has characteristic $p = 2$. But we shall be satisfied with the estimates that result from using the most obvious algorithms for arithmetic in finite fields.

**2.** If we happen to know the number $N$ of points on our elliptic curve $E$, and if $k > N$, then since $NP = O$ we can replace $k$ by its least nonnegative residue modulo $N$ before computing $kP$; in this case we can replace the time estimate by $O(log^4 q)$ (recall that $N \le q+1+2\sqrt{q} = O(q)$). There is an algorithm due to René Schoof which computes $N$ in $O(log^8 q)$ bit operations.

**Imbedding plaintexts.** We shall want to encode our plaintexts as points on some given elliptic curve $E$ defined over a finite field $\mathbf{F}_q$. We want to do this in a simple systematic way, so that the plaintext $m$ (which we may regard as an integer in some range) can readily be determined from knowledge of the coordinates of the corresponding point $P_m$. Notice that this "encoding" is not the same thing as encryption. Later we shall discuss ways to encrypt the plaintext points $P_m$. But an authorized user of the system must be able to recover $m$ after deciphering the ciphertext point.

There are two remarks that should be made here. In the first place, there is no polynomial time (in $log\, q$) *deterministic* algorithm known for writing down a large number of points on an arbitrary elliptic curve $E$ over $\mathbf{F}_q$. However, there are probabilistic algorithms for which the chance of failure is very small, as we shall see below. In the second place, it is not enough to generate random points of $E$: in order to encode a large number of possible messages $m$, we need a systematic way to generate points that are related to $m$ in some way, for example, the $x$-coordinate has a simple relationship to the integer $m$.

Here is one possible probabilistic method to imbed plaintexts as points on an elliptic curve $E$ defined over $\mathbf{F}_q$, where $q = p^r$ is assumed to be large (and odd; see Exercise 8 below for $q = 2^r$). Let $\kappa$ be a large enough integer so that we are satisfied with a failure probability of 1 out of $2^\kappa$ when we attempt to imbed a plaintext message unit $m$; in practice $\kappa = 30$ or at worse $\kappa = 50$ should suffice. We suppose that our message units $m$ are integers $0 \le m < M$. We also suppose that our finite field is chosen so that $q > M\kappa$. We write the integers from 1 to $M\kappa$ in the form $m\kappa + j$, where $1 \le j \le \kappa$, and we set up a 1-to-1 correspondence between such integers and a set of elements of $\mathbf{F}_q$. For example, we write such an integer as an $r$-digit integer to the base $p$, and take the $r$ digits, considered as elements of $\mathbf{Z}/p\mathbf{Z}$, as the coefficients of a polynomial of degree $r-1$ corresponding to an element of $\mathbf{F}_q$. That is, the integer $(a_{r-1}a_{r-2}\cdots a_1 a_0)_p$ corresponds to the polynomial $\sum_{i=0}^{r-1} a_j X^j$, which, considered modulo some fixed degree-$r$ irreducible polynomial over $\mathbf{F}_p$, gives an element of $\mathbf{F}_q$.

Thus, given $m$, for each $j = 1, 2, \ldots, \kappa$ we obtain an element $x$ of $\mathbf{F}_q$ corresponding to $m\kappa + j$. For such an $x$, we compute the right side of the equation

$$y^2 = f(x) = x^3 + ax + b,$$

and try to find a square root of $f(x)$ using the method explained at the end of § II.2. (Although the algorithm was given for the prime field $\mathbf{F}_p$, it carries over word for word to any finite field $\mathbf{F}_q$. In order to use it we must have a nonsquare $g$ in the field, which can easily be found by a probabilistic algorithm.) If we find a $y$ such that $y^2 = f(x)$, we take $P_m = (x, y)$. If it turns out that $f(x)$ is a nonsquare, then we increment $j$ by 1 and try again with the corresponding $x$. Provided we find an $x$ for which $f(x)$ is a square before $j$ gets bigger than $\kappa$, we can recover $m$ from the point $(x, y)$ by the formula $m = \left[(\tilde{x} - 1)/\kappa\right]$, where $\tilde{x}$ is the integer corresponding to $x$ under the 1-to-1 correspondence between integers and elements of $\mathbf{F}_q$. Since $f(x)$ is a square for approximately 50% of all $x$, there is only about a $2^{-\kappa}$ probability that this method will fail to produce a point $P_m$ whose $x$-coordinate corresponds to an integer $\tilde{x}$ between $m\kappa+1$ and $m\kappa+\kappa$. (More precisely, the probability that $f(x)$ is a square is essentially equal to $N/2q$; but $N/2q$ is very close to $1/2$.)

**Discrete log on $E$.** In § IV.3 we discussed public key cryptosystems based on the discrete logarithm problem in the multiplicative group of a finite field. Now we do the same in the group (under addition of points) of an elliptic curve $E$ defined over a finite field $\mathbf{F}_q$.

**Definition.** If $E$ is an elliptic curve over $\mathbf{F}_q$ and $B$ is a point of $E$, then the *discrete log problem* on $E$ (to the base $B$) is the problem, given a point $P \in E$, of finding an integer $x \in \mathbf{Z}$ such that $xB = P$ if such an integer $x$ exists.

It is likely that the discrete log problem on elliptic curves will prove to be more intractible than the discrete log problem in finite fields. The strongest techniques developed for use in finite fields do not seem to work on elliptic curves. This is especially true in the case of characteristic 2. As explained in Odlyzko's survey article cited in the references, special methods for solving the discrete log problem in $\mathbf{F}_{2^r}^*$ make it relatively easy to compute discrete logs, and hence break the cryptosystems discussed in § IV.3, unless $r$ is chosen to be rather large. It seems that the analogous systems using elliptic curves defined over $\mathbf{F}_{2^r}$ (see below) will be secure with significantly smaller values of $r$. Since there are practical reasons (relating to both computer hardware and software) for preferring to do arithmetic over the fields $\mathbf{F}_{2^r}$, the public key cryptosystems discussed below may turn out to be more convenient in applications than the systems based on the discrete log problem in $\mathbf{F}_q^*$.

Until 1990, the only discrete log algorithms known for an elliptic curve were the ones that work in any group, irrespective of any particular structure. These are exponential time algorithms, provided that the order of the group is divisible by a large prime factor. But then Menezes, Okamoto, and Vanstone found a new approach to the discrete log problem on an elliptic curve $E$ defined over $\mathbf{F}_q$. Namely, they used the Weil pairing (see §III.8 of Silverman's textbook cited in the references to §1) to imbed the group $E$ into the multiplicative group of some extension field $\mathbf{F}_{q^k}$. This imbedding reduces the discrete log problem on $E$ to the discrete log problem in $\mathbf{F}_{q^k}^*$.

However, in order for the Weil pairing reduction to help, it is essential for the extension degree $k$ to be small. Essentially the only elliptic curves for which $k$ is small are the so-called "supersingular" elliptic curves, the most familiar examples of which are curves of the form $y^2 = x^3 + ax$ when the characteristic $p$ of $\mathbf{F}_q$ is $\equiv -1 \pmod 4$, and curves of the form $y^2 = x^3 + b$ when $p \equiv -1 \pmod 3$. The vast majority of elliptic curves, however, are nonsupersingular. For them, the reduction almost never leads to a subexponential algorithm (see my paper in *Journal of Cryptology* cited in the references).

Thus, a key advantage of elliptic curve cryptosystems is that no subexponential algorithm is known that breaks the system, provided that we avoid supersingular curves and also curves whose order has no large prime factor.

We now describe analogs of the public key systems in § IV.3 based on the discrete log problem on an elliptic curve $E$ defined over a finite field $\mathbf{F}_q$.

**Analog of the Diffie–Helman key exchange.** Suppose that Aïda and Bernardo want to agree upon a key which will later be used in conjunction with a classical cryptosystem. They first publicly choose a finite field $\mathbf{F}_q$ and an elliptic curve $E$ defined over it. Their key will be constructed from a random point $P$ on the elliptic curve. For example, if they have a random point $P \in E$, then taking the $x$-coordinate of $P$ gives a random element of $\mathbf{F}_q$, which can then be converted to a random $r$-digit base-$p$ integer (where $q = p^r$) which serves as the key to their classical cryptosystem. (Here we're using the word "random" in an imprecise sense; all we mean is that its choice is arbitrary and unpredictable in a large set of admissible keys.) Their task is to choose the point $P$ in such a way that all of their communication with one another is public and yet no one other than the two of them knows what $P$ is.

Aïda and Bernardo first publicly choose a point $B \in E$ to serve as their "base." $B$ plays the role of the generator $g$ in the finite–field Diffie–Hellman system. However, we do not want to insist that $B$ be a generator of the group of points on $E$. In fact, the latter group may fail to be cyclic. Even if it is cyclic, we want to avoid the effort of verifying that $B$ is a generator (or even determining the number $N$ of points, which we do not need to know in what follows). We would like the subgroup generated by $B$ to be large, preferably of the same order of size as $E$ itself. This question will be discussed later. For now, let us suppose that $B$ is a fixed publicly known point on $E$ whose order is very large (either $N$ or a large divisor of $N$).

To generate a key, first Aïda chooses a random integer $a$ of order of magnitude $q$ (which is approximately the same as $N$), which she keeps secret. She computes $aB \in E$, which she makes public. Bernardo does the same: he chooses a random $b$ and makes public $bB \in E$. The secret key they use is then $P = abB \in E$. Both users can compute this key. For example,

Aïda knows $bB$ (which is public knowledge) and her own secret $a$. However, a third party knows only $aB$ and $bB$. Without solving the discrete logarithm problem — finding $a$ knowing $B$ and $aB$ (or finding $b$ knowing $B$ and $bB$) — there seems to be no way to compute $abB$ knowing only $aB$ and $bB$.

**Analog of Massey–Omura.** As in the finite-field situation, this is a public key cryptosystem for transmitting message units $m$, which we now suppose have been imbedded as points $P_m$ on some fixed (and publicly known) elliptic curve $E$ over $\mathbf{F}_q$ (where $q$ is large). We also suppose that the number $N$ of points on $E$ has been computed (and is also publicly known). Each user of the system secretly selects a random integer $e$ between 1 and $N$ such that $g.c.d.(e, N) = 1$ and, using the Euclidean algorithm, computes its inverse $d = e^{-1}$ *mod* $N$, i.e., an integer $d$ such that $de \equiv 1$ *mod* $N$. If Alice wants to send the message $P_m$ to Bob, first she sends him the point $e_A P_m$ (where the subscript $A$ denotes the user Alice). This means nothing to Bob, who, knowing neither $d_A$ nor $e_A$, cannot recover $P_m$. But, without attempting to make sense of this point, he multiplies it by *his* $e_B$, and sends $e_B e_A P_m$ back to Alice. The third step is for Alice to unravel the message part of the way by multiplying the point $e_B e_A P_m$ by $d_A$. Since $NP_m = O$ and $d_A e_A \equiv 1$ *mod* $N$, this gives the point $e_B P_m$, which Alice returns to Bob, who can read the message by multiplying the point $e_B P_m$ by $d_B$.

Notice that an eavesdropper would know $e_A P_m$, $e_B e_A P_m$ and $e_B P_m$. If (s)he could solve the discrete log problem on $E$, (s)he could determine $e_B$ from the first two points and then compute $d_B = e_B^{-1}$ *mod* $N$ and $P_m = d_B(e_B P_m)$.

**Analog of ElGamal.** This is another public key cryptosystem for transmitting messages $P_m$. As in the key exchange system above, we start with a fixed publicly known finite field $\mathbf{F}_q$, elliptic curve $E$ defined over it, and base point $B \in E$. (We do not need to know the number of points $N$.) Each user chooses a random integer $a$, which is kept secret, and computes and publishes the point $aB$.

To send a message $P_m$ to Björn, Aniuta chooses a random integer $k$ and sends the pair of points $(kB,\ P_m + k(a_B B))$ (where $a_B B$ is Björn's public key). To read the message, Björn multiplies the first point in the pair by his secret $a_B$ and subtracts the result from the second point:

$$P_m + k(a_B B) - a_B(kB) = P_m.$$

Thus, Aniuta sends a disguised $P_m$ along with a "clue" $kB$ which is enough to remove the "mask" $ka_B B$ if one knows the secret integer $a_B$. An eavesdropper who can solve the discrete log problem on $E$ can, of course, determine $a_B$ from the publicly known information $B$ and $a_B B$.

**The choice of curve and point.** There are various ways of choosing an elliptic curve and (in the Diffie–Hellman and ElGamal set-up) a point $B$ on it.

**Random selection of $(E, B)$.** Once we choose our large finite field $\mathbf{F}_q$, we can choose both $E$ and $B = (x, y) \in E$ at the same time as follows. (We shall assume that the characteristic is $> 3$, so that elliptic curves are given by equation (1) in §1; one makes the obvious modifications if $q = 2^r$ or $3^r$.) First let $x, y, a$ be three random elements of $\mathbf{F}_q$. Then set $b = y^2 - (x^3 + ax)$. Check that the cubic $x^3 + ax + b$ does not have multiple roots, which is equivalent to: $4a^3 + 27b^2 \neq 0$. (If this condition is not met, make another random choice of $x, y, a$.) Set $B = (x, y)$. Then $B$ is a point on the elliptic curve $y^2 = x^3 + ax + b$.

If you need to know the number $N$ of points, there are several techniques now available for computing $N$. The first polynomial time algorithm to compute $\#E$ was discovered by René Schoof. Schoof's algorithm is even deterministic. It is based on the idea of finding the value of $\#E$ modulo $l$ for all primes $l$ less than a certain bound. This is done by examining the action of the "Frobenius" (the $p$-th power map) on points of order $l$.

In Schoof's original paper the bound for running time was essentially $O(\log^8 q)$, which is polynomial but quite unpleasant. At first it looked like the algorithm was not practical. However, since then many people have worked on speeding up Schoof's algorithm (V. Miller, N. Elkies, J. Buchmann, V. Müller, A. Menezes, L. Charlap, R. Coley, and D. Robbins). In addition, A. O. L. Atkins has developed a somewhat different method that, while not guaranteed to work in polynomial time, functions extremely well in practice. As a result of all of these efforts it has become feasible to compute the order of an arbitrary elliptic curve over $\mathbf{F}_q$ if $q$ is, say, a 50-digit or even a 100-digit prime power. Some of the methods for computing the number of points on an elliptic curve are discussed in the references listed at the end of the section.

It should also be remarked that, even though one does not have to know $N$ in order to implement the Diffie–Helman or the ElGamal system, in practice one wants to be confident in its security, which depends upon $N$ having a large prime factor. If $N$ is a product of small primes, then the method of Pohlig–Silver–Hellman (see §IV.3) can be used to solve the discrete log problem. Note that the Pohlig–Silver–Hellman method carries over to the discrete log problem in any finite abelian group (unlike the index–calculus algorithm also discussed in §IV.3, which depends upon the specific nature of $\mathbf{F}_q^*$). Thus, one has to know that $N$ is not a product of small primes, and it is not likely that you will know this unless you have the actual value of $N$.

**Reducing a global $(E, B)$ modulo $p$.** We now mention a second way to determine a pair consisting of an elliptic curve and a point on it. We first choose once and for all a "global" elliptic curve and a point of infinite order on it. Thus, let $E$ be an elliptic curve defined over the field of rational numbers (or, more generally, we could use an elliptic curve defined over a number field), and let $B$ be a point of infinite order on $E$.

**Example 2.** It turns out that the point $B = (0, 0)$ is a point of infinite order on the elliptic curve $E:\ y^2 + y = x^3 - x$, and in fact generates the entire group of rational points on $E$.

**Example 3.** It turns out that the point $B = (0,0)$ is a point of infinite order on $E: \ y^2 + y = x^3 + x^2$, and generates the entire group of rational points.

Next, we choose a large prime $p$ (or, if our elliptic curve is defined over an extension field $K$ of $\mathbf{Q}$, then we choose a prime ideal of $K$) and consider the *reduction* of $E$ and $B$ modulo $p$. More precisely, for all $p$ except for some small primes the coefficients in the equation for $E$ have no $p$ in their denominators, so we may consider the coefficients in this equation modulo $p$. If we make a change of variables taking the resulting equation over $\mathbf{F}_p$ to the form $y^2 = x^3 + ax + b$, the cubic on the right has no multiple roots (except in the case of a few small primes $p$), and so gives an elliptic curve (which we shall denote $E$ *mod* $p$) over $\mathbf{F}_p$. The coordinates of $B$ will also reduce modulo $p$ to give a point (which we shall denote $B$ *mod* $p$) on the elliptic curve $E$ *mod* $p$.

When we use this second method, we fix $E$ and $B$ once and for all, and then get many different possibilities by varying the prime $p$.

**Order of the point** $B$**.** What are the chances that a "random" point $B$ on a "random" elliptic curve is a generator? Or, in the case of our second method of selecting $(E, B)$, what are the chances, as $p$ varies, that the point $B$ reduces modulo $p$ to a generator of $E$ *mod* $p$? This question is closely analogous to the following question concerning the multiplicative groups of finite fields: Given an integer $b$, what are the chances, as $p$ varies, that $b$ is a generator of $\mathbf{F}_p^*$? The question has been studied both in the finite–field and elliptic–curve situations. For further discussion, see the paper by Gupta and Murty cited in the references.

As mentioned before, for the security of the above cryptosystems it is not really necessary for $B$ to be a generator. What is needed is for the cyclic subgroup generated by $B$ to be a group in which the discrete log problem is intractible. This will be the case — i.e., all known methods for solving the discrete logarithm problem in an arbitrary abelian group will be very slow — provided that the order of $B$ is divisible by a very large prime, say, having order of magnitude almost as large as $N$.

One way to guarantee that our choice of $B$ is suitable — and, in fact, that $B$ generates the elliptic curve — is to choose our elliptic curve and finite field so that the number $N$ of points is itself a prime number. If we do that, then every point $B \neq O$ will be a generator. Thus, if we use the first method described above, then for a fixed $\mathbf{F}_q$ we might keep choosing pairs $(E, B)$ until we find one for which the number of points on $E$ is a prime number (as determined by one of the primality tests discussed in § V.1). If we use the second method, then for a fixed global elliptic curve $E$ over $\mathbf{Q}$ we keep choosing primes $p$ until we find a prime for which the number of points on $E$ *mod* $p$ is a prime number. How long are we likely to have to wait? This question is analogous to the following question about the groups $\mathbf{F}_p^*$: is $(p-1)/2$ prime, i.e., is any element $\neq \pm 1$ either a generator or the square of a generator (see Exercise 13 of § II.1)? Neither the elliptic curve nor the finite field question has been definitively answered, but it is conjectured in both cases that the probability that a chosen $p$ has the desired property is $O(1/log\, p)$.

**Remark.** In order for $E$ *mod* $p$ to have any chance of being of prime order $N$ for large $p$, $E$ must be chosen so as to have trivial torsion, i.e., to have no points except $O$ of finite order. Otherwise, $N$ will be divisible by the order of the torsion subgroup.

## *Exercises*

1. Give a probabilistic algorithm for finding a nonsquare in $\mathbf{F}_q$.
2. Describe a polynomial time *deterministic* algorithm for imbedding plaintexts $m$ as points on an elliptic curve in the following cases:
   (a) $E$ has equation $y^2 = x^3 - x$ and $q \equiv 3\ mod\ 4$.
   (b) $E$ has equation $y^2 + y = x^3$ and $q \equiv 2\ mod\ 3$.
3. Let $E$ be the elliptic curve $y^2 + y = x^3 - x$ defined over the field of $p = 751$ elements. (A change of variables of the form $y' = y + 376$ will convert this equation to the form (1) of §1.) This curve contains $N = 727$ points. Suppose that the plaintext message units are the decimal digits 0—9 and the letters A—Z with numerical equivalents 10—35, respectively. Take $\kappa = 20$.
   (a) Use the method in the text to write the message "STOP007" as a sequence of seven points on the curve.
   (b) Translate the sequence of points $(361, 383)$, $(241, 605)$, $(201, 380)$, $(461, 467)$, $(581, 395)$ into a reply message.
4. Let $E$ be an elliptic curve defined over $\mathbf{Q}$, and let $p$ be a large prime, in particular, large enough so that reducing the equation $y^2 = x^3 + ax + b$ modulo $p$ gives an elliptic curve over $\mathbf{F}_p$. Show that (a) if the cubic $x^3 + ax + b$ splits into linear factors modulo $p$, then $E$ *mod* $p$ is not cyclic; (b) if this cubic has a root modulo $p$, then the number $N$ of elements on $E$ *mod* $p$ is even.
5. Let $E$ be the elliptic curve in Example 5 of §1. Let $q = 2^r$, and let $N_r$ be the number of $\mathbf{F}_{2^r}$-points on $E$.
   (a) Show that $N_r$ is never prime for $r > 1$.
   (b) When $4|r$, find conditions that are equivalent to $N_r$ being divisible by an $(r/4)$-bit or $(r/4 + 1)$-bit prime.
6. Let $E$ be an elliptic curve defined over $\mathbf{F}_p$, and let $N_r$ denote the number of $\mathbf{F}_{p^r}$-points on $E$.
   (a) Prove that if $p > 3$, then $N_r$ is never prime for $r > 1$.
   (b) Give a counterexample to part (a) when $p = 2$ and when $p = 3$.
7. (a) Find an elliptic curve $E$ defined over $\mathbf{F}_4$ which has only one $\mathbf{F}_4$-point (the point at infinity $O$).
   (b) Show that the number of $\mathbf{F}_{4^r}$-points on the curve in part (a) is the square of the Mersenne number $2^r - 1$.

(c) Find a very simple formula for the double of an $\mathbf{F}_{4^r}$-point on this elliptic curve.
(d) Prove that, if $2^r - 1$ is a Mersenne prime, then every $\mathbf{F}_{4^r}$-point (except $O$) has exact order $2^r - 1$.

8. Let $r$ be odd, and let $K$ denote the field $\mathbf{F}_{2^r}$. For $z \in K$ let $g(z)$ denote $\sum_{j=0}^{(r-1)/2} z^{2^{2j}}$, and let $tr(z)$ (called the "trace") denote $\sum_{j=0}^{r-1} z^{2^j}$.
(a) Prove that $tr(z) \in \mathbf{F}_2$; $tr(z_1 + z_2) = tr(z_1) + tr(z_2)$; $tr(1) = 1$; and $g(z) + g(z)^2 = z + tr(z)$.
(b) Prove that $tr(z) = 0$ for exactly half of the elements of $K$ and $tr(z) = 1$ for the other half.
(c) Describe a probabilistic algorithm for generating $\mathbf{F}_{2^r}$-points on the elliptic curve $y^2 + y = x^3 + ax + b$.

9. Let $E$ be the elliptic curve $y^2 = x^3 + ax + b$ with $a, b \in \mathbf{Z}$. Let $P \in E$. Let $p > 3$ denote a prime that does not divide either $4a^3 + 27b^2$ or the denominator of the $x$- or $y$-coordinate of $P$. Show that the order of $P$ *mod* $p$ on the elliptic curve $E$ *mod* $p$ is the smallest positive integer $k$ such that either (1) $kP = O$ on $E$; or (2) $p$ divides the denominator of the coordinates of $kP$.

10. Let $E$ be the elliptic curve $y^2 + y = x^3 - x$ defined over $\mathbf{Q}$, and let $P = (0,0)$. By computing $2^j P$ for $j = 1, 2, \ldots$, find an example of a prime $p$ such that $E$ *mod* $p$ is *not* generated by $P$ *mod* $p$. (Note: it can be shown that the point $P$ *does* generate the group of rational points of $E$.)

11. Use the elliptic curve analog of ElGamal to send the message in Exercise 3(a) with $E$ and $p$ as in Exercise 3 and $B = (0,0)$. Suppose that your correspondent's public key is the point $(201, 380)$ and your sequence of random $k$'s (one used to send each message unit) is 386, 209, 118, 589, 312, 483, 335. What sequence of 7 pairs of points do you send?

    Note that in this exercise we used a rather small value of $p$; a more realistic example of the sort one would encounter in practice would require working with numbers of several dozen decimal digits.

# References for § VI.2

1. G. Agnew, R. Mullin, and S. A. Vanstone, An implementation of elliptic curve cryptosystems over $\mathbf{F}_{2^{155}}$, *IEEE J. Selected Areas in Communications* **11** (1993), 804–813.
2. R. Gupta and M. R. Murty, "Primitive points on elliptic curves," *Compositio Math.* **58** (1986), 13–44.
3. N. Koblitz, "Elliptic curve cryptosystems," *Math. Comp.* **48** (1987).
4. N. Koblitz, "Primality of the number of points on an elliptic curve over a finite field," *Pacific J. Math.* **131** (1988), 157–165.

5. N. Koblitz, Constructing elliptic curve cryptosystems in characteristic 2, *Advances in Cryptology — Crypto '90*, Springer-Verlag, 1991, 156–167.
6. N. Koblitz, Elliptic curve implementation of zero-knowledge blobs, *J. Cryptology* **4** (1991), 207–213.
7. N. Koblitz, CM-curves with good cryptographic properties, *Advances in Cryptology — Crypto '91*, Springer-Verlag, 1992, 279–287.
8. H. W. Lenstra, Jr., "Elliptic curves and number-theoretic algorithms," Report 86–19, Mathematisch Instituut, Universiteit van Amsterdam, 1986.
9. A. Menezes, *Elliptic Curve Public Key Cryptosystems*, Kluwer Acad. Publ., 1993.
10. A. Menezes, T. Okamoto, and S. A. Vanstone, Reducing elliptic curve logarithms to logarithms in a finite field, *IEEE Transactions on Information Theory IT-39* (1993), 1639–1646.
11. A. Menezes, S. Vanstone, and R. Zuccherato, Counting points on elliptic curves over $\mathbf{F}_{2^m}$, *Math. Comp.* **60** (1993), 407–420.
12. V. Miller, "Use of elliptic curves in cryptography," *Abstracts for Crypto 85*, 1985.
13. A. M. Odlyzko, "Discrete logarithms in finite fields and their cryptographic significance," *Advances in Cryptology, Proc. Eurocrypt 84*, Springer-Verlag, 1985, 224–314.
14. R. Schoof, "Elliptic curves over finite fields and the computation of square roots mod $p$," *Math. Comp.* **44** (1985), 483–494.

## 3 Elliptic curve primality test

The elliptic curve primality test, due to S. Goldwasser, J. Kilian and (in another variant) A. O. L. Atkin, is an analog of the following primality test of Pocklington based on the group $(\mathbf{Z}/n\mathbf{Z})^*$:

**Proposition 6.3.1.** *Let $n$ be a positive integer. Suppose that there is a prime $q$ dividing $n-1$ which is greater than $\sqrt{n}-1$. If there exists an integer $a$ such that* (i) $a^{n-1} \equiv 1 \pmod{n}$; *and* (ii) $g.c.d.(a^{(n-1)/q} - 1, n) = 1$, *then $n$ is prime.*

**Proof.** If $n$ is not prime, then there is a prime $p \le \sqrt{n}$ which divides $n$. Since $q > p - 1$, it follows that $g.c.d.(q, p-1) = 1$, and hence there exists an integer $u$ such that $uq \equiv 1 \pmod{p-1}$. Then $a^{(n-1)/q} \equiv a^{uq(n-1)/q} = a^{u(n-1)} \equiv 1 \pmod{p}$ by condition (i), and this contradicts condition (ii).

**Remarks.** This is an excellent test provided that $n-1$ is divisible by a prime $q > \sqrt{n} - 1$, and we have been able to find $q$ (and prove that it's prime). Otherwise, we're out of luck. (This is not quite true — there's a more general version which can be used whenever we have a large divisor of $n-1$ in fully factored form, see Exercise 2 below.)

Note that this primality test is probabilistic only in the sense that a randomly chosen $a$ may or may not satisfy condition (ii) (of course, if it fails to satisfy (i), then $n$ is not prime). But once such an $a$ is found (and $a = 2$ will usually work), then the test shows that $n$ is definitely a prime. Unlike the primality tests in §V.1 (the Solovay–Strassen and Miller–Rabin tests), the conclusion of Pocklington's test is a certainty: $n$ is a prime, not a "probable prime."

The elliptic curve primality test is based on an analogous proposition, where we suppose that we have an equation $y^2 = x^3 + ax + b$ considered modulo $n$. That is, $a$ and $b$ are integers modulo $n$, and we let $E$ denote the set of all integers $x, y \in \mathbf{Z}/n\mathbf{Z}$ which satisfy the equation, along with a symbol $O$, which we call the "point at infinity." If $n$ is prime (as is almost certainly the case — since in practice we are only considering numbers $n$ which have already passed some of the probable prime tests in §V.1), then $E$ is an elliptic curve with identity element $O$.

Before stating the analog of Proposition 6.3.1 for $E$, we note that, even without knowing that $n$ is prime, we can apply the formulas in §1 to add elements of $E$. One of three things happens when we add two points (or double a point): (1) we get a well-defined point, (2) if the points are of the form $(x, y)$ and $(x, -y)$ modulo $n$, then we get the point at infinity, (3) the formulas are undefined, because we have a denominator which is not invertible modulo $n$. But case (3) means that $n$ is composite, and we can find a nontrivial divisor by taking the *g.c.d.* of $n$ with the denominator. So without loss of generality in what follows we may assume that case (3) never occurs.

It can be shown that for $P$ an element of $E$ modulo $n$, even if $n$ is composite the answer our algorithm gives for $mP$ does not depend on the particular manner in which we successively add and double points. (This is not *a priori* obvious.) However, this fact will not be needed below. It suffices to let $mP$ denote *any* point which is obtained working modulo $n$ with the formulas in §1.

Just as we can add points modulo $n$ without knowing that $n$ is prime, similarly, given an algorithm for computing the number of points on an elliptic curve (such as Schoof's method), we can apply it to our set $E$ modulo $n$. We will either obtain some number $m$ — which if $n$ is prime is guaranteed to be the number of points on the *elliptic curve* $E$ — or else encounter an undefined expression whose denominator has a nontrivial common factor with $n$. As in the case of the addition of points, without loss of generality we may assume that the latter never happens.

Such an $m$ will play the role of $n - 1$ in Proposition 6.3.1 — notice that $n - 1$ is the order of $(\mathbf{Z}/n\mathbf{Z})^*$ if $n$ is prime.

We are now ready to state the elliptic curve analog of Pocklington's criterion.

**Proposition 6.3.2.** *Let $n$ be a positive integer. Let $E$ be the set given by an equation $y^2 = x^3 + ax + b$ modulo $n$, as above. Let $m$ be an integer.*

*Suppose that there is a prime $q$ dividing $m$ which is greater than $\left(n^{1/4}+1\right)^2$. If there exists a point $P$ of $E$ such that* (i) $mP = O$; *and* (ii) $(m/q)P$ *is defined and not equal to $O$, then $n$ is prime.*

**Proof** (compare with the proof of Proposition 6.3.1). If $n$ is not prime, then there is a prime $p \leq \sqrt{n}$ which divides $n$. Let $E'$ be the elliptic curve given by the same equation as $E$ but considered modulo $p$, and let $m'$ be the order of the group $E'$. By Hasse's Theorem, we have $m' \leq p+1+2\sqrt{p} = (\sqrt{p}+1)^2 \leq \left(n^{1/4}+1\right)^2 < q$, and hence $g.c.d.(q, m') = 1$, and there exists an integer $u$ such that $uq \equiv 1 \ (mod\ m')$. Let $P' \in E'$ be the point $P$ considered modulo $p$. Then in $E'$ we have $(m/q)P' = uq(m/q)P' = umP' = O$, by (i), since $mP'$ is obtained using the same procedure as $mP$, only working modulo $p|n$ rather than modulo $n$. But this contradicts (ii), since if $(m/q)P$ is defined and $\neq O$ modulo $n$, then the same procedure working modulo $p$ rather than modulo $n$ will give $(m/q)P' \neq O$. This completes the proof.

This proposition leads to an algorithm for proving that an integer $n$ (which we may suppose is already known to be a "probable prime") is definitely prime. We proceed as follows. We randomly select three integers $a, x, y$ modulo $n$ and set $b \equiv y^2 - x^3 - ax \ (mod\ n)$. Then $P = (x, y)$ is an element of E, where $E$ is given by $y^2 = x^3 + ax + b$. We use Schoof's algorithm (or another method for counting the number of points on an elliptic curve) to find a number $m$ which, if $n$ is prime, is equal to the number of points on the elliptic curve $E$ over $\mathbf{F}_n$. If we cannot write $m$ in the form $m = kq$, where $k \geq 2$ is a small integer and $q$ is a "probable prime" (i.e., it passes a test as in §V.1), then we choose another random triple $a, x, y$ and start again. Suppose we finally obtain an elliptic curve for which $m$ has the desired form. Then we use the formulas in §VI.1 (working modulo $n$) to compute $mP$ and $kP$. If we ever obtain an undefined expression — either in computing a multiple of $P$ or in applying Schoof's algorithm — then we immediately find a nontrivial factor of $n$. We may assume that this doesn't happen. If $mP \neq O$, then we know that $n$ is composite (because if $n$ were prime, then the group $E$ would have order $m$, and any element of $E$ would be killed by multiplication by $m$). If $kP = O$ (which is highly unlikely), we are out of luck, and must start again with another triple. But if $mP = O$ and $kP \neq O$, then by Proposition 6.3.2 we know that $n$ is prime, provided that the large factor $q$ of $m$ is really a prime (we only know it to be a "probable prime"). This reduces the problem to proving primality of $q$, which has magnitude at most about $n/2$. We then start over with $n$ replaced by $q$. Thus, we obtain a recursive procedure with $t$ repetitions of the primality test, where $t$ is no more than about $\log_2 n$. When we're done, we have obtained a number $q_t$ which we know to be prime, from which it follows that the previous $q_{t-1}$ was really a prime (not just a "probable prime"), from which it follows that the same is true of $q_{t-2}$, and so on, until $q_1 = q$, and finally $n$ itself is truly a prime. This concludes the description of the elliptic curve primality test.

There are two difficulties with this test, one practical and the other theoretical. In the first place, although Schoof's algorithm takes time polynomial in $\log n$, in practice it is quite cumbersome. Some progress has been made recently in supplementing and streamlining it, but even so it is rather unpleasant to have to count the number of points on a large number of $E$ until we finally find one for which $m$ has the desired form $m = kq$. In order to deal with this problem, A. O. L. Atkin developed a variant of the elliptic curve primality test using carefully constructed elliptic curves with complex multiplication, for which it is much easier to compute the number of points on their reduction modulo $n$. For more information on Atkin's method, see the article by Lenstra and Lenstra in the references below.

The second difficulty is theoretical. In order to find an elliptic curve $E$ over $\mathbf{F}_n$ (assuming that $n$ is prime) whose number of points is "almost prime" (i.e., of the form $m = kq$ for $k$ small and $q$ prime), we have to know something about the distribution of primes (rather, of "near primes") in the interval from $p+1-2\sqrt{p}$ to $p+1+2\sqrt{p}$ which, by Hasse's Theorem, is known to contain $m$. Because the length of this interval is relatively small, there is no theorem which guarantees that we have a high probability of finding such an $E$ after only polynomially many tries (polynomial in $\log n$). However, there is a very plausible conjecture which would guarantee this, and for practical purposes there should be no problem. But if one wants a provably polynomial time probabilistic algorithm, one has to work much harder: such a primality test was developed by Adleman and Huang using two-dimensional abelian varieties, which are a generalization of elliptic curves to 2 dimensions. However, their algorithm is completely impractical, as well as very complicated.

## *Exercises*

1. (a) In Pocklington's primality test, if $n$ is prime, $n-1$ is divisible by a prime $q$ as in Proposition 6.3.1, and $a$ is chosen at random in $(\mathbf{Z}/n\mathbf{Z})^*$, then what is the probability that $a$ will satisfy the conditions of the proposition?
(b) In the elliptic curve primality test, if $n$ is prime, one has an elliptic curve of order divisible by a prime $q$ as in Proposition 6.3.2, and $P$ is a random point on it, then what is the probability that $P$ will satisfy the conditions of the proposition?
2. Generalize Pocklington's primality test to the case when one knows an integer $s$ dividing $n-1$ which is greater than $\sqrt{n}-1$ and for which one knows all primes $q|s$. Condition (ii) is required to hold for all $q|s$.
3. (a) (Pépin's primality test for Fermat numbers.) Prove that a Fermat number $n = 2^{2^k} + 1$ is a prime if and only if there exists an integer $a$ such that $a^{2^{2^k-1}} \equiv -1 \bmod n$. Prove that if $n$ is a prime, then 50% of all $a \in (\mathbf{Z}/n\mathbf{Z})^*$ have this property. Also prove that $a$ can always be chosen to be 3, or 5, or 7, if $k > 1$.

(b) Prove that a Mersenne number $n = 2^p - 1$ is a prime if and only if there exists a point $P = (x, y)$ on the curve $E : y^2 \equiv x^3 + x \bmod n$ such that (1) $2^{p-1}P$ can be computed without encountering non-invertible denominators mod $n$, and (2) $2^{p-1}P$ has $y$-coordinate zero. To do this, first prove that, if $n = 2^p - 1$ is prime, then the group of points on $E \bmod n$ is cyclic of order $2^p$, and 50% of all $P \in E \bmod n$ have the properties (1)–(2) above. Explain how one can generate random points $P \in E \bmod n$. You may use any algorithm that assumes that $b^{n-1} \equiv 1 \bmod n$ (i.e., that $n$ is a pseudoprime to various bases $b$), because if you ever encounter a $b$ for which this fails, your test ends with the conclusion that $n$ must be composite.

Note that this is a probabilistic primality test in the sense that, if $n$ is a prime, there is no guarantee of when a suitable $P$ will turn up. However, once such a $P$ is found, then the test ensures that $n$ *must* be prime. In this respect it is different from the pseudoprime tests in § V.1. For a generalization which can test primality of any odd $n$, see W. Bosma's paper cited below.

## References for § VI.3


1. L. Adleman and M. Huang, "Recognizing primes in random polynomial time," *Proc. 19th Annual ACM Symposium on Theory of Computing*, 1987, 462–469.
2. W. Bosma, "Primality testing using elliptic curves," Report 85–12, Mathematisch Instituut, Universiteit van Amsterdam, 1985.
3. S. Goldwasser and J. Kilian, "Almost all primes can be quickly certified," *Proc. 18th Annual ACM Symposium on Theory of Computing*, 1986, 316–329.
4. A. K. Lenstra and H. W. Lenstra, Jr., "Algorithms in number theory," Technical Report 87–008, University of Chicago, 1987.
5. F. Morain, "Implementation of the Goldwasser–Kilian–Atkin primality testing algorithm," INRIA report 911, 1988.
6. H. Pocklington, "The determination of the prime and composite nature of large numbers by Fermat's theorem," *Proc. Cambridge Philos. Soc.*, **18** (1914–16), 29–30.
7. R. Schoof, "Elliptic curves over finite fields and the computation of square roots mod $p$," *Math. Comp.* **44** (1985), 483–494.


## 4 Elliptic curve factorization

A key reason for the increasing interest in elliptic curves on the part of cryptographers is the recent ingenious use of elliptic curves by H. W. Lenstra to obtain a new factorization method that in many respects is better than the earlier known ones. The improvement in efficiency is not significant enough in practice to pose a threat to the security of cryptosystems based on the assumed intractability of factoring (its time estimate has the same form that we encountered in § V.3); nevertheless, the discovery of an improvement using an unexpected new device serves as a warning that one should never be too complacent about the supposed imperviousness of the factoring problem to dramatic breakthroughs. The purpose of this final section is to describe Lenstra's method.

Before proceeding to Lenstra's elliptic curve factorization algorithm, we give a classical factoring technique which is analogous to Lenstra's method.

**Pollard's $p-1$ method.** Suppose that we want to factor the composite number $n$, and $p$ is some (as yet unknown) prime factor of $n$. If $p$ happens to have the property that $p-1$ has no large prime divisor, then the following method is virtually certain to find $p$.

The algorithm proceeds as follows:

1. Choose an integer $k$ that is a multiple of all or most integers less than some bound $B$. For example, $k$ might be $B!$, or it might be the least common multiple of all integers $\leq B$.
2. Choose an integer $a$ between 2 and $n-2$. For example, $a$ could equal 2, or 3, or a randomly chosen integer.
3. Compute $a^k$ *mod* $n$ by the repeated squaring method.
4. Compute $d = g.c.d.(a^k - 1, n)$ using the Euclidean algorithm and the residue of $a^k$ modulo $n$ from step 3.
5. If $d$ is not a nontrivial divisor of $n$, start over with a new choice of $a$ and/or a new choice of $k$.

To explain when this algorithm will work, suppose that $k$ is divisible by all positive integers $\leq B$, and further suppose that $p$ is a prime divisor of $n$ such that $p-1$ is a product of small prime powers, all less than $B$. Then it follows that $k$ is a multiple of $p-1$ (because it is a multiple of all of the prime powers in the factorization of $p-1$), and so, by Fermat's Little Theorem, we have $a^k \equiv 1 \mod p$. Then $p|g.c.d.(a^k - 1, n)$, and so the only way we could fail to get a nontrivial factor of $n$ in step 4 is if it so happens that $a^k \equiv 1 \mod n$.

**Example 1.** We factor $n = 540143$ by this method, choosing $B = 8$ (and hence $k = 840$, which is the least common multiple of $1, 2, \ldots, 8$) and $a = 2$. We find that $2^{840}$ *mod* $n$ is 53047, and $g.c.d.(53046, n) = 421$. This leads to the factorization $540143 = 421 \cdot 1283$.

The main weakness of the Pollard method is clear if we attempt to use it when all of the prime divisors $p$ of $n$ have $p-1$ divisible by a relatively large prime (or prime power).

**Example 2.** Let $n = 491389$. We would be unlikely to find a nontrivial divisor until we chose $B \geq 191$. This is because it turns out that $n =$

$383 \cdot 1283$. We have $383 - 1 = 2 \cdot 191$ and $1283 - 1 = 2 \cdot 641$ (both 191 and 641 are primes). Except for $a \equiv 0, \pm 1 \ mod\ 383$, all other $a$'s have order modulo 383 either 191 or 382; and except for $a \equiv 0, \pm 1 \ mod\ 1283$, all other $a$'s have order modulo 1283 either 641 or 1282. So unless $k$ is divisible by 191 (or 641), we are likely to find again and again that $g.c.d.(a^k - 1, n) = 1$ in step 4.

The basic dilemma with Pollard's $p-1$ method is that we are pinning our hopes on the group $(\mathbf{Z}/p\mathbf{Z})^*$ (more precisely, the various such groups as $p$ runs through the prime divisors of $n$). For a fixed $n$, these groups are fixed. If all of them happen to have order divisible by a large prime, we are stuck.

The key difference in Lenstra's method, as we shall see, is that, by working with elliptic curves over $\mathbf{F}_p = \mathbf{Z}/p\mathbf{Z}$, we suddenly have a whole gaggle of groups to use, and we can realistically hope always to find one whose order is not divisible by a large prime or prime power.

We start our description of Lenstra's algorithm with some comments about reducing points on elliptic curves modulo $n$, where $n$ is a composite integer (unlike in §2, where we worked modulo prime numbers and in finite fields).

**Elliptic curves — reduction modulo $n$.** For the remainder of the section we let $n$ denote an odd composite integer and $p$ an (as yet unknown) prime factor of $n$. We shall suppose that $p > 3$. For any integer $m$ and any two rational numbers $x_1, x_2$ with denominators prime to $m$, we shall write $x_1 \equiv x_2 \ mod\ m$ if $x_1 - x_2$, written in lowest terms, is a fraction with numerator divisible by $m$. For any rational number $x_1$ with denominator prime to $m$ there is a unique integer $x_2$ (called the "least nonnegative residue") between 0 and $m-1$ such that $x_1 \equiv x_2 \ mod\ m$. Sometimes we shall write "$x_1 \ mod\ m$" to denote this least nonnegative residue.

Suppose that we have an equation of the form $y^2 = x^3 + ax + b$ with $a, b \in \mathbf{Z}$ and a point $P = (x, y)$ which satisfies it. In practice, the curve $E$ together with the point $P$ will be generated in some "random" way, for example, by choosing three random integers $a, x, y$ in some range and then setting $b = y^2 - x^3 - ax$. We shall assume that the cubic has distinct roots, i.e., $4a^3 + 27b^2 \neq 0$; this is almost certain if the coefficients were chosen in the random way described. For simplicity, in what follows we shall also suppose that $4a^3 + 27b^2$ has no common factor with $n$; in other words, $x^3 + ax + b$ has no multiple roots modulo $p$ for any prime divisor $p$ of $n$. In practice, once we have made a choice of $a$ and $b$, we can check this by computing $g.c.d.(4a^3 + 27b^2, n)$. If this is $> 1$, then either $n | 4a^3 + 27b^2$ (in which case we must make another choice of $a$ and $b$) or else we have obtained a nontrivial divisor of $n$ (in which case we're done). So we shall suppose that $g.c.d.(4a^3 + 27b^2, n) = 1$.

Now suppose that we want to find the multiple $kP$, using the repeated doubling method described in § VI.2. This can be done in $O(log\ k)$ steps, each involving a doubling or an addition of two distinct points. There are many ways to go about this. For example, $k$ can be written in binary as $a_0 + a_1 \cdot 2 + \cdots + a_{m-1}2^{m-1}$, then $P$ can be successively doubled, with $2^j P$ added to the partial sum whenever the corresponding bit $a_j$ is 1. Alternately, $k$ could be factored first into a product of primes $\ell_j$, and then one could successively compute $\ell_1(P)$, $\ell_2(\ell_1 P)$, and so on, where $\ell_1, \ell_2, \ldots$ are the primes in the factorization (listed, say, in non-decreasing order). Here each multiple $\ell_j P_j$, where $P_j = \ell_{j-1}\ell_{j-2}\cdots\ell_1 P$, is computed by writing $\ell_j$ in binary and using repeated doublings.

We shall suppose that some such technique has been chosen to compute multiples $kP$.

We shall consider the point $P$ and all of its multiples modulo $n$. This means that we let $P \ mod \ n = (x \ mod \ n, y \ mod \ n)$, and, every time we compute some multiple $kP$, we really compute only the reduction of the coordinates modulo $n$. In order to be able to work modulo $n$, there is a nontrivial condition that must hold whenever we perform a doubling step or add two different points. Namely, all denominators must be prime to $n$.

**Proposition VI.3.1.** *Let $E$ be an elliptic curve with equation $y^2 = x^3 + ax + b$, where $a, b \in \mathbf{Z}$ and g.c.d.$(4a^3 + 27b^2, n) = 1$. Let $P_1$ and $P_2$ be two points on $E$ whose coordinates have denominators prime to $n$, where $P_1 \neq -P_2$. Then $P_1 + P_2 \in E$ has coordinates with denominators prime to $n$ if and only if there is no prime $p|n$ with the following property: the points $P_1$ mod $p$ and $P_2$ mod $p$ on the elliptic curve $E$ mod $p$ add up to the point at infinity $O$ mod $p \in E$ mod $p$. Here $E$ mod $p$ denotes the elliptic curve over $\mathbf{F}_p$ obtained by reducing modulo $p$ the coefficients of the equation $y^2 = x^3 + ax + b$.*

**Proof.** First suppose that $P_1 = (x_1, y_1)$, $P_2 = (x_2, y_2)$, and $P_1 + P_2 \in E$ all have coordinates with denominators prime to $n$. Let $p$ be any prime divisor of $n$. We must show that $P_1 \ mod \ p + P_2 \ mod \ p \neq O \ mod \ p$. If $x_1 \not\equiv x_2 \ mod \ p$, then, according to the description of the addition law on $E \ mod \ p$, we immediately conclude that $P_1 \ mod \ p + P_2 \ mod \ p$ is not the point at infinity on $E \ mod \ p$. Now suppose that $x_1 \equiv x_2 \ mod \ p$. First, if $P_1 = P_2$, then the coordinates of $P_1 + P_2 = 2P_1$ are found by the formula (5) of §1, and $2P_1 \ mod \ p$ is found by the same formula with each term replaced by its residue modulo $p$. We must show that the denominator $2y_1$ is not divisible by $p$. If it were, then, because the denominator of the $x$-coefficient of $2P_1$ is not divisible by $p$, it would follow that the numerator $3x_1^2 + a$ would be divisible by $p$. But this would mean that $x_1$ is a root modulo $p$ of both the cubic $x^3 + ax + b$ and its derivative, contradicting our assumption that there are no multiple roots modulo $p$. Now suppose that $P_1 \neq P_2$. Since $x_2 \equiv x_1 \ mod \ p$ and $x_2 \neq x_1$, we can write $x_2 = x_1 + p^r x$ with $r \geq 1$ chosen so that neither the numerator nor denominator of $x$ is divisible by $p$. Because we have assumed that $P_1 + P_2$ has denominator not divisible by $p$, we can use the formula (4) of §1 to conclude that $y_2$ is of the form $y_1 + p^r y$. On the other hand,

$$\begin{aligned} y_2^2 &= (x_1 + p^r x)^3 + a(x_1 + p^r x) + b \\ &\equiv x_1^3 + ax_1 + b + p^r x(3x_1^2 + a) = y_1^2 + p^r x(3x_1^2 + a) \bmod p^{r+1}. \end{aligned} \tag{1}$$

But since $x_2 \equiv x_1 \bmod p$ and $y_2 \equiv y_1 \bmod p$, it follows that $P_1 \bmod p = P_2 \bmod p$, and so $P_1 \bmod p + P_2 \bmod p = 2P_1 \bmod p$, which is $O \bmod p$ if and only if $y_1 \equiv y_2 \equiv 0 \bmod p$. If the latter congruence held, then $y_2^2 - y_1^2 = (y_2 - y_1)(y_2 + y_1)$ would be divisible by $p^{r+1}$ (i.e., its numerator would be), and so the congruence (1) would imply that $3x_1^2 + a \equiv 0 \bmod p$. This is impossible, because the polynomial $x^3 + ax + b$ modulo $p$ has no multiple roots, and so $x_1$ cannot be a root both of this polynomial and its derivative modulo $p$. We conclude that $P_1 \bmod p + P_2 \bmod p \neq O \bmod p$, as claimed.

Conversely, suppose that for all prime divisors $p$ of $n$ we have $P_1 \bmod p + P_2 \bmod p \neq O \bmod p$. We must show that the coordinates of $P_1 + P_2$ have denominators prime to $n$, i.e., that the denominators are not divisible by $p$ for any $p|n$. Fix some $p|n$. If $x_2 \not\equiv x_1 \bmod p$, then the formula (4) of §1 shows that there are no denominators divisible by $p$. So suppose that $x_2 \equiv x_1 \bmod p$. Then $y_2 \equiv \pm y_1 \bmod p$; but since $P_1 \bmod p + P_2 \bmod p \neq O \bmod p$, we must have $y_2 \equiv y_1 \not\equiv 0 \bmod p$. First, if $P_2 = P_1$, then the formula (5) of §1 together with the fact that $y_1 \not\equiv 0 \bmod p$ shows that the coordinates of $P_1 + P_2 = 2P_1$ have denominators prime to $p$. Finally, if $P_2 \neq P_1$, we again write $x_2 = x_1 + p^r x$ with $x$ not divisible by $p$, and we use the congruence (1) above to write $(y_2^2 - y_1^2)/(x_2 - x_1) \equiv 3x_1^2 + a \bmod p$. Since $p$ does not divide $y_2 + y_1 \equiv 2y_1 \bmod p$, it follows that there is no $p$ in the denominator of $\frac{y_2^2 - y_1^2}{(y_2+y_1)(x_2-x_1)} = \frac{y_2 - y_1}{x_2 - x_1}$, and hence, by formula (4) of §1, there is no $p$ in the denominator of the coordinates of $P_1 + P_2$. This completes the proof.

**Lenstra's method.** We are given a composite odd integer $n$ and want to find a nontrivial factor $d|n$, $1 < d < n$. We start by taking some elliptic curve $E : y^2 = x^3 + ax + b$ with integer coefficients along with a point $P = (x, y)$ on it. The pair $(E, P)$ is probably generated in some random way, although we could choose to use some deterministic method which is capable of generating many such pairs (as in Example 4 below). We attempt to use $E$ and $P$ to factor $n$, as will be presently explained; if our attempt fails, we take another pair $(E, P)$, and continue in this way until we find a factor $d|n$. If the probability of failure is $\rho < 1$, then the probability that $h$ successive choices of $(E, P)$ all fail is $\rho^h$, which is very small for $h$ large. Thus, with a very high probability we will factor $n$ in a reasonable number of tries.

Once we have a pair $(E, P)$, we choose an integer $k$ which is divisible by powers of small primes ($\leq B$) which are less than some bound $C$. That is, we set

$$k = \prod_{\ell \leq B} \ell^{\alpha_\ell}, \tag{2}$$

where $\alpha_\ell = \left[\log C / \log \ell\right]$ is the largest exponent such that $\ell^{\alpha_\ell} \leq C$. We then attempt to compute $kP$, working all the time modulo $n$. This computation is uneventful and useless, unless we run into the following difficulty: when attempting to find the inverse of $x_2 - x_1$ in the formula (4) of §1 or the inverse of $2y_1$ in (5), we encounter a number that is *not* prime to $n$. According to Proposition VI.3.1, this will happen when we have some multiple $k_1P$ (a partial sum encountered along the way in our computation of $kP$) which for some $p|n$ has the property $k_1(P \textit{ mod } p) = O \textit{ mod } p$, i.e., the point $P \textit{ mod } p$ in the group $E \textit{ mod } p$ has order dividing $k_1$. In the process of using the Euclidean algorithm to try to find the inverse modulo $n$ of a denominator which is divisible by $p$, we instead find the *g.c.d.* of $n$ with that denominator. That *g.c.d.* will be a proper divisor of $n$, unless it is $n$ itself, i.e., unless the denominator is divisible by $n$. That would mean, by Proposition VI.3.1, that $k_1P \textit{ mod } p = O \textit{ mod } p$ for *all* prime divisors $p$ of $n$ — something which is highly unlikely if $n$ has two or more very large prime divisors. Thus, it is virtually certain that as soon as we try to compute $k_1P$ modulo $n$ for a $k_1$ which is a multiple of the order of $P \textit{ mod } p$ for some $p|n$, we will obtain a proper divisor of $n$.

Notice the similarity with Pollard's $p-1$ method. Instead of the group $(\mathbf{Z}/p\mathbf{Z})^*$, we are using the group $E \textit{ mod } p$. However, this time, if our $E$ proves to be a bad choice — i.e., for each $p|n$ the group $E \textit{ mod } p$ has order divisible by a large prime (and so $kP \textit{ mod } p$ is not likely to equal $O \textit{ mod } p$ for $k$ given by (2)) — all we have to do is throw it away and pick out another elliptic curve $E$ together with a point $P \in E$. We did not have such an option in the Pollard method.

**The algorithm.** Let $n$ be a positive odd composite integer. We now describe Lenstra's probabilistic method for factoring $n$.

We suppose we have a method for generating pairs $(E, P)$ consisting of an elliptic curve $y^2 = x^3 + ax + b$ with $a, b \in \mathbf{Z}$ and a point $P = (x, y) \in E$. Given such a pair, we go through the procedure about to be described. If that procedure fails to yield a nontrivial factor of $n$, then we generate a new pair $(E, P)$ and repeat the process.

Before working with our $E$ modulo $n$, we must verify that it is in fact an elliptic curve modulo any $p|n$, i.e., that the cubic on the right has distinct roots modulo $p$. This holds if and only if the discriminant $4a^3 + 27b^2$ is prime to $n$. Thus, if $\textit{g.c.d.}(4a^3 + 27b^2, n) = 1$, we may proceed. Of course, if this *g.c.d.* is strictly between 1 and $n$, we have a divisor of $n$, and we're done. If this *g.c.d.* equals $n$, then we must choose a different elliptic curve.

Next, we suppose that we have chosen two positive integer bounds $B$, $C$. Here $B$ is a bound for the prime divisors of the integer $k$ by which we multiply the point $P$. If $B$ is large, then there is a greater probability that our pair $(E, P)$ has the property that $kP \textit{ mod } p = O \textit{ mod } p$ for some $p|n$; on the other hand, the larger $B$ the longer it will take to compute $kP \textit{ mod } p$. So $B$ must be chosen in some way which we estimate minimizes the running time. $C$, roughly speaking, is a bound for the prime divisors $p|n$ for which we are at all likely to obtain a relation $kP \textit{ mod } p = O \textit{ mod } p$. We then choose $k$ to be given by (2), i.e., $k$ is the product of all prime powers $\leq C$

which are powers of primes $\leq B$. Then Hasse's Theorem tells us that, if $p$ is such that $p+1+2\sqrt{p} < C$ and the order of $E \bmod p$ is not divisible by any prime $> B$, then $k$ is a multiple of this order and so $kP \bmod p = O \bmod p$.

**Example 3.** Suppose we choose $B = 20$, and we want to factor a 10–decimal–digit integer $n$ which may be a product of two 5–digit primes (i.e., not be divisible by any prime of fewer than 5 digits). Then choose $C = 100700$ and $k = 2^{16} \cdot 3^{10} \cdot 5^7 \cdot 7^5 \cdot 11^4 \cdot 13^4 \cdot 17^4 \cdot 19^3$.

We now return to the description of the algorithm. Working modulo $n$, attempt to compute $kP$ as follows. Use the repeated doubling method to compute $2P$, $2(2P)$, $2(4P)$, ..., $2^{\alpha_2}P$, then $3(2^{\alpha_2})P$, $3(3 \cdot 2^{\alpha_2}P)$, ..., $3^{\alpha_3}2^{\alpha_2}P$, and so on, until finally you have $\prod_{\ell \leq B} \ell^{\alpha_\ell} P$. (Multiply successively by the prime factors $\ell$ of $k$ from smallest to largest.) In these computations, whenever you have to divide modulo $n$, you use the Euclidean algorithm to find the inverse modulo $n$. If at any stage the Euclidean algorithm fails to provide an inverse, then either you find a nontrivial divisor of $n$ or you obtain $n$ itself as the *g.c.d.* of $n$ and the denominator. In the former case, the algorithm has been successfully completed. In the latter case, you must go back and choose another pair $(E, P)$. If the Euclidean algorithm always provides an inverse — and so $kP$ modulo $n$ is actually calculated — then you must also go back and choose another pair $(E, P)$. This completes the description of the algorithm.

**Example 4.** Let us use the family of elliptic curves $y^2 = x^3 + ax - a$, $a = 1, 2, \ldots$, each of which contains the point $P = (1, 1)$. Before using an $a$ for a given $n$, we must verify that the discriminant $4a^3 + 27a^2$ is prime to $n$. Let us try to factor $n = 5429$ with $B = 3$ and $C = 92$. (In this example and the exercises below we illustrate the method using small values of $n$. Of course, in practice the method becomes valuable only for much, much larger $n$.) Here our choice of $C$ is motivated by our desire to find a prime factor $p$ which could be almost as large as $\sqrt{n} \approx 73$; for $p = 73$ the bound on the number of $\mathbf{F}_p$-points on an elliptic curve is $74 + 2\sqrt{73} < 92$. Using (2), we choose $k = 2^6 \cdot 3^4$. For each value of $a$, we successively multiply $P$ by 2 six times and then by 3 four times, working modulo $n$, on the elliptic curve $y^2 = x^3 + ax - a$. When $a = 1$ we find that the multiplication proceeds smoothly, and it turns out that $3^4 2^6 P \bmod p$ is a finite point on $E \bmod p$ for all $p|n$. So we try $a = 2$. Then we find that when we try to compute $3^2 2^6 P$, we obtain a denominator whose g.c.d. with $n$ is the proper factor 61. That is, the point $(1, 1)$ has order dividing $3^2 2^6$ on the curve $y^2 = x^3 + 2x - 2$ modulo 61. (See Exercise 5 below.) Thus, our second attempt succeeds. By the way, if we try $a = 3$ we find that the method gives the other prime factor 89 when we try to compute $3^4 2^6 P$. (Usually, but not always, the method gives the smallest prime factor.)

**Running time.** The central issue in estimating the running time is to compute, for a fixed $p$ and a given choice of bound $B$ (which is chosen in some optimal manner), the probability that a randomly chosen elliptic curve modulo $p$ has order $N$ not divisible by any prime $> B$. Now the orders $N$ of all elliptic curves modulo $p$ are known to be distributed fairly uniformly in the interval $p+1-2\sqrt{p} \le N \le p+1+2\sqrt{p}$ where Hasse's Theorem tells us they all fall (except that the density of $N$'s drops off near the endpoints of this interval). Thus, the probability is roughly equal to the chance that a randomly chosen integer of size approximately $p$ is not divisible by any prime $> B$. We already saw in our heuristic time estimate in § V.3 that this probability is approximately $u^{-u}$, where $u = log\, p/log\, B$. This leads to an estimate of the form $O(e^{C\sqrt{r\, log\, r}})$, where $r$ is the number of bits in $n$. For a detailed derivation of an estimate for the running time, see Lenstra's article.

More precisely, suppose that $n$ is a positive integer which is not a prime power and is not divisible by 2 or 3. Assuming a plausible conjecture about the distribution of integers not divisible by any prime $> B$ in a small interval around $p$, Lenstra proves the following probabilistic time estimate for the number of bit operations required to produce a nontrivial divisor of $n$:

$$e^{\sqrt{(2+\epsilon) log\, p\, log\, log\, p}}, \tag{3}$$

where $p$ is the smallest prime factor of $n$ and $\epsilon$ approaches zero for large $p$. Since always $p < \sqrt{n}$, it follows from (3) that we also have the estimate

$$e^{\sqrt{(1+\epsilon) log\, n\, log\, log\, n}}. \tag{4}$$

The estimate (4) has exactly the same form as the (conjectural) time estimates for the best general factoring methods known. However, Lenstra's method has certain advantages over its competitors:

1. It is the only method which is substantially faster if $n$ is divisible by a prime which is much smaller than $\sqrt{n}$.
2. For this reason, it can be used in combination with other factoring methods when the factorization of certain auxiliary numbers is required. (For example, in the continued fraction method in § V.4, we needed the complete factorization of $b_i^2\ mod\ n$ if it is a product of relatively small primes.)
3. It has a very small storage requirement, unlike most of its competitors.

But perhaps the most exciting feature of Lenstra's factorization algorithm is the use for the first time of elliptic curves, which are among the most richly structured and intensively studied objects in modern number theory and algebraic geometry. This shows that new factoring techniques might be found using unexpected constructions from hitherto unrelated branches of mathematics.

## *Exercises*

1. Use Pollard's method with $k = 840$ and $a = 2$ to try to factor $n = 53467$. Then try with $a = 3$.
2. Suppose that only one of the prime divisors $p$ of $n$ has the property that $p-1$ has no large prime factors. Suppose that in Pollard's algorithm

you take a value of $k$ which is not quite a multiple of $p-1$, and try various values of $a$. Estimate in terms of $k$ and $p-1$ the probability that you obtain the factor $d=p$ in step 4.

3. For the following values of $p$ and $B$, find (using a computer if necessary) the fraction of the integers between $p+1-2\sqrt{p}$ and $p+1+2\sqrt{p}$ which have no prime divisors greater than $B$: (a) $p=109$, $B=3$; (b) $p=109$, $B=19$; (c) $p=1009$, $B=19$; (d) $p=1009$, $B=97$; (e) $p=9973$, $B=97$.
4. Each of the values of $n$ in Exercise 5 of § V.4 has a factor $p<100$. In each case (a)–(k) find this factor by Lenstra's elliptic curve method, choosing $B=5$, $C=120$, $P=(1,1)$, and $E: y^2=x^3+ax-a$ with $a=1,2,\ldots$ (taking $a$'s for which the discriminant is prime to $n$). In each case, what is the first value of $a$ for which you find the factor, and what is the value of $k_1$ for which the factor appears as $g.c.d.(\text{denominator}, n)$ in your computation of $k_1P$?
5. With $k$ given by equation (2), suppose that you find a factor of $n$ in the process of computing $k_1P$ modulo $n$, where $k_1$ is a partial product in (2). (Recall that we compute $kP$ by successively multiplying by the $\ell$'s, proceeding in order of increasing $\ell$.) Prove that $k_1P \ mod\ p = O\ mod\ p$ for some $p|n$, i.e., rule out the possibility that you obtained a denominator not prime to $n$ in the computation of $\ell$ times $(k_1/\ell)P$ during one of the stages of the repeated doubling method before the last step.
6. (a) Suppose that for any $a\in\mathbf{Z}$ you have an efficient way of generating a point $P=(x,y)$ such that $y^2\equiv x^3+ax\ mod\ n$. Explain why it would *not* be a good idea to use the elliptic curves $y^2=x^3+ax$ with various $a$'s to factor $n$.

   (b) Same question for the family of elliptic curves $y^2=x^3+b$ with various $b$'s.
7. Suppose you want to increase very slightly the probability that the order of $E\ mod\ p$ for some $p|N$ is a product of small prime factors by ensuring in advance that 4 divides this order. Describe how to do this.

## References for § VI.4


1. H. W. Lenstra, Jr., "Factoring integers with elliptic curves," *Annals of Math.* (2) **126** (1987), 649–673.
2. P. Montgomery, Speeding the Pollard and elliptic curve methods of factorization, *Math. Comp.* **48** (1987), 243–264.
3. J. M. Pollard, "Theorems on factorization and primality testing," *Proc. Cambridge Philos. Soc.* **76** (1974), 521–528.

# Answers to Exercises

**§ I.1.**

1. $(112111)_3$.
2. $(260\frac{12}{126})_7$.
3. 10001100101; $1101\frac{1010}{1011}$.
4. MPJNS; $\text{LIKE}\frac{\text{IT}}{\text{WE}}$ (in other words, JQVXHJ=WE·LIKE+IT).
5. (a) 10.101101111110000; (b) C.SRO.
6. If $b^f - 1$ is a multiple of $d$, then the fraction can be written in the form $a/(b^f - 1)$, where $a$ is an integer of at most $f$ digits. Then use the formula for the sum of a geometric progression with initial term $a \cdot b^{-f}$ and ratio $b^{-f}$. Conversely, given a pure period-$f$ expansion $x$, you find that $b^f x$ differs from $x$ by an $f$-digit integer $a$, and this means that $x = a/(b^f - 1)$.
7. (a) $(\text{BAD})_{16}$; (b) no division is required: for example, to go from binary to hexadecimal simply start from the right and break off the digits in blocks of four; each four-tuple can be viewed as a hexadecimal digit (or replaced by one of the symbols 0—9, A—F).
8. (1) Look at the top and bottom bit and also at whether there's a borrow; (2) if both bits are the same and there is no borrow, or if the top bit is 1, the bottom bit is 0 and there is a borrow, then put down 0 and move on; (3) if the top bit is 1, the bottom bit is 0 and there is no borrow, then put down 1 and move on; (4) if the top bit is 0, the bottom bit is 1 and there is a borrow, then put down 0, put a borrow in the next column, and move on; (5) if both bits are the same and there is a borrow, or if the top bit is 0, the bottom bit is 1 and there is no borrow, then put down 1, put a borrow in the next column, and move on.

9. (a) One needs $n-1$ multiplications; in each case the partial product $3^j$ has at most $O(n)$ digits and 3 has 2 digits, so there are $O(n)$ bit operations; thus, the total is $O(n^2)$. (b) Here the partial product has $O(n\,log\,n)$ digits, so each multiplication takes $O(n\,log^2 n)$ bit operations; the total is $O(n^2 log^2 n)$.
10. $O(n^2 log^2 N)$.
11. (a) $O(n\,log^2 n)$; (b) $O(log^2 n)$.
12. $O(rsn(log^2 m + log\,n))$.
13. (a) The product of $O(n/log\,n)$ numbers each with $O(log\,n)$ digits has $O(n/log\,n)\cdot O(log\,n) = O(n)$ digits. (b) $O(n\,log\,n)$; (c) $O(n^2)$.
14. (a) $O(\sqrt{n}\,log^2 n)$; (b) $O(\sqrt{n}\,log\,n)$.
15. $O(m\,log\,n)$.
16. Suppose that $n$ has $k+1$ bits. As a first approximation to $m = [\sqrt{n}]$ take a 1 followed by $[k/2]$ zeros. Find the digits of $m$ from left to right after the 1 by each time trying to change the zero to 1, and if the square of the resulting $m$ is larger than $n$, putting it back to 0.

§ **I.2.**

1. (b) A simple counterexample: let $b = -a$.
2. 16 divisors: 1, 3, 5, 7, 9, 15, 21, 27, 35, 45, 63, 105, 135, 189, 315, 945.
3. (a) When $a|n$ write $n = ab$ and let $a \longleftrightarrow b$. (b) Given $n = ab$ with $a \geq b$, set $s = (a+b)/2$ and $t = (a-b)/2$. Conversely, given $n = s^2 - t^2$, set $a = s+t$, $b = s-t$ to get the reverse correspondence. (c) $473^2 - 472^2$, $159^2 - 156^2$, $97^2 - 92^2$, $71^2 - 64^2$, $57^2 - 48^2$, $39^2 - 24^2$, $33^2 - 12^2$, $31^2 - 4^2$.
4. (b) $100! = 2^{97}\cdot 3^{48}\cdot 5^{24}\cdot 7^{16}\cdot 11^9\cdot 13^7\cdot 17^5\cdot 19^5\cdot 23^4\cdot 29^3\cdot 31^3\cdot 37^2\cdot 41^2\cdot 43^2\cdot 47^2\cdot 53\cdot 59\cdot 61\cdot 67\cdot 71\cdot 73\cdot 79\cdot 83\cdot 89\cdot 97$. (c) The formula is $(n - S_p(n))/(p-1)$. To prove this, write $n = d_{k-1}p^{k-1} + \cdots + d_1 p + d_0$, and note that for each $j$: $[n/p^j] = d_{k-1}p^{k-1-j} + \cdots + d_{j+1}p + d_j$. Then use the formula in part (a).
6. (a) $1 = 11\cdot 19 - 8\cdot 26$; (b) $17 = 1\cdot 187 - 5\cdot 34$; (c) $1 = 205\cdot 160 - 39\cdot 841$; (d) $13 = 65\cdot 2171 - 54\cdot 2613$.
7. For example, here's a comparison between the two ways in the case of part (d):

| | |
|---|---|
| $2613 = 2171 + 442$ | $2613 = 2171 + 442$ |
| $2171 = 4\cdot 442 + 403$ | $2171 = 5\cdot 442 - 39$ |
| $442 = 403 + 39$ | $442 = 11\cdot 39 + 13$ |
| $403 = 10\cdot 39 + 13$ | $39 = 3\cdot 13.$ |
| $39 = 3\cdot 13.$ | |

8. (b)

$$\begin{aligned}
&g.c.d.(101000110101,\ 100001111011)\\
&= g.c.d.(110111010,\ 100001111011)\\
&= g.c.d.(11011101,\ 100001111011) = g.c.d.(11011101,\ 11110011110)\\
&= g.c.d.(11011101,\ 1111001111) = g.c.d.(11011101,\ 1011110010)\\
&= g.c.d.(11011101,\ 101111001) = g.c.d.(11011101,\ 10011100)\\
&= g.c.d.(11011101,\ 100111) = g.c.d.(10110110,\ 100111)\\
&= g.c.d.(1011011,\ 100111) = g.c.d.(110100,\ 100111)\\
&= g.c.d.(1101,\ 100111) = g.c.d.(1101,\ 11010)\\
&= g.c.d.(1101,\ 1101) = 1101.
\end{aligned}$$

(c) Consider the product $ab$, and show that every two steps must decrease the product of the two numbers whose g.c.d. you're taking at least by a factor of 2. Thus, there are $O(log\, a)$ steps. Each step is at most a subtraction, so takes $O(log\, a)$ bit operations. (Notice that no division or multiplication is involved.) (d) It doesn't give a way of expressing the g.c.d. as an integer combination of the original two numbers. However, it can be modified so as to do this: see "Extending the Binary GCD Algorithm" by G. H. Norton in *Algebraic Algorithms and Error Correcting Codes*, Springer-Verlag, 1986, 363–372.

9. $O(log\, a\, log\, b + log^3 b)$.

10. (a) The remainders decrease at the slowest rate when all of the quotients are 1. (b) Write $\begin{pmatrix} 1 & 1 \\ 1 & 0 \end{pmatrix} = BAB^{-1}$, where $A = \begin{pmatrix} \alpha & 0 \\ 0 & \alpha' \end{pmatrix}$ is the diagonal matrix made up from the eigenvalues and $B$ is a matrix whose columns are eigenvectors, e.g., $B = \begin{pmatrix} \alpha & \alpha' \\ 1 & 1 \end{pmatrix}$. (c) Since $\sqrt{5}\, a \geq \sqrt{5}\, f_{k+2} = \alpha^{k+2} - {\alpha'}^{k+2} > \alpha^{k+2} - 1$, it follows that $k < \big(log(1+\sqrt{5}\, a)/log\, \alpha\big) - 2$; we can also get the simpler estimate $k < log\, a/\log \alpha$. The latter estimate is equal to $1.44042\cdots log_2 a$, while the estimate in the proof of Proposition I.2.1 is $2\, log_2 a$.

11. (b) In the sum of $(log\, r_i)(1 + log\, q_{i+1})$, use the inequalities $r_i \leq b$ and $\prod q_{i+1} \leq a$. Conclude that the sum is bounded by $O\big((log\, b)(log\, a + log\, a)\big)$.

12. (a) $x^4 + x^2 + 1 = (x^2)(x^2+1) + 1$; $\;1 = 1(x^4 + x^2 + 1) - x^2(x^2+1)$. (b) $x^4 - 4x^3 + 6x^2 - 4x + 1 = (x-3)(x^3 - x^2 + x - 1) + (2x^2 - 2)$, $x^3 - x^2 + x - 1 = (\frac{1}{2}x - \frac{1}{2})(2x^2 - 2) + (2x - 2)$, $\;2x^2 - 2 = (x+1)(2x-2)$, so the g.c.d. is $x - 1$; $\;x - 1 = (-\frac{1}{4}x + \frac{1}{4})f + (\frac{1}{4}x^2 - x + \frac{5}{4})g$.

13. $g.c.d.(f, f') = x^2 - x - 1$, and the multiple roots are the golden ratio and its conjugate $(1 \pm \sqrt{5})/2$.

14. (a) $5+6i = 2i(3-2i)+1$; $\;1 = 1(5+6i) - 2i(3-2i)$. (b) $8-19i = 2(7-11i) + (-6+3i)$, $\;7-11i = (-2+i)(-6+3i) + (-2+i)$, $\;-6+3i = 3(-2+i)$, so $-2+i$ is the g.c.d.; $-2+i = (-3+2i)(7-11i) + (2-i)(8-19i)$.

15. (a) $12^2 + 25^2$; (b) $54^2 + 31^2$; (c) $116^2 + 159^2$.

§ **I.3.**

1. (a) $x = 6 + 7n$, $n$ any integer; (b) no solution; (c) same as (a); (d) $219 + 256n$; (e) $36 + 100n$; (f) $636 + 676n$.
2. 0, 1, 4, 9.
3. 3, B.
4. The difference between $n = 10^{k-1}d_{k-1} + \cdots + 10d_1 + d_0$ and the sum of the digits $d_{k-1} + \cdots + d_1 + d_0$ is a sum of multiples of numbers of the form $10^j - 1$, which is divisible by 9.
5. Prove separately that it is divisible by 2, 3 and 5.
6. Let $x$ and $y$ be the two digits. Then 72 — and hence both 8 and 9 — divide the cost $1000x + 60 + y$ cents. Thus, $8|60 + y$, which means that $y = 4$, and then $9|1000x + 64$, which is $\equiv x + 1$ *mod* 9. So $x = 8$. Thus each tile cost \$1.12.
7. (a) For example, suppose that $m = 2p^\alpha$. Since $m|(x^2-1) = (x+1)(x-1)$, we must have $\alpha$ powers of $p$ appearing in both $x+1$ and $x-1$ together. But since $p \geq 3$, it follows that $p$ cannot divide both $x + 1$ and $x - 1$ (which are only 2 apart from one another), and so all of the $p$'s must divide one of them. If $p^\alpha|x + 1$, this means that $x \equiv -1$ *mod* $p^\alpha$; if $p^\alpha|x-1$, then $x \equiv 1$ *mod* $p^\alpha$. Finally, since $2|x^2-1$ it follows that $x$ must be odd, i.e., $x \equiv 1 \equiv -1$ *mod* 2. Thus, by Property 5 of congruences, either $x \equiv 1$ *mod* $2p^\alpha$ or $x \equiv -1$ *mod* $2p^\alpha$. (b) First, if $m \geq 8$ is a power of 2, it's easy to show that $x = m/2 + 1$ gives a contradiction to part (a). Next, suppose that $m$ is not a prime power (or twice a prime power), and $p^\alpha||m$. Set $m' = m/p^\alpha$. Use the Chinese Remainder Theorem to find an $x$ which is $\equiv 1$ *mod* $p^\alpha$ and $\equiv -1$ *mod* $m'$. Show that this $x$ contradicts part (a).
8. Pair every integer from 1 to $p - 1$ with its multiplicative inverse. According to Exercise 7(a), only 1 and $-1$ are their own inverses. Thus, when the $p - 1$ numbers are multiplied, each pair containing two numbers which are each other's inverses must cancel, leaving just 1 and $-1$.
9. Of course, 4 has the desired property, but it is not a 3-digit number. By the last part of the Chinese Remainder Theorem, any other number which leaves the right remainders must differ from 4 by a multiple of $7 \cdot 9 \cdot 11 = 693$. The only 3-digit possibility is $4 + 693 = 697$.
10. One can apply the Chinese Remainder Theorem to the congruences $x \equiv 1$ *mod* 11, $x \equiv 2$ *mod* 12, $x \equiv 3$ *mod* 13. Alternately, one can observe that obviously $-10$ leaves the right remainders, and then proceed as in Exercise 9 to get $-10 + 11 \cdot 12 \cdot 13 = 1706$.
11. (a) 1973; (b) 63841; (c) 58837.
12. The quotient leaves remainders of 5, 1, 4 when divided by 9, 10, 11, and so (by the Chinese Remainder Theorem) is of the form $851+990m$. Similarly, the divisor is of the form $817 + 990n$. Since the divisor has 3 digits, $n = 0$. Since the product has 6 digits, also $m = 0$. Thus, the answer is 851.

13. The most time-consuming parts of implementing the Chinese Remainder Theorem are: (i) computing $M$; (ii) computing $M_i = M/m_i$ for each of the $r$ different $i$'s; (iii) finding the inverse of $M_i$ modulo $m_i$ for each $i$; (iv) multiplying out $a_iM_iN_i$ in the formula for $x$ for each $i$; (v) dividing the resulting $x$ by $M$ to get the least nonnegative value. We use $O(log\,B)$ for the number of bits in the $m_i$ or $a_i$ or $N_i$, and $O(r\,log\,B)$ for the number of bits in $M$ or the $M_i$. This gives $O(r^2 log^2 B)$ for the number of bit operations to do (i)–(ii), (iv)–(v). In (iii), we need $O(r^2 log^2 B)$ bit operations to reduce each of the $M_i$ modulo the corresponding $m_i$ before taking the inverses, and then $O(r\,log^3 B)$ bit operations to find all $r$ inverses by the Euclidean algorithm. This gives the combined estimate $O\big(r\,log^2 B(r + log\,B)\big)$. Whether the $r^2 log^2 B$ term or the $r\,log^3 B$ term dominates depends on the relative size of $r$ and $log\,B$ (i.e., the number of equations and the number of bits in our moduli).
14. $38^{1+2+2^3+2^6} \equiv 38 \cdot 2 \cdot 16 \cdot 63 \equiv 79 \; mod \; 103$.
15. If we use the $O(k^2)$ estimate for the time to perform one multiplication of $k$-bit integers (as we have been doing), then there is no saving of time. In fact, the very last multiplication already uses time $O((n\,log\,b)^2)$, which is the estimate we get by multiplying $b$ by itself $n$ times. The difficulty is that, unlike in modular arithmetic, in the repeated squaring method we end up dealing with pairs of very large integers, and this offsets the advantage of having far fewer multiplications to perform. But if we were to use a more clever way of multiplying two $k$-bit integers, for example, if we used an algorithm requiring only $O(k\,log\,k\,log\,log\,k)$ bit operations, then it would save time to use the repeated squaring method.
16. (a) Repeated squaring requires $O(log^3 p)$ bit operations whereas a time estimate of $O(log^2 p)$ can be proved for the Euclidean algorithm. (b) Repeated squaring still requires time $O(log^3 p)$, but after we perform the first step of the Euclidean algorithm — dividing $p$ by $a$ (which requires $O(log\,p\,log\,a)$ bit operations) — the rest of the Euclidean algorithm takes $O(log^2 a)$ bit operations. So the Euclidean algorithm is faster, especially for $a$ very small compared to $p$.
17.

| $n$ | 90 | 91 | 92 | 93 | 94 | 95 | 96 | 97 | 98 | 99 | 100 |
|---|---|---|---|---|---|---|---|---|---|---|---|
| $\varphi(n)$ | 24 | 72 | 44 | 60 | 46 | 72 | 32 | 96 | 42 | 60 | 40 |

18. There is no $n$ for which $\varphi(n)$ is an odd number greater than 1; $\varphi(n) = 1$ for $n = 1, 2$; $\varphi(n) = 2$ for $n = 3, 4, 6$; $\varphi(n) = 4$ for $n = 5, 8, 10, 12$; $\varphi(n) = 6$ for $n = 7, 9, 14, 18$; $\varphi(n) = 8$ for $n = 15, 16, 20, 24, 30$; $\varphi(n) = 10$ for $n = 11, 22$; $\varphi(n) = 12$ for $n = 13, 21, 26, 28, 36, 42$. To prove, for example, that these are all of the $n$ for which $\varphi(n) = 12$, compare the possible factorizations of 12 (with 1 allowed as a factor but not 3) with the formula $\varphi(\prod p^\alpha) = \prod(p^\alpha - p^{\alpha-1})$. One has $1 \cdot 2 \cdot 6$, $1 \cdot 12$, $2 \cdot 6$, and 12. The first gives $2 \cdot 3 \cdot 7$, the second gives $2 \cdot 13$, the third gives $(3 \; or \; 4) \cdot 7$ and $4 \cdot 9$, and the fourth gives 13.

19. $n$ cannot be a prime, since if it were $\varphi(n) = n - 1$. By assumption, $n$ is not the square of a prime. If it were not a product of two distinct primes, then it would be a product of three or more primes (not necessarily distinct). Let $p$ be the smallest. Then $p \le n^{1/3}$, and we have $\varphi(n) \le n(1 - \frac{1}{p}) \le n(1 - n^{-1/3}) = n - n^{2/3}$, a contradiction.
20. Show that the square of any odd number is $\equiv 1$ *mod* 8, and then use induction just as in the first paragraph of the proof of Proposition I.3.5.
21. (a) Notice that 360 is a multiple of $\varphi(p^\alpha)$ for each $p^\alpha \| m$. By the remark just before Example 3 in the text, this means that $6647^{362} \equiv 6647^2 \equiv 44182609$ *mod* $m$. (Here we're also using the fact that $g.c.d.(6647, m) = 1$, which follows because $6647 = 17^2 \cdot 23$.) (b) Raise $a$ to the 359th power modulo $m$ by the repeated squaring method. Since $m = (101100111)_2$, we find that there are 8 squarings plus 5 multiplications (of at most 63-bit integers), in each case combined with a division (at worst of a 126-bit integer by a 63-bit integer). Thus, the number of bit operations is at most $13 \times 63 \times 63 + 13 \times 64 \times 63 = 104013$.
22. (a) Show that, if $x = j \cdot \frac{n}{d}$, then $x$ generates $S_d$ if and only if $g.c.d.(x, d) = 1$. Notice that $j$ runs through $0, 1, \ldots, d-1$. (b) Partition the set $\mathbf{Z}/n\mathbf{Z}$ into subsets according to which $S_d$ an element generates. The subset corresponding to a given $S_d$ has $\varphi(d)$ elements, according to part (a).
23. (a) Expand each term in the product in a geometric series: $(1 + \frac{1}{p} + \frac{1}{p^2} + \frac{1}{p^3} + \cdots)$. In expanding all the parentheses, the denominators will be all possible expressions of the form $p_1^{\alpha_1} p_2^{\alpha_2} \cdots p_r^{\alpha_r}$. According to the Fundamental Theorem, every positive integer $n$ occurs exactly once as such an expression. Hence, the product is equal to the harmonic series $\sum_{n=1}^{\infty} \frac{1}{n}$, which we know diverges. (b) First prove that for $x \le \frac{1}{2}$ we have $x > -\frac{1}{2} log(1 - x)$ (look at the graph of *log*). Apply this when $x = \frac{1}{p}$, and compare $\sum \frac{1}{p}$ with the *log* of the product in part (a). (c) For any sequence of prime numbers $n$ approaching infinity we have $\frac{\varphi(n)}{n} = 1 - \frac{1}{n} \longrightarrow 1$; for any sequence of $n$'s which are divisible by increasingly many of the successive primes (for example, take $n_j = j!$), we have $\frac{\varphi(n)}{n} = \prod_{p|n}(1 - \frac{1}{p}) \longrightarrow \prod_{\text{all } p}(1 - \frac{1}{p}) = 0$ by part (a).
24. (a) Give $p_i$ and the residue of $N$ modulo $p_i$ to the $i$-th lieutenant general, and use the Chinese Remainder Theorem. (b) Choose each $p_i > \sqrt[k]{N}$ but much smaller than $\sqrt[k-1]{N}$.

§ **I.4.**

3. Use the same argument as in the proof of the last proposition to conclude that $b^d \equiv \pm 1$ *mod* $m$. But since $(b^d)^{a/d} \equiv -1$ *mod* $m$, it follows that $b^d \equiv -1$ *mod* $m$ and $a/d$ is odd.
4. Use Exercise 3 with $a = n$ and $c = (p-1)/2$.
5. (a) $2^8 + 1 = 257$; (b) use Exercise 4; (c) $m = 97 \cdot 257 \cdot 673$.
6. $2 \cdot 11^2 \cdot 13 \cdot 4561$, $2^5 \cdot 5 \cdot 7 \cdot 13 \cdot 41 \cdot 73 \cdot 6481$.
7. $2^4 \cdot 3^2 \cdot 7 \cdot 13 \cdot 31 \cdot 601$.

8. $3^2 \cdot 41 \cdot 271$, $3^3 \cdot 7 \cdot 11 \cdot 13 \cdot 37$, $3^2 \cdot 11 \cdot 73 \cdot 101 \cdot 137$.
9. $7 \cdot 23 \cdot 89 \cdot 599479$; $7^2 \cdot 127 \cdot 337$ (this example shows that a prime $p|b^d - 1$ in Proposition I.4.3 may divide $b^n - 1$ to a greater power than it divides $b^d - 1$).
10. $7 \cdot 31 \cdot 151$, $3^2 \cdot 7 \cdot 11 \cdot 31 \cdot 151 \cdot 331$, $3^2 \cdot 5^2 \cdot 7 \cdot 11 \cdot 13 \cdot 31 \cdot 41 \cdot 61 \cdot 151 \cdot 331 \cdot 1321$.
11. (a) Apply side by side the Euclidean algorithm to find $g.c.d.(a^m - 1, a^n - 1)$ and to find $g.c.d.(m, n)$. Notice that at each stage the remainder in the first Euclidean algorithm is $a^r - 1$, where $r$ is the remainder in the second Euclidean algorithm. For example, in the first step one divides $a^m - 1$ by $a^n - 1$ to get $a^r - 1$, where $r$ is the remainder when $m$ is divided by $n$. (b) By part (a) and the Chinese Remainder Theorem, no two numbers between 0 and $\prod(2^{m_i} - 1)$ have the same set of remainders. This product is greater than $2^{r\ell/2} > 2^{2k} > ab$. For the time estimate, one has $r$ multiplications of at most $\ell$-bit integers, which take $O(r\ell^2) = O(k\ell)$ bit operations. This is better by a factor of $r$ than the usual multiplication of $a$ and $b$ (which takes time $O(k^2)$).

## § II.1.

1.

| | | | | | | | |
|---|---|---|---|---|---|---|---|
| prime $p$ | 2 | 3 | 5 | 7 | 11 | 13 | 17 |
| smallest generator | 1 | 2 | 2 | 3 | 2 | 2 | 3 |
| number of generators | 1 | 1 | 2 | 2 | 4 | 4 | 8 |

2. (a) If $g^{p-1} \equiv 1 \ mod\ p^2$, then replace $g$ by $(p+1)g$ and show that then one has $g^{p-1} = 1 + g_1 p$ with $g_1$ prime to $p$. Now if $g^j \equiv 1 \ mod\ p^\alpha$, first show that $p-1|j$, i.e., $j = (p-1)j_1$, and so $(1+g_1p)^{j_1} \equiv 1 \ mod\ p^\alpha$. But show that $(1+g_1p)^{j_1} = 1 + j_1 g_1 p + higher\ powers\ of\ p$, and that then $p^{\alpha-1}$ must divide $j_1$. (b) For the first part, see Exercise 20 of § I.3; the proof of the second part (which reduces to showing that $5^j$ cannot be $\equiv 1 \ mod\ 2^\alpha$ unless $2^{\alpha-2}|j$) is similar to part (a).
3. $5^6$.
4. 2 for $d = 1$: $X$, $X+1$; 1 for $d = 2$: $X^2+X+1$; 2 for $d = 3$: $X^3+X^2+1$, $X^3+X+1$; 3 for $d = 4$: $X^4+X^3+1$, $X^4+X+1$, $X^4+X^3+X^2+X+1$; 6 for $d = 5$: $X^5 + X^3 + 1$, $X^5 + X^2 + 1$, $X^5 + X^4 + X^3 + X^2 + 1$, $X^5+X^4+X^3+X+1$, $X^5+X^4+X^2+X+1$, $X^5+X^3+X^2+X+1$; 9 for $d = 6$: $X^6+X^5+1$, $X^6+X^3+1$, $X^6+X+1$, $X^6+X^5+X^4+X^2+1$, $X^6+X^5+X^4+X+1$, $X^6+X^5+X^3+X^2+1$, $X^6+X^5+X^2+X+1$, $X^6+X^4+X^3+X+1$, $X^6+X^4+X^2+X+1$.
5. 3 for $d = 1$: $X$, $X \pm 1$; 3 for $d = 2$: $X^2+1$, $X^2 \pm X - 1$; 8 for $d = 3$: $X^3 + X^2 \pm (X-1)$, $X^3 - X^2 \pm (X+1)$, $X^3 \pm (X^2-1)$, $X^3 - X \pm 1$; 18 for $d = 4$; 48 for $d = 5$; 116 for $d = 6$.
6. $(p^f - p^{f/\ell})/f$.
7. (a) $g.c.d. = 1 = X^2 g + (X+1)f$; (b) $g.c.d. = X^3 + X^2 + 1 = f + (X^2+X)g$; (c) $g.c.d. = 1 = (X-1)f - (X^2 - X + 1)g$; (d) $g.c.d. = X + 1 = (X-1)f - (X^3 - X^2 + 1)g$; (e) $g.c.d. = X + 78 = (50X + 20)f + (51X^3 + 26X^2 + 27X + 4)g$.

8. Since $g.c.d.(f, f') = X^2+1$, the multiple roots are $\pm\alpha^2$, where $\alpha$ is the generator of $\mathbf{F}_9^*$ in the text.
9. (a) Raising $0 = \alpha^2 + b\alpha + c$ to the $p$-th power and using the fact that $b^p = b$ and $c^p = c$, we obtain $0 = (\alpha^p)^2 + b\alpha^p + c$. (b) The polynomial's two distinct roots are then $\alpha$ and $\alpha^p$. Then $a$ is minus the sum of the roots, and $b$ is the product of the roots. (c) $(c\alpha + d)^{p+1} = (c\alpha^p + d)(c\alpha + d)$, and then multiply out and use part (b). (d) $(2 + 3i)^{5(19+1)+1} = (2^2 + 3^2)^5(2 + 3i) = 14(2 + 3i) = 9 + 4i$.
10. In each division of polynomials (first $f$ by $g$, then $r_j$ by $r_{j+1}$), after first finding the inverse modulo $p$ of the leading coefficient of $r_{j+1}$ (which takes $O(log^3p)$ bit operations), one needs to perform $O(d^2)$ multiplications in the field (i.e., of integers modulo $p$), each taking $O(log^2p)$ bit operations. Thus, each division takes $O(log^3p+d^2log^2p)$ bit operations, and so the entire Euclidean algorithm takes $O(d)\cdot O(log^2p(log\,p+d^2)) = O(d\,log^2p(log\,p+d^2))$ operations. (This can be simplified to $O(d\,log^3p)$ if $d$ is constrained not to grow faster than $\sqrt{log\,p}$, and to $O(d^3log^2p)$ if $p$ is constrained not to grow faster than $e^{d^2}$.)
11. (a) Let $\alpha$ be a root of $X^2+X+1 = 0$; then the three successive powers of $\alpha$ are $\alpha$, $\alpha + 1$, and 1. (b) Let $\alpha$ be a root of $X^3 + X + 1 = 0$; then the seven successive powers of $\alpha$ are $\alpha$, $\alpha^2$, $\alpha + 1$, $\alpha^2 + \alpha$, $\alpha^2 + \alpha + 1$, $\alpha^2+1$, 1. (c) Let $\alpha$ be a root of $X^3 - X - 1 = 0$; then the 26 successive powers of $\alpha$ are $\alpha$, $\alpha^2$, $\alpha+1$, $\alpha^2+\alpha$, $\alpha^2+\alpha+1$, $\alpha^2-\alpha+1$, $-\alpha^2-\alpha+1$, $-\alpha^2 - 1$, $-\alpha + 1$, $-\alpha^2 + \alpha$, $\alpha^2 - \alpha - 1$, $-\alpha^2 + 1$, $-1$, followed by the same 13 elements with all +'s and −'s reversed. (d) Let $\alpha$ be a root of $X^2 - X + 2 = 0$; then the 24 successive powers of $\alpha$ are $\alpha$, $\alpha - 2$, $-\alpha - 2$, $2\alpha + 2$, $-\alpha + 1$, 2, then the same six elements multiplied by 2, then multiplied by $-1$, then multiplied by $-2$, giving all 24 powers of $\alpha$.
12. $O(f2^f)$, since for each of the $O(2^f)$ powers of $\alpha$ one has to multiply the previous expression by $\alpha$ and, if $\alpha^f$ occurs, add the lower degree polynomial which equals $\alpha^f$ to the result of increasing the lower powers of $\alpha$ by 1 in the previous expression; all of this takes only $O(f)$ bit operations.
13. (a) $p = 2$ and $2^f - 1$ is a "Mersenne" prime (see Example 1 and Exercise 2 of § I.4); (b) besides the cases in part (a), also when $p = 3$ and $(3^f - 1)/2$ is a prime (as in part (a), this requires that $f$ itself be prime, but that is not sufficient, as the example $f = 5$ shows), and when $p$ is of the form $2p' + 1$ with $p'$ a prime and $f = 1$. It is not known, incidentally, whether there are infinitely many prime fields with any of the conditions in (a)–(b) (but it is conjectured that there are). Primes $p'$ for which $p = 2p' + 1$ is also prime are called "Germain primes" after Sophie Germain, who in 1823 proved that the first case of Fermat's Last Theorem holds if the exponent is such a prime.
14. Choose a sequence $n_j$ for which $\varphi(n_j)/n_j \longrightarrow 0$ as $j \longrightarrow \infty$ (see Exercise 23 of § I.3) with none of the $n_j$ divisible by $p$, and let $f_j$ be

the *order* of $p$ modulo $n_j$ (the smallest power of $p$ that is $\equiv 1 \; mod \; n_j$).

15. All polynomials in which $X^j$ occurs with nonzero coefficient only if $p|j$.
16. Reduce to the case when $j = d$ by showing that, if $\sigma^j(a) = a$ and $\sigma^f(a) = a$, we have $\sigma^d(a) = a$ (see the proof of Proposition I.4.2). Notice that the field $\mathbf{F}_{p^d}$, which is the splitting field of $X^{p^d} - X$, is contained in $\mathbf{F}_q$, because any root $a$ of this polynomial also satisfies $X^q = X$ (to see this, raise both sides of $a^{p^d} = a$ to the $p^d$-th power $f/d$ times).
17. Show that $b' = b^{(p^n-1)/(p^d-1)}$ is in $\mathbf{F}_{p^d}$ by showing that it is fixed under $\sigma^d$ (i.e., raising to the $p^d$-th power); show that it is a generator by showing that all of the powers $(b')^j$, $j = 0, \ldots, p^d - 2$ are distinct (this follows from the fact that the first $p^n - 1$ powers of $b$ are distinct).

**§ II.2.**

1. The sets of residues are: for $p = 3$, $\{1\}$; for $p = 5$, $\{1, 4\}$; for $p = 7$, $\{1, 2, 4\}$; for $p = 13$, $\{1, 3, 4, 9, 10, 12\}$; for $p = 17$, $\{1, 2, 4, 8, 9, 13, 15, 16\}$; for $p = 19$, $\{1, 4, 5, 6, 7, 9, 11, 16, 17\}$.
2. (b) From part (a) and Propositions II.2.2 and II.2.4 you know that $\left(\frac{2}{p}\right) = 1 \equiv 2^{(p-1)/2} \; mod \; p$. This means that the $((p-1)/2^\ell)$-th power of 2 is $\equiv -1 \; mod \; p$ for some $\ell \geq 2$. Since $2^{2^k} \equiv -1 \; mod \; p$, you can show that $g.c.d.((p-1)/2^\ell, \, 2^k) = 2^k$ and this immediately gives $p \equiv 1 \; mod \; 2^{k+\ell}$. (c) The only prime which is $\equiv 1 \; mod \; 64$ and $< \sqrt{65537}$ is 193, which does not divide 65537.
3. $g.c.d.(84, 1330) = 14$.
4. Write $\left(\frac{-2}{p}\right) = \left(\frac{-1}{p}\right)\left(\frac{2}{p}\right)$, and consider the four possible cases of $p \; mod \; 8$.
5. $\left(\frac{91}{167}\right) = \left(\frac{7}{167}\right)\left(\frac{13}{167}\right) = -\left(\frac{167}{7}\right)\left(\frac{167}{13}\right) = -\left(\frac{-1}{7}\right)\left(\frac{-2}{13}\right) = -(-1)(-1) = -1$
6. (a) 14; (b) 9; (c) $9\alpha$.
7. $a^3 - a$ (see the proof of Proposition II.2.4); 6, 60, 4080, 24, 210, 336.
8. Since $q \equiv 1 \; mod \; p$, there is a primitive $p$-th root of unity $\xi$ in $\mathbf{F}_q$. Then $G = \sum_{j=1}^{p-1}\left(\frac{j}{p}\right)\xi^j$ has square $\left(\frac{-1}{p}\right)p$ (see the lemma in the proof of Proposition II.2.5).
9. (a) $\left(\frac{-1}{p}\right)\sum_{j=1}^{p-1}\left(\frac{j}{p}\right)a^j$; 6, 45, 3126, 906 (in the last case use: $1093 = (3^7 - 1)/2$). (b) Let $G = \sum_{j=1}^{p-1}\left(\frac{j}{p}\right)2^j$. Then the least positive square root of $\left(\frac{-1}{p}\right)p$ modulo $2^p - 1$ is $g$ if $p \equiv 5 \; mod \; 8$; $-g$ if $p \equiv 3 \; mod \; 8$; $p + g$ if $p \equiv 7 \; mod \; 8$; $p - g$ if $p \equiv 1 \; mod \; 8$.
10. (a) $\left(\frac{1801}{8191}\right) = \left(\frac{8191}{1801}\right) = \left(\frac{987}{1801}\right) = \left(\frac{3}{1801}\right)\left(\frac{7}{1801}\right)\left(\frac{47}{1801}\right) = \left(\frac{1}{3}\right)\left(\frac{2}{7}\right)\left(\frac{15}{47}\right) = 1 \cdot 1 \cdot \left(\frac{3}{47}\right)\left(\frac{5}{47}\right) = -\left(\frac{2}{3}\right)\left(\frac{2}{5}\right) = -1$. (b) $\left(\frac{987}{1801}\right) = \left(\frac{1801}{987}\right) = \left(\frac{2 \cdot 407}{987}\right) = -(-1)\left(\frac{987}{407}\right) = \left(\frac{173}{407}\right) = \left(\frac{407}{173}\right) = \left(\frac{61}{173}\right) = \left(\frac{173}{61}\right) = \left(\frac{51}{61}\right) = \left(\frac{61}{51}\right) = \left(\frac{2 \cdot 5}{51}\right) = -\left(\frac{5}{51}\right) = -\left(\frac{51}{5}\right) = -1$.
11. (a) 1; (b) 1; (c) 1; (d) 1; (e) 1; (f) 1; (g) $-1$.
12. (a) $\left(\frac{-3}{p}\right) = \left(\frac{-1}{p}\right)\left(\frac{3}{p}\right) = (-1)^{(p-1)/2}(-1)^{(3-1)(p-1)/4}\left(\frac{p}{3}\right) = \left(\frac{p}{3}\right)$, which $= 1$ if and only if $p \equiv 1 \; mod \; 3$. (b) $\left(\frac{3}{2^p - 1}\right) = -\left(\frac{2^p - 1}{3}\right) = -\left(\frac{1}{3}\right) = -1$.

13. last decimal digit being 1 or 9.
14. Any power of a residue is a residue, so none of the nonresidues can occur as a power, and that means a residue cannot be a generator.
15. (a) Since $p-1$ is a power of 2, the order of any element $g$ is a power of 2. If $-1 = \left(\frac{g}{p}\right) \equiv g^{(p-1)/2}$ *mod* $p$, then this order cannot be less than $p-1$. (b) If $k > 1$ and $p = 2^{2^k}+1$, then $p \equiv 2$ *mod* 5 (since the exponent of 2 is a multiple of 4). Then $\left(\frac{5}{p}\right) = \left(\frac{p}{5}\right) = -1$. (c) Similar to part (b): since the exponent of 2 is not divisible by 3, it follows that the power of 2 is $\equiv 2$ or 4 modulo 7; hence $p \equiv 3$ or 5 *mod* 7, and $\left(\frac{7}{p}\right) = \left(\frac{p}{7}\right) = -1$.
16. (a) We have $(a+bi)^{p+1} = (a^p + b^p i^p)(a+bi) = (a-bi)(a+bi) = a^2+b^2$. **Claim:** If $(a+bi)^m \in \mathbf{F}_p$, then $p+1|m$. To prove the claim, let $d = g.c.d.(m, p+1)$. Using the same argument as in the proof of Proposition I.4.2, we see that $(a+bi)^d \in \mathbf{F}_p$. But since $p+1$ is a power of 2, if $d < p+1$ we find that $(a+bi)^{(p+1)/2}$ is an element of $\mathbf{F}_p$ whose square is $a^2+b^2$. But $a^2+b^2$ is not a residue (by Exercise 14). Hence, $d = p+1$ and $p+1|m$. Now that the claim has been proved, suppose that $n = n'(p+1)$ is such that $(a+bi)^n = 1$ (note that $p+1|n$ by the claim). Then $(a^2+b^2)^{n'} = 1$, and so $p-1|n'$ because $a^2+b^2$ is a generator of $\mathbf{F}_p^*$. (b) Show that 17 and 13 are generators of $\mathbf{F}_{31}^*$.
17. In both cases you get $O(log^3 p)$. But note that Proposition II.2.2 applies only for $\left(\frac{a}{n}\right)$ when $n = p$ is prime, whereas the method in part (a) applies generally for any positive odd $n$. Also notice that the time for part (a) can be reduced to $O(log^2 p)$ by the method used in Exercise 11 of §I.2.
18. (a) Solve by completing the square; show that the number of solutions is the same as for the equation $x^2 \equiv D$ *mod* $p$. There is 1 solution if $D = 0$, none if $D$ is a nonresidue, and 2 if $D$ is a residue. (b) 0, 0, 2, 1, 2; (c) 2, 2, 1, 0, 0.
19. $n = 3$; $p - 1 = 2^5 \cdot 65$; $r \equiv a^{33} \equiv 203$ *mod* $p$ (we compute $302^{33}$ by the repeated squaring method, successively squaring 5 times and multiplying the result by 302); also by the repeated squaring method we compute $b \equiv n^{65} \equiv 888$ *mod* $p$; one takes $j = 2^2$, i.e., $\sqrt{302}$ *mod* $p \equiv b^4 r \equiv 1292$ *mod* $p$.
20. (a) Use induction on $\alpha$. To go from $\alpha - 1$ to $\alpha$, suppose you have an $(\alpha-1)$-digit base-$p$ integer $\tilde{x}$ such that $\tilde{x}^2 \equiv a$ *mod* $p^{\alpha-1}$. To determine the last digit $x_{\alpha-1} \in \{0, 1, \ldots, p-1\}$ of $x = \tilde{x} + x_{\alpha-1}p^{\alpha-1}$, write $\tilde{x}^2 = a + bp^{\alpha-1}$ for some integer $b$, and then work modulo $p^\alpha$ as follows: $x^2 = (\tilde{x} + x_{\alpha-1}p^{\alpha-1})^2 \equiv \tilde{x}^2 + 2x_0x_{\alpha-1}p^{\alpha-1} = a + p^{\alpha-1}(b + 2x_0x_{\alpha-1})$. So it suffices to choose $x_{\alpha-1} \equiv -(2x_0)^{-1}b$ *mod* $p$ (note that $2x_0$ is invertible because $p$ is odd, and $a \equiv x_0^2$ *mod* $p$ is prime to $p$). (b) Use the Chinese remainder theorem to find an $x$ which is congruent modulo each $p^\alpha$ to the square root found in part (a).
21. (a) If $(*)$ were true for $b_1$ and for $b_1b_2$, then dividing the two congru-

ences would give (∗) for $b_2$ (since both sides are multiplicative). Next, suppose (∗) were false for some $b$. Then the set of $b$'s obtained by multiplying $b$ by all the elements for which (∗) is true would consist of elements for which (∗) is false. (b) For example, take $b = 1+n/p$, where $p^2|n$. Then $(\frac{b}{n}) = 1$, but $b^j \equiv 1$ only when $p|j$, which is not the case for $j = (n-1)/2$. (c) Show that $(\frac{b}{n}) = -1$ but that $b^{(n-1)/2} \equiv 1 \ mod\ n/p$ and hence one could not have $b^{(n-1)/2} \equiv -1$ modulo $n/p$, let alone modulo $n$. Next, let $a_1$ be any nonresidue modulo $p$, and let $a_2 = 1$. Use the Chinese Remainder Theorem to find a solution $b$ to: $x \equiv a_1 \ mod\ p$, $x \equiv a_2 \ mod\ n/p$.

22. $b^2 = (t+\alpha)^p(t+\alpha) = (t+\alpha^p)(t+\alpha) = (t-\alpha)(t+\alpha) = t^2 - \alpha^2 = a$, where the third equality comes from the fact that $\alpha = \sqrt{t^2-a}$ has conjugate $\alpha^p = -\sqrt{t^2-a}$; note that $b$ must be in $\mathbf{F}_p$, since $a$ has two square roots in $\mathbf{F}_p$ by assumption, and so its square roots in $\mathbf{F}_{p^2}$ are actually in $\mathbf{F}_p$.
23. Let $b$ be the least positive residue of $n^{(p-1)/4}$ modulo $p$; then $b$ is a square root of $-1$ modulo $p$, i.e., $p|b^2+1$. Now compute $c + di = g.c.d.(p, b+i)$ (see Exercise 14 of § I.2).

§ **III.1.**

1. "We sewed a smile on a horse's ass, and a year later it was elected President."
2. Use the fact that "X" occurs most frequently in the ciphertext to find that $b = 19$. The message is: WEWERELUCKYBECAUSEOFTEN THEFREQUENCYMETHODNEEDSLONGERCIPHERTEXT.
3. THRPXDH.
4. SUCCESSATLAST.
5. AGENT 006 IS DEAD 007.
6. You find 9 possibilities for $a'$ and $b'$: $a' = 1, 4, 7, 10, 13, 16, 19, 22, 25$, and $b' = 21, 6, 18, 3, 15, 0, 12, 24, 9$, respectively. Since you have no more information to go on, simply try all nine possibilities; it turns out that only the third one $P \equiv 7C + 18 \ mod\ 27$ gives a meaningful plaintext. The plaintexts of the nine tranformations are, respectively: "I DY IB RIF", "I PS IH RIX", "I AM IN RIO", "I MG IT RIF", "I YA IZ RIX", "I JV IE RIO", "I VP IK RIF", "I GJ IQ RIX", "I SD IW RIO".
7. (a) $N$; (b) $N\varphi(N) = N^2 \prod_{p|N}(1-\frac{1}{p})$; (c) 312, 486, 812, 240.
8. (a) If $a \neq 1$, then the congruence $(a-1)P \equiv -b \ mod\ N$ has exactly one solution in the field $\mathbf{F}_N = \mathbf{Z}/N\mathbf{Z}$. (b) $P = 0$ is always fixed; for $N$ even (so $a$ must be odd) the congruence $(a-1)P \equiv 0 \ mod\ N$ at least has the two solutions $P = 0$ and $P = N/2$. (c) Any example with $N$ even and $b$ odd; more generally, any example in which $b$ is not divisible by $g.c.d.(a-1, N)$.
9. $N^2\varphi(N^2) = N^4 \prod_{p|N}(1-\frac{1}{p})$; 210,912; 354,294; 682,892; 216,000.

10. (a) $a' = 435,\ \ b' = 64$; "FOUNDTHEGOLD"; (b) $a = 115,\ \ b = 76$; "AWOFUWAE."
11. (a) You cannot find the key from the first two congruences; but subtracting the third from the first gives $139a' \equiv 247 \ mod\ 900$, and then $a' = 73,\ \ b' = 768$; "ARE YOU JOKING?"; (b) $a = 37,\ \ b = 384$; "FWU ORI DCCUVGA ."
12. "CCCP", which is Russian for USSR.
13. $P \equiv 37P + 384\ mod\ 900$ leads to $3P \equiv 43\ mod\ 75$; none.
14. (a) The product of $I \equiv P + b_1\ mod\ N$ and $C \equiv I + b_2\ mod\ N$ is $C \equiv P + b\ mod\ N$ with $b = b_1 + b_2$. (b) The product is the linear transformation with $a = a_1 \cdot a_2$. (c) The product is the affine transformation with $a = a_1 \cdot a_2$ and $b = a_2 \cdot b_1 + b_2$.
15. $P \equiv 642C + 187\ mod\ 853$; "DUMB IDEA ."
16. First compute $I \equiv 201C + 250\ mod\ 881$ and then $P \equiv 331I + 257\ mod\ 757$; "NO RETREAT."

## § III.2.

1. The key-word for enciphering is "SPY." The plaintext (with blanks and punctuation inserted for readability) is: "I had asked that a cable from Washington to New Delhi summarizing the results of the aid consortium be repeated to me through the Toronto Consulate. It arrived in code; no facilities existed for decoding. They brought it to me at the airport — a mass of numbers. I asked if they assumed I could read it. They said no. I asked how they managed. They said when something arrived in code, they phoned Washington and had the original message read to them." (John Kenneth Galbraith, *Ambassador's Journal*, quoted by G. E. Mellen in "Cryptology, computers and common sense," vol. III of *Computers and Security.*)
2. (a) $\begin{pmatrix} 3 & 2 \\ 1 & 1 \end{pmatrix}$; (b) $\begin{pmatrix} 19 & 10 \\ 23 & 16 \end{pmatrix}$; (c) $\begin{pmatrix} 11 & 11 \\ 24 & 1 \end{pmatrix}$; (d) $\begin{pmatrix} 820 & 0 \\ 0 & 801 \end{pmatrix}$; (e) $\begin{pmatrix} 127 & 303 \\ 546 & 353 \end{pmatrix}$.
3. (a) $\binom{6}{1}$; (b) none (since multiplying the second congruence by 2 and subtracting from the first gives $6y \equiv 8\ mod\ 9$, which would mean $3|8$); (c) $\binom{6}{1}, \binom{3}{4}, \binom{0}{7}$; (d) $\binom{0}{0}, \binom{6}{3}, \binom{3}{6}$.
4. (a) $\binom{9}{21}$; (b) $\binom{0}{0}$; (c) any vector with $y = x$, i.e., $\binom{0}{0}, \binom{1}{1}, \binom{2}{2}$, etc.; (d) any vector of the form $\binom{n}{15+n}$; (e) none.
5. (a) $\binom{787}{759}$; (b) $\binom{626}{233}$; (c) $\binom{0}{0}$; (d) $\binom{0}{0}, \binom{101}{505}, \binom{202}{1010}, \binom{303}{404}, \binom{404}{909}, \binom{505}{303}, \binom{606}{808}, \binom{707}{202}, \binom{808}{707}, \binom{909}{101}, \binom{1010}{606}$; (e) add $\binom{31}{800}$ to any of the 11 vectors in part (d) and reduce mod 1111.
6. Use mathematical induction, proving the assertion for $n = 1, 2, \ldots, b$ by inspection and then proving that the assertion for $n$ implies the assertion for $n + b$. Namely, compute:

$$\begin{aligned}\begin{pmatrix} f_{n+b+1} & f_{n+b} \\ f_{n+b} & f_{n+b-1}\end{pmatrix} &= \begin{pmatrix} 1 & 1 \\ 1 & 0\end{pmatrix}^{n+b} = \begin{pmatrix} 1 & 1 \\ 1 & 0\end{pmatrix}^{b}\begin{pmatrix} 1 & 1 \\ 1 & 0\end{pmatrix}^{n} \\ &= \begin{pmatrix} f_{b+1} & f_b \\ f_b & f_{b-1}\end{pmatrix}\begin{pmatrix} f_{n+1} & f_n \\ f_n & f_{n-1}\end{pmatrix} \\ &\equiv \begin{pmatrix} c & 0 \\ 0 & c\end{pmatrix}\begin{pmatrix} f_{n+1} & f_n \\ f_n & f_{n-1}\end{pmatrix} \\ &= \begin{pmatrix} cf_{n+1} & cf_n \\ cf_n & cf_{n-1}\end{pmatrix} \quad mod\ a,\end{aligned}$$

where $c \in (\mathbf{Z}/a\mathbf{Z})^*$, and use the induction assumption. (It can be proved that for *any* integer $a$ there is an integer $b$ such that $a|f_n \iff b|n$, and that if $a = p^\alpha$ is a power of a prime $p \neq 5$, then $b$ is a divisor of $p^{\alpha-1}(p^2-1)$; the proof uses a little algebraic number theory in the real quadratic field generated by the golden ratio — note that the golden ratio and its conjugate are the eigenvalues of the matrix in the definition of Fibonacci numbers.)

7. $A^{-1} = \begin{pmatrix} 23 & 7 \\ 18 & 5\end{pmatrix}$, "SENATORTOOK."
8. $A^{-1} = \begin{pmatrix} 22 & 16 \\ 21 & 17\end{pmatrix}$, "MEET AT NOON."
9. $A^{-1} = \begin{pmatrix} 22 & 20 \\ 28 & 8\end{pmatrix}$, "WHY NO GO? MARIA"; $A = \begin{pmatrix} 3 & 7 \\ 4 & 1\end{pmatrix}$, "JMLD W EFWJV."
10. "СЛАВА КПСС", which is Russian for GLORY TO THE CPSU (Communist Party of the Soviet Union).
11. The product cryptosystem has enciphering matrix $A_2A_1$.
12. "?CVK"; first apply $\begin{pmatrix} 18 & 28 \\ 19 & 20\end{pmatrix}$ to the ciphertext vector, working modulo 29, and then apply $\begin{pmatrix} 15 & 15 \\ 22 & 3\end{pmatrix}$ to the resulting vector, working modulo 26; "STOP."
13. By Proposition 3.2.1 (namely, (b) false implies (c) false), there exists a nonzero vector which the matrix $A$ takes to $\binom{0}{0}$. That plaintext digraph-vector can be added to any plaintext digraph-vector without changing the corresponding ciphertext.
14. Here the ciphertext is

$$\begin{pmatrix} 18 & 6 & 11 & 10 & 29 & 14 & 16 & 11 & 14 & 10 & 11 & 21 \\ 26 & 13 & 8 & 3 & 10 & 25 & 11 & 8 & 12 & 20 & 27 & 24\end{pmatrix}$$

and the last three columns of plaintext are $\begin{pmatrix} 10 & 17 & 0 \\ 0 & 11 & 27\end{pmatrix}$. The determinant of the matrix formed by the first two of the latter three columns is 20 *mod* 30, which is not invertible modulo 30 but is invertible modulo 3. The determinant of the matrix formed by the second and third columns is 9 *mod* 30, which is not invertible modulo 30 but is invertible modulo 10. Working with the first two columns modulo 3 gives $A^{-1}\ mod\ 3 = \begin{pmatrix} 10 & 17 \\ 0 & 11 \end{pmatrix} \cdot \begin{pmatrix} 10 & 11 \\ 20 & 27 \end{pmatrix}^{-1} = \begin{pmatrix} 1 & 2 \\ 0 & 2 \end{pmatrix} \cdot \begin{pmatrix} 1 & 2 \\ 2 & 0 \end{pmatrix}^{-1}$ $= \begin{pmatrix} 1 & 0 \\ 1 & 1 \end{pmatrix}$. Similarly, working with the last two columns modulo 10 gives $A^{-1} \equiv \begin{pmatrix} 4 & 9 \\ 5 & 8 \end{pmatrix}$. By the Chinese Remainder Theorem there is a unique matrix $A^{-1}$ modulo 30 that satisfies these two congruences: $A^{-1} = \begin{pmatrix} 4 & 9 \\ 25 & 28 \end{pmatrix}$. The plaintext is "GIVE THE PLANS TO KARLA."

15. Here the ciphertext is $\begin{pmatrix} 10 & 22 & 26 & 0 & 10 & 1 & 5 & 17 \\ 21 & 27 & 19 & 28 & 9 & 27 & 21 & 26 \end{pmatrix}$ and the first three columns of plaintext are $\begin{pmatrix} 2 & 8 & 0 \\ 29 & 29 & 29 \end{pmatrix}$. In attempting to use $A^{-1} = PC^{-1}$; note that the matrix formed from the first two digraphs of $C$ has determinant whose *g.c.d.* with 30 is 6. Using the 1st and 3rd digraphs improves the situation: $det\begin{pmatrix} 10 & 26 \\ 21 & 19 \end{pmatrix} = 4$, and $g.c.d.(4, 30) = 2$. Use this matrix for $C$ and work modulo 15 to find that $A^{-1} = \begin{pmatrix} 2 & 2 \\ 8 & 4 \end{pmatrix} + 15A_1$, where $A_1 \in M_2(\mathbf{Z}/2\mathbf{Z})$. Use the fact that $A^{-1}\begin{pmatrix} 10 & 22 & 26 \\ 21 & 27 & 19 \end{pmatrix} = \begin{pmatrix} 2 & 8 & 0 \\ 29 & 29 & 29 \end{pmatrix}$ and the fact that $det(A^{-1})$ is odd to show that either $A^{-1} = \begin{pmatrix} 17 & 2 \\ 8 & 19 \end{pmatrix}$ or $\begin{pmatrix} 17 & 2 \\ 23 & 19 \end{pmatrix}$. The first possibility gives the plaintext message "C.I.A. WILLLHTLA;" the second possibility gives "C.I.A. WILL HELP."
16. Use the Chinese Remainder Theorem.
17. $(p^2 - 1)(p^2 - p)$.
18. The determinant has no common factor with $p^\alpha$ if and only if it has no common factor with $p$; $p^{4\alpha-3}(p^2 - 1)(p - 1)$.
19. $N^4 \prod_{p|N}(1 - \frac{1}{p})(1 - \frac{1}{p^2})$; 157248, 682080, 138240.
20. $N^{(k^2)} \prod_{p|N}\left((1 - \frac{1}{p})(1 - \frac{1}{p^2}) \cdots (1 - \frac{1}{p^k})\right)$.
21. $N^6 \prod_{p|N}(1 - \frac{1}{p})(1 - \frac{1}{p^2})$; 106,299,648; 573,629,280; 124,416,000.
22. (a) $(p^2 - 1)(p^2 - p)$; (b) $p^2 - p$.
23. (a) $A_0 = \begin{pmatrix} 21 & 27 \\ 18 & 27 \end{pmatrix}$; (b) $\begin{pmatrix} 1 \\ 1 \end{pmatrix}$; (c) six (this agrees with Exercise 22(b), where $p = 3$); they are: $A = \begin{pmatrix} a & 7 \\ c & 7 \end{pmatrix}$, where $\binom{a}{c} = \binom{21}{28}, \binom{21}{8}, \binom{1}{18}, \binom{1}{8}, \binom{11}{18}$, or $\binom{11}{28}$.
24. (a) $g.c.d.(det(A - I),\ N) = 1$, where $det(A - I) = (a - 1)(d - 1) - bc$ (apply the (a)$\Longleftrightarrow$(c) part of Proposition 3.2.1 with $A$ replaced by $A - I = \begin{pmatrix} a-1 & b \\ c & d-1 \end{pmatrix}$). (b) Let $\mathbf{F}_N$ be the field $\mathbf{Z}/N\mathbf{Z}$. The digraphs are a 2-dimensional vector space, of which the fixed digraphs form a subspace. Any subspace that contains more than the zero-vector must either be 1-dimensional, in which case it has $N$ elements, or else contain all digraphs, in which case $A = I$.
25. (a) $P = A'C + B'$; $A' = \begin{pmatrix} 14 & 781 \\ 821 & 206 \end{pmatrix}$, $B' = \begin{pmatrix} 322 \\ 202 \end{pmatrix}$; "HIT ARMY

BASE! HEADQUARTERS" (b) $C = AP+B$, $A = \begin{pmatrix} 103 & 30 \\ 10 & 7 \end{pmatrix}$, $B = \begin{pmatrix} 301 \\ 412 \end{pmatrix}$; "!NJUFYKTEGOUL IB!VFEXU!JHALGQGJ?"

26. $29^8(29^2-1)(29^2-29) = 341,208,073,352,438,880$.
27. 91,617,661,629,000,000.
28. $A^{-1} = \begin{pmatrix} 18 & 21 & 19 \\ 13 & 18 & 3 \\ 3 & 19 & 11 \end{pmatrix}$, "SENDROSESANDCAVIARJAMESBOND."

## § IV.1.

1. $\binom{m}{2} = m(m-1)/2$ for classical; $m$ for public key; 499500 versus 1000 when $m = 1000$.
2. Here is one possible method. The investors and stockbrokers use a system with $\mathcal{P} = \mathcal{C}$. Then user A sends a message to user B by taking each message unit $P$ and transmitting $f_B f_A^{-1}(P)$. Each message includes an identification number. Then user B must immediately send an acknowledgment message which includes the identification number of the message received from A. User B transforms each message unit $P$ of the acknowledgment message to $f_A f_B^{-1}(P)$ before transmitting it (this is completely analogous to A's double enciphering of the original message). If A does not receive an acknowledgment message very soon after sending his message, he repeats the message until he does. Later, after the stock loses money or for some reason there is a dispute about who sent what message, the stockbroker can prove that a message was sent by A, because no one except A (and the judge) has the information necessary to produce a message that can be read by applying $f_A f_B^{-1}$. Similarly, A can prove that a message with a given identification number was received by B (since no one else could have sent the acknowledgment message), and so B can be required to produce the message for the judge.
3. A public key cryptosystem is agreed upon which uses random integers (subject to some conditions, perhaps) to form enciphering and deciphering keys according to some algorithm. The computer is then programmed to generate random integers which it then uses to form a pair of keys $K = (K_E, K_D)$. The computer transmits $K_D$ (*not* $K_E$) to the outside world and keeps $K_E$ (*not* $K_D$) to itself. Thus, anyone at all can read its messages, but no one at all can create a message that can be deciphered using the deciphering algorithm with key $K_D$. (This is the reverse of the usual situation in public key cryptography, where anyone can send a message but only the user with the secret key can read it.) It is possible for the scientists working jointly to program the computer to generate random numbers in a way that no one can predict or duplicate once the computer is "on its own." (Note the profound realism of this example, which assumes that the two countries have infinite mistrust of each other and at the same time infinite trust of computers.)

4. Björn chooses at random an element $p \in \mathcal{P}$, computes $c = f(p)$ and sends Aniuta $c$. Aniuta then computes the two preimages $p_1$ and $p_2$ and sends only one of them, say $p_1$, to Björn. If $p_1 \neq p$, then Björn can name both preimages $p_1$ and $p_2 = p$, in which case we say that Björn wins; otherwise, Aniuta wins. If Aniuta wins, she has to produce the second preimage, which Björn can verify does in fact satisfy $f(p_2) = c$ (otherwise, Aniuta could cheat by choosing an improper key, for which each $c$ has only one preimage). (Aniuta would have no interest in choosing a key for which each $c$ has more than two preimages, since that would just lessen her chances of sending Björn the preimage that he already knows.)

§ **IV.2.**

1. (a) BH A 2AUCAJEAR0; (b) $2047 = 23 \cdot 89$ (see Example 1 in § I.4), $d_A = 411$; (c) since $\varphi(23)$ and $\varphi(89)$ have small least common multiple 88, any inverse of 179 modulo 88 will work as $d_A$ (e.g., 59).
2. $n_A$ is the product of the Mersenne prime 8191 and the Fermat prime 65537 — a flamboyantly bad choice; $d_A = 201934721$; "DUMPTHE-STOCK."
3. (a) STOP PAYMENT; (b) (i) 6043; (ii) $n = 113 \cdot 191$.
4. On the third try $t = 152843, 152844, 152845$ you find that $t^2 - n = 804^2$, and so $p = 152845 + 804 = 153649$, $q = 152845 - 804 = 152041$.
5. To show that one cannot feasibly compute the companion element in $\mathcal{P}$ that has the same image as a given element, we suppose that a person who knows only $K_E$ (i.e., knows $n$ but not its factorization) obtained a second pair $\pm x_2$ with the same square modulo $n$ as $\pm x_1$. Then show that $g.c.d.(x_1 + x_2, n)$ is either $p$ or $q$. In other words, finding a *single* pair of companion elements of $(\mathbf{Z}/n\mathbf{Z})^*/\pm 1$ is tantamount to factoring $n$.
6. It suffices to prove that $a^{de} \equiv a \bmod p$ for any integer $a$ and each prime divisor $p$ of $n$. This is obvious if $p|a$; otherwise use Fermat's Little Theorem (Proposition I.3.2).
7. If $m/2 \equiv (p-1)/2 \bmod p-1$, then $a^{m/2} \equiv (\frac{a}{p})$, which is $+1$ half the time and $-1$ half the time. In case (ii), use the Chinese Remainder Theorem to show that the probability that an element in $(\mathbf{Z}/n\mathbf{Z})^*$ is a residue modulo $p$ and the probability that it is a residue modulo $q$ are independent of one another, i.e., the situation in case (ii) is like two independent tosses of a coin.

§ **IV.3.**

1. (a) 24, 30, 11, 13; (b) 1, $\alpha^2 + \alpha$, $\alpha$, $\alpha + 1$.
2. (i) To justify moving the $a$ to the left, notice that if $x < \varphi(3^\alpha)$ is the solution of $2^x a \equiv 1 \bmod 3^\alpha$, then $\varphi(3^\alpha) - x$ is the solution of the original congruence. If $a \equiv 2 \bmod 3$, then solve the problem $2^x(2a) \equiv 1 \bmod 3^\alpha$, in which we do have $2a \equiv 1 \bmod 3$, and then $x+1$ is the solution of the original congruence. If $a \equiv 1 \bmod 3$, then the solution $x$ must be even,

because $2^{odd} \equiv 2 \ mod\ 3$. (iii) To show that $(*)_j$ holds after choosing $x_{j-2} = (1-a_{j-1})/3^{j-1}$, you compute the left side of $(*)_j$ modulo $3^j$ as follows: it equals $a_{j-1}g_{j-1}^{x_{j-2}} \equiv (1-3^{j-1}x_{j-2})g_{j-1}^{x_{j-2}}$, and then show that $(1+3)^{3^{j-2}x_{j-2}} \equiv 1+3^{j-1}x_{j-2} \ mod\ 3^j$ (use the binomial expansion). Thus, the left side of $(*)_j$ is $\equiv (1-x_{j-2}^2 3^{2(j-1)}) \equiv 1 \ mod\ 3^j$. Finally, to estimate the number of bit operations, note that each time step (iii) is performed one does a couple of multiplications and reductions (divisions) with integers having $O(\alpha)$ bits, i.e., each step takes $O(\alpha^2)$ bit operations; thus, the whole thing takes $O(\alpha^3)$ bit operations.

3. (a) To make your computation of $(g^b)^a$ in $\mathbf{F}_{31^2}$ easier, use the fact that $(c+di)^{32} = c^2+d^2$; you find that $A+Bi = 26+28i$; (b) $20+13i$; (c) $P \equiv 6C+18 \ mod\ 31$; (d) YOU'RE JOKING!
4. (a) $K_E = 1951280$, its least nonnegative residue modulo $26^4$ is $7\cdot 26^3 + 0\cdot 26^2 + 13\cdot 26 + 6$; but you have to add 1 to this in order to get an invertible enciphering matrix $\begin{pmatrix} 7 & 0 \\ 13 & 7 \end{pmatrix}$; (b) $\begin{pmatrix} 15 & 0 \\ 13 & 15 \end{pmatrix}$, DONOTPAY.
5. The $f_A$'s must commute, i.e., $f_A f_B = f_B f_A$ for all pairs of users $A$ and $B$; you need to use it with a good signature scheme (as explained in the text); and it must not be feasible to determine the key for $f_A$ from the knowledge of pairs $(P,\ f_A(P))$. For example, a translation map $f_A(P) \equiv P+b$ or a linear map $f_A(P) \equiv aP$ has the first property but not the last one, since knowing any pair $(P,\ P+b)$ (or $(P,\ aP)$) immediately enables anyone to find $b$ (or $a$). The example in the text satisfies this property because of our assumption that the discrete log problem cannot be solved in a reasonable length of time.
6. $P = 6229 =$ "GO!"
7. (a) First replace $x$ by $p-1-x$ so as to reduce to the equivalent congruence $g^x a \equiv 1 \ mod\ p$. Set $l = 2^k$, and $x = x_0 + 2x_1 + \cdots + 2^{l-1}x_{l-1}$. Define $g_j = g^{2^j} \ mod\ p$ and $a_j = g^{x_0+2x_1+\cdots+2^{j-1}x_{j-1}}a \ mod\ p$ (with $a_0$ taken to be $a$). At the $j$-th step, compute $a_{j-1}^{2^{k-j}} = \pm 1$, and set $x_{j-1} = 0$ if it is $+1$ and $x_{j-1} = 1$ if it is $-1$; also compute $g_j = g_{j-1}^2$, and $a_j = g_{j-1}^{x_{j-1}}$. When $j = l$, you're done. (b) $O(log^4 p)$. (c) $k = 7912$.
8. THEYREFUSEOURTERMS.
9. To find $x$, Alice converts the congruence $g^S \equiv y^r r^x \equiv g^{ar+kx}$ to the congruence $S \equiv ar + kx \ mod\ p-1$, which has solution $x = k^{-1}(S - ar) \ mod\ p-1$. Bob knows $p$, $g$, and $y = y_A$, and so can verify that $g^S \equiv y^r r^x \ mod\ p$ once he is sent the pair $(r,\ x)$ along with $S$. Finally, someone who can solve the discrete log problem can determine $a$ from $g$ and $y$, and hence forge the signature by finding $x$.
10. 107.
11. (a) $9/128 = 7.03\%$, $160/1023 = 15.64\%$; (b) $70/2187 = 3.20\%$, $1805/29524 = 6.11\%$. (See the corollary to Proposition II.1.8.)
12. (a) Neglect terms beyond the leading power of $p$. Then the number of monic polynomials is $(p^{n+1}-1)/(p-1) \approx p^n$. The number of products of degree $< n$ can be neglected. The number $n_f$ of irreducible monic

polynomials of degree $f$ is $\frac{1}{f}(p^f - \sum_{d<f,\ d|f} dn_d) \approx \frac{p^f}{f}$. The number of products of degree $n$ is then the following sum taken over all partitions $n = \sum_{d=1}^{m} i_d d$ $(i_d \geq 0)$:

$$\sum \binom{n_1 + i_1 - 1}{i_1} \cdots \binom{n_m + i_m - 1}{i_m}$$
$$\approx p^n \sum \frac{1}{2^{i_2} 3^{i_3} \cdots m^{i_m} i_1! i_2! \cdots i_m!}.$$

Thus,

$$P(n, m) = \sum \Big( \prod_{d=1}^{m} d^{i_d} i_d! \Big)^{-1}.$$

This is obviously $> 0$; to see that $P(n, m) < 1$, notice that there are approximately $p^n/n$ monic irreducible polynomials of degree $n$, and so the probability that a monic polynomial fails to factor as desired is at least $1/n$. (b) $\sum_{i+2j=n,\ 0 \leq i,j} (2^j i! j!)^{-1}$. (c) $P(3, 2) = 2/3$, $P(4, 2) = 5/12$, $P(5, 2) = 13/60$, $P(6, 2) = 19/180$, $P(7, 2) = 29/630$.

**§ IV.4.**

1. (a) yes, 1; (b) yes, 0; (c) no, 2; (d) no, 0; (e) yes, 1; (f) no, 1.
2. (a) Use induction on $k$. (b) To show the second part, let $v_i$ be strictly greater than $1 + v_{i-1} + \cdots + v_0$, and set $V = v_i - 1$.
3. Use induction.
4. (a) INTERCEPTCONVOY; (b) 89, 3, 25, 11, 41, 60, 65.
5. FORMULA STOLEN!
6. BRIBE HIM!

**§ IV.5.**

1. $2^T$ to 1.
2. (a) The numbers $e$ and $x + e$ modulo $N$ that Vivales receives in steps (2) and (3) are in the range from 0 to $N - 1$; so after a large number of trials Vivales will get a good idea of the magnitude of $N$. (b) Let $N'$ be a very large multiple of $N$, and replace $N$ by $N'$ in steps (1) and (3).
3. The values Vivales receives in step (3) are upper bounds for $x$. The values Clyde sends in step (3) are not bounded from below, unlike the values $x + e$ that Pícara sends.
4. Pícara would have $y$ as her public key; signing a document would consist of convincing the recipient that she knows its discrete log $x$.
5. Knowing the factorization enables one to take square roots, using the method at the end of §II.2 along with the Chinese Remainder Theorem (see also Exercise 5 of §IV.2). Conversely, suppose you have an algorithm to take square roots. Then choose a random number $x$, and apply the algorithm to the least nonnegative residue of $x^2$ *mod* $n$. The result will be $x'$ such that $x'^2 \equiv x^2$ $(mod\ n)$. There is a 50% chance

that $x' \not\equiv \pm x$ (*mod* $n$), in which case you immediately obtain a nontrivial factor, i.e., $g.c.d.(x'+x, n)$. By repeating the procedure $T$ times, you have probability $1 - 2^{-T}$ of factoring $n$.

6. Yes. Suppose that another person Pícara$_2$ playing the role of Pícara intercepts the message $(b^{y_1}, b^{y_2}, \alpha_1, \alpha_2)$ that Pícara sent to Vivales, and wants to fool Vivales into believing that she also knows the factorization of $n$ (or the 3-coloring, or the discrete logarithm, etc.). Suppose also that Vivales will not accept from Pícara$_2$ a repetition of the exact same four-tuple that Pícara sent. Without knowing Pícara's secret random integers $y_1, y_2$ or her messages $m_1, m_2$ or the discrete logarithm of either $\beta_1$ or $\beta_2$, Pícara$_2$ has no way to construct a different four-tuple that gives Vivales the impression that she knows the factorization.
7. Pícara randomly selects $0 \le x' < N$, and sends Vivales $y' = b^{x'}$. Then the two messages for oblivious transfer are $m_1 = x'$ and $m_2 = x + x'$ (*mod* $N$). Vivales verifies either $b^{x'} = y'$ or else $b^{x+x'} = yy'$. If the procedure is repeated $T$ times, then the odds against Pícara being lucky (i.e., being able to fool Vivales into thinking she knows the discrete log of $y$) are $2^T$ to 1.
8. Vivales can easily get Pícara to betray the factorization of $n$, as follows. He randomly chooses integers $z$ until he finds a $z$ whose Jacobi symbol modulo $n$ is $-1$. He then sends Pícara $y = z^2$ *mod* $n$. Pícara replies with the value $x^2$ of a square root of $y$ *mod* $n$ which is different from $\pm z$. Vivales can now find a nontrivial factor of $n$, namely, $g.c.d.(x^2 + z, n)$.
9. The proof of zero knowledge transmission using a simulator Clyde will not work. Another problem is that Pícara would have to be certain that every $y_i$ had been produced by the trusted Center, and not by Vivales pretending to be the trusted Center.

**§ V.1.**

1. (a) 4, 11; (b) 8, 13; (c) see part (d); (d) Show that $n - 1 \equiv p - 1$ *mod* $2p - 2$, so that $b^{n-1} \equiv 1$ *mod* $p$, and $b^{n-1} \equiv b^{(2p-1-1)/2} \equiv (\frac{b}{2p-1})$ *mod* $2p - 1$. Then $b^{n-1} \equiv 1$ *mod* $p(2p-1)$ if and only if $(\frac{b}{2p-1}) = 1$.
2. (a) Use the fact that $n = n'p = n'(p - 1 + 1) \equiv n'$ *mod* $p - 1$. (b) Use part (a) with $n' = 3$ to conclude that $p$ would have to be a divisor of $2^2 - 1$, $5^2 - 1$, $7^2 - 1$. (c) $p$ would have to be a divisor of $2^4 - 1$, $3^4 - 1$, $7^4 - 1$. (d) Any smaller $n$ would be the product of 2 primes greater than 5 (by part (c)). Then check 49 and 77.
3. Divide the congruence (1) with $n = p^2$ by the congruence $b^{p^2-p} \equiv 1$ *mod* $p^2$, which always holds by Euler's theorem (Proposition I.3.5).
4. (a) 217; (b) 341.
5. (a) First suppose that $n$ is a pseudoprime to the base $b$. Since $n - 1 = pq - 1 \equiv q - 1$ *mod* $p - 1$, you have $b^{q-1} \equiv 1$ *mod* $p$; but since $b^{p-1} \equiv 1$ *mod* $p$ always by Fermat's little theorem, and since $d$ is an integer linear combination of $p - 1$ and $q - 1$, it follows that $b^d \equiv 1$ *mod* $p$.

Interchanging the roles of $p$ and $q$ gives $b^d \equiv 1 \, mod \, q$, and so $b^d \equiv 1 \, mod \, n$. The converse is similar (actually, easier). There are $d^2$ bases in $(\mathbf{Z}/n\mathbf{Z})^*$. (b) four: $\pm 1$, $\pm(4p+1)$. (c) $d^2/\varphi(341) = 100/300 = \frac{1}{3}$.

7. (a) See part (b). (b) Since $N - 1 = b(b^{n-1}-1)/(b-1)$, where the numerator is divisible by $n$ (because $n$ is a pseudoprime to the base $b$) and the denominator is prime to $n$, it follows that $n|N-1$. Since $b^n \equiv 1 \, mod \, N$ (namely, $(b-1)N = b^n - 1$), we have $b^{N-1} \equiv 1 \, mod \, N$. One must also show that $N$ is composite, but this is easy if we use the fact that $n$ is composite by assumption (see the corollary to Proposition I.4.1). The fact that $N$ is odd (whether $b$ is odd or even) follows by writing $N$ in the form $b^{n-1} + b^{n-2} + \cdots + b + 1$. (c) Start with 341, 91, or 217, respectively, and use part (b) to find a sequence of larger and larger pseudoprimes. Note that the condition $g.c.d.(b-1, n) = 1$ always holds when $b = 2, 3, 5$. (d) 15 is a pseudoprime to the base 4, but $N = (4^{15}-1)/3$ is not. (To see the latter, note that 4 has order 15 in $(\mathbf{Z}/N\mathbf{Z})^*$, but $N - 1 = 4(4^{14}-1)/3$ is not divisible by 3, let alone 15.)

8. (a) $n = \left(\frac{b^p-1}{b-1}\right)\left(\frac{b^p+1}{b+1}\right)$ (b) Note that $n$ is odd (see the answer to 7(b) above), and so $2|n-1$. Next, since $(n-1)(b^2-1) = b^2(b^{2(p-1)}-1) \equiv 0 \, mod \, p$ and $p$ does not divide $(b+1)(b-1) = b^2 - 1$, it follows that $p|n-1$. (c) Since $n$ is an odd composite number, $b^{2p} \equiv 1 \, mod \, n$, and $2p|n-1$, it follows that $n$ is a pseudoprime to the base $b$. Since there are infinitely many primes greater than $b+1$, in this way we get infinitely many pseudoprimes to the base $b$.

9. (a) $3^{2046} \equiv 1013 \, mod \, 2047$, so (1) fails for $b = 3$. (b) If composite, they will still be pseudoprimes to the base 2. To see this for $n = 2^{2^k} + 1$, we note that $2^{2^k} \equiv -1 \, mod \, n$, and then $2^{n-1} \equiv 1 \, mod \, n$ can be obtained from this by repeated squaring. For $n = 2^p - 1$, we have $n - 1 = 2(2^{p-1}-1) \equiv 0 \, mod \, p$, and so $2^p = n + 1 \equiv 1 \, mod \, n$ implies $2^{n-1} \equiv 1 \, mod \, n$. Using (2) with $b = 2$ also won't work, since both sides will be 1, even if the number is composite. Using (3) with $b = 2$ also won't work: for a Fermat number this follows because $2^{2^k} \equiv -1 \, mod \, n$, and for a Mersenne number it follows by Proposition V.1.5.

10. Expand the parentheses to show that $n - 1$ is divisible by $36m$, and hence by $6m$, $12m$, and $18m$.

12. We suppose $p < q$. The technique to answer (a)–(b) is given in part (c). (a) $561 = 3{\cdot}11{\cdot}17$; (b) $1105 = 5{\cdot}13{\cdot}17$; $2465 = 5{\cdot}17{\cdot}29$; $10585 = 5{\cdot}29{\cdot}73$. (c) Suppose $p < q$. Since $q - 1|rpq - 1 \equiv rp - 1 \, mod \, q - 1$, we must have $rp - 1 = a(q-1)$ for some $a$, $1 < a < r$. Also $p - 1|rq - 1$, and so $p-1|a(rq-1) = r(aq) - a = r(a+rp-1) - a \equiv (r-1)(a+r) \, mod \, p-1$. Thus, with $r$ fixed and for each fixed $a$ from 2 to $r-1$, there are only finitely many possibilities for $p$, namely, the primes such that $p - 1$ is a divisor of $(r-1)(a+r)$. Then each prime $p$ uniquely determines $q$, because $rp - 1 = a(q-1)$. Of course, not all $a$ and $p$ lead to a

Carmichael number (for example, $a$ might not divide $rp-1$).

13. Any Carmichael number not listed in Exercise 12(a)–(b) must be at least a product of three distinct primes all $\geq 7$.

14. $n=21$, $b=8$.

16. (a) By Exercise 1(d), we need only look at the $b$ for which $b^{p-1} \equiv (\frac{b}{2p-1}) = 1$ *mod* $2p-1$. Since $n-1 \equiv p-1$ *mod* $2p-2$, we have $b^{(n-1)/2} \equiv b^{(p-1)/2}$ *mod* $p$ and *mod* $2p-1$, i.e., $b^{(n-1)/2} \equiv b^{(p-1)/2}$ *mod* $n$. Now $(\frac{b}{n}) = (\frac{b}{2p-1})(\frac{b}{p}) = (\frac{b}{p}) \equiv b^{(p-1)/2}$ *mod* $p$, so condition (2) holds if and only if $b^{(p-1)/2} \equiv (\frac{b}{p})$ *mod* $2p-1$. This holds for exactly half of all $b$ for which $b^{p-1} \equiv 1$ *mod* $2p-1$ (since in $(\mathbf{Z}/(2p-1)\mathbf{Z})^*$ such $b$ must be a power $g^j$ of a generator $g$ such that $\frac{p-1}{2}j \equiv 0$ *mod* 4 if $(\frac{b}{p}) = 1$, $\frac{p-1}{2}j \equiv 2$ *mod* 4 if $(\frac{b}{p}) = -1$). (b) $n = p(2p-1)$ where $p \equiv 3$ *mod* 4 (by Proposition V.1.5).

17. Compute $n$ modulo $72m$: $n \equiv 36m^2+36m+1$. Thus, $\frac{n-1}{2} \equiv 18m(m+1)$ *mod* $36m$. If $m$ is odd, this means that we always have $b^{(n-1)/2} \equiv 1$ *mod* $n$ (because $p-1|36m$ for each $p|n$), and so (2) holds if and only if $(\frac{b}{n}) = 1$, i.e., 50% of the time. If $m$ is even, we still have $b^{(n-1)/2} \equiv 1$ *mod* $6m+1$ and *mod* $18m+1$, while $b^{(n-1)/2} \equiv b^{6m} \equiv (\frac{b}{12m+1})$ *mod* $12m+1$. Thus, in that case (2) holds if and only if $(\frac{b}{12m+1}) = 1$ (so that $b^{(n-1)/2} \equiv 1$ *mod* $n$) and also $(\frac{b}{n}) = 1$, i.e., 25% of the time.

18. (a) $O(log^3 n\, log\, m)$; (b) $O(log^5 n)$.

19. (a) $N$ is composite because $n$ is composite (by the corollary to Proposition I.4.1); then proceed as in Exercise 9 to see that $2^{(N-1)/2} = 2^{2^{n-1}-1} \equiv 1$ *mod* $N$. But since $N \equiv -1$ *mod* 8, we also have $(\frac{2}{N}) = 1$. Thus, $N$ is an Euler pseudoprime; by Proposition V.1.5, it is also a strong pseudoprime. (b) Use the same argument as in Exercise 7(c).

20. If the first possibility in (3) holds, then obviously $(b^k)^t \equiv 1$ *mod* $n$. Now suppose that $b^{2^r t} \equiv -1$ *mod* $n$. Write $k = 2^i j$ with $j$ odd. If $i > r$, then $(b^k)^t \equiv 1$ *mod* $n$; if $i \leq r$, then $(b^k)^{2^{r-i}t} = (b^{2^r t})^j \equiv (-1)^j = -1$ *mod* $n$.

21. (a) Show that the necessary and sufficient conditions on $b$ are: $(\frac{b}{17}) = 1$, $(\frac{b}{561}) = 1$. These conditions both hold 25% of the time, i.e., for 80 bases in $(\mathbf{Z}/561\mathbf{Z})^*$. (b) Since $b^{70} \equiv 1$ *mod* 3 and *mod* 11, it follows that 561 is a strong pseudoprime to the base $b$ if and only if $b^{35} \equiv \pm 1$ *mod* 561, i.e., if and only if either (i) $b \equiv 1$ *mod* 3, $b \equiv 1$ *mod* 17, $(\frac{b}{11}) = 1$, or else (ii) $b \equiv -1$ *mod* 3, $b \equiv -1$ *mod* 17, $(\frac{b}{11}) = -1$. There are 10 such bases, 5 in case (i) and 5 in case (ii), by the Chinese Remainder Theorem. The 8 nontrivial bases $b \neq \pm 1$ are: 50, 101, 103, 256, 305, 458, 460, 511.

22. Use Exercise 7(a) of § I.3, which says that the only square roots of 1 are $\pm 1$.

23. (a) $8^2 \equiv 18^2 \equiv -1$ *mod* 65; $14^2 \equiv 1$ *mod* 65, but $14^1 \not\equiv \pm 1$ *mod* 65. (b) The case when $n$ is a prime power follows from the previous exercise, so

suppose that $n$ is not a prime power. First, if $p|n$ with $p \equiv 3 \; mod \; 4$, then no integer raised to an even power gives $-1 \; mod \; n$ (since $-1$ is not a quadratic residue modulo $p$); hence, in this case the strong pseudoprime condition can be stated: $b^t \equiv \pm 1 \; mod \; n$. This condition obviously has the multiplicative property. Next, suppose that $n = p_1^{\alpha_1} \cdots p_r^{\alpha_r}$ where $p_j \equiv 1 \; mod \; 4$ for $1 \leq j \leq r$. Let $\pm a_j$ be the two square roots of $-1$ modulo $p_j^{\alpha_j}$ (a square root modulo $p_j$ can be lifted to a square root modulo $p_j^{\alpha_j}$; see Exercise 20 of § II.2). Then any $b$ which satisfies $b \equiv \pm a_j \; mod \; p_j^{\alpha_j}$ (for any choice of the $\pm$) is a base to which $n$ is a strong pseudoprime, since then $b^{2t} \equiv (-1)^t \equiv -1 \; mod \; n$. Choose $b_1$ by taking all of the $\pm a_j$ equal to $a_j$, and choose $b_2$ by taking any of the $2^r - 2$ possible choices of sign other than all positive or all negative. Then show that for $b = b_1 b_2$ one has $b^{2t} \equiv 1 \; mod \; n$ and $b^t \equiv b \not\equiv \pm 1 \; mod \; n$.

24. (a) In that case you obtain a number $c$ other than $\pm 1$ whose square is 1; then $g.c.d.(c+1, n)$ is a nontrivial factor of $n$. (b) Choose $p$ and $q$ so that $p-1$ and $q-1$ do not have a large common divisor (see Exercise 5 above).

§ **V.2.**

1. $g.c.d.(x_5 - x_3, n) = g.c.d.(21 - 63, 91) = 7$; $91 = 7 \cdot 13$.
2. $g.c.d.(x_6 - x_3, n) = g.c.d.(2839 - 26, 8051) = 97$; $8051 = 83 \cdot 97$.
3. $g.c.d.(x_9 - x_7, n) = g.c.d.(869 - 3397, 7031) = 79$; $7031 = 79 \cdot 89$.
4. $g.c.d.(x_6 - x_3, n) = g.c.d.(630 - 112, 2701) = 37$; $2701 = 37 \cdot 73$.
5. (a) Prove by induction on $k$ that for $1 \leq k \leq r$ there is a $1/r$ probability that $x_0, \ldots, x_{k-1}$ are distinct and $x_k$ is equal to one of the earlier $x_j$. For $k = 1$ there is a $1/r$ probability that $f(x_0) = x_0$. The induction step is as follows. By the induction assumption, the probability that none of the earlier $k$'s was the first for which $x_k = x_j$ for some $j < k$ is $1 - \frac{k-1}{r} = \frac{r-(k-1)}{r}$. Assuming this to be the case, there are $r - (k-1)$ possible values for $f(x_{k-1})$, since a bijection cannot take $x_{k-1}$ to any of the $k-1$ values $f(x_j)$, $0 \leq j \leq k-2$. Of the $r - (k-1)$ possible values, one is $x_0$, and all the others are distinct from $x_0, x_1, \ldots, x_{k-1}$. Thus, there is a $1/(r-(k-1))$ chance that the value is one of the earlier $x_j$ (namely, if this is the case, note that $j = 0$). The probability that *both* things happen — none of the earlier $k$'s was the first for which $x_k = x_0$ but our present $k$ has $x_k = x_0$ — is the product of the individual probabilities, i.e., $\frac{r-(k-1)}{r} \cdot \frac{1}{r-(k-1)} = \frac{1}{r}$. (b) Since all of the values from 1 to $r$ are equally probable, the average is $\frac{1}{r}\sum_{k=1}^{r} k = \frac{1}{r}(r(r+1)/2) = (r+1)/2$.
6. Suppose that $a$ has no common factor with $n$ (otherwise, we would immediately find a factor of $n$ by computing $g.c.d.(a, n)$ and we would have no need of the rho method at all). Then $f(x) = ax+b$ is a bijection of $\mathbf{Z}/r\mathbf{Z}$ to itself (for any $r|n$), and so the expected number of steps

before we get a repetition modulo $r$ is of the order of $r/2$ (by Exercise 5(b)) rather than $\sqrt{r}$, i.e., it is much worse.

7. (a) $2^k \equiv 2^\ell$ *mod* $r-1$; (b) $\ell = s$ and $k = s+m$, where $m$ is the *order* of 2 modulo $t$, i.e., the smallest positive integer such that $2^m \equiv 1$ *mod* $t$. $m$ is also the *period* of the repeating binary expansion of $1/t$, as we see by writing $2^m - 1 = ut$ and then $1/t = u\sum_{i=1}^{\infty} 2^{-mi}$. (c) $k$ can easily have order almost as large as $r$, e.g., if $r-1$ is twice a prime and 2 happens to be a generator modulo that prime (in which case $s = 1$, $m = (r-3)/2$).

§ **V.3.**

1. (a) (using $t = [\sqrt{n}] + 1 = 93$) $89 \cdot 97$; (b) (using $t = [\sqrt{n}] + 4 = 903$) $823 \cdot 983$; (c) (using $t = [\sqrt{n}] + 6 = 9613$) $9277 \cdot 9949$; (d) (using $t = [\sqrt{n}] + 1 = 9390$) $9343 \cdot 9437$; (e) (using $t = [\sqrt{n}] + 8 = 75$) $43 \cdot 107$.
2. In the factorization $n = ab$ with $a > b$, if $a < \sqrt{n} + \sqrt[4]{n}$, then $b = n/a > n/(\sqrt{n} + \sqrt[4]{n}) > \sqrt{n} - \sqrt[4]{n}$. On the other hand, if we start with $b > \sqrt{n} - \sqrt[4]{n}$, then we must have $a < \sqrt{n} + \sqrt[4]{n} + 2$, because otherwise we would have $n = ab > (\sqrt{n} + \sqrt[4]{n} + 2)(\sqrt{n} - \sqrt[4]{n}) = n + \sqrt{n} - 2\sqrt[4]{n} > n$ (as soon as $n > 15$; we check Exercise 2 separately for the first few $n$). Thus, in either case $a - b < 2(\sqrt[4]{n} + 1)$. But if Fermat factorization fails to work for the first value of $t$, then the $s$ and $t$ corresponding to the factorization $n = ab$ satisfy: $t > \sqrt{n} + 1$, and so $s = \sqrt{t^2 - n} > \sqrt{(\sqrt{n}+1)^2 - n} = \sqrt{2\sqrt{n}+1} > \sqrt{2}\sqrt[4]{n}$, which contradicts the relationship $s = (a-b)/2 < \sqrt[4]{n} + 1$ as soon as $n > 33$.
3. (a) We would have $t^2 - s^2 = kn \equiv 2$ *mod* 4; but modulo 4 the difference of two squares cannot be 2. (b) We would have $t^2 - s^2 = 4n \equiv 4$ *mod* 8, which can hold only if both $s$ and $t$ are even; but then $(t/2)^2 - n = (s/2)^2$, and so simple Fermat factorization would have worked equally well.
4. (a) (using $t = [\sqrt{3n}] + 1 = 455$) $149 \cdot 463$; (b) (using $t = [\sqrt{3n}] + 2 = 9472$) $3217 \cdot 9293$; (c) (using $t = [\sqrt{5n}] + 1 = 9894$) $1973 \cdot 9923$; (d) (using $t = [\sqrt{5n}] + 2 = 9226$) $1877 \cdot 9067$.
5. $B = \{2, 3\}$; the vectors are $\{0, 1\}$ and $\{0, 1\}$; $b = 52 \cdot 53$ *mod* $n = 55$, $c = 2 \cdot 3^2 = 18$; *g.c.d.*$(55 + 18, 2701) = 73$; $2701 = 37 \cdot 73$.
6. $B = \{-1, 2, 3, 61\}$; the vectors are $\{1, 0, 0, 0\}$, $\{1, 0, 0, 1\}$, and $\{0, 0, 0, 1\}$; $b = 68 \cdot 152 \cdot 153$ *mod* $n = 1555$, $c = 2 \cdot 3 \cdot 61 = 366$; *g.c.d.*$(1555 + 366, 4633) = 113$; $4633 = 41 \cdot 113$.
7. (a) Estimate the difference by taking the sum of the "triangular regions" between the graph of $\log x$ and the Riemann sum rectangles. (b) Compare $\int_1^n \log x\, dx$ with the sum of the areas of the trapezoids whose tops join the points $(j, \log j)$, and show that the total area between the curve and the trapezoids is bounded by a constant. (c) $\lim_{y \longrightarrow \infty}(\frac{1}{y}\log y! - (\log y - 1)) = 0$, so $\log y - 1$ is the answer.
8. (a) $(1 - 2^{-n})(1 - 2^{-n+1}) \cdots (1 - 2^{-n+k-1})$; (b) 0.298.

9. The term from the rho method becomes $3.2\times 10^{12}$ times as great, while the term from the factor base method becomes $2.6\times 10^{6}$ times as great.
10. (a) For $s < s_0$, we have $h(s) \geq f(s) > f(s_0) = \frac{1}{2}h(s_0)$, and for $s > s_0$, we have $h(s) \geq g(s) > g(s_0) = \frac{1}{2}h(s_0)$. (b) Apply part (a) to $log(f(s))$ and $log(g(s))$.

**§ V.4.**

1. (a) $\frac{1}{1+}\frac{1}{1+}\frac{1}{44}$; (b) $\frac{1}{1+}\frac{1}{1+}\frac{1}{1+}\frac{1}{1+}\frac{1}{1+}\frac{1}{1+}\frac{1}{1+}\frac{1}{1+}\frac{1}{1+1}$; (c) $1+\frac{1}{7+}\frac{1}{1+}\frac{1}{2+}\frac{1}{4}$.
2. (a) Since $a+\frac{1}{x} = x$, it follows that $x$ is the positive root of $x^2 - ax - 1 = 0$, i.e., $x = (a+\sqrt{a^2+4})/2$. (b) Since the $a_i$'s are 1, the recurrence relation for the numerators and denominators of the convergents are the same as for the Fibonacci numbers.
3. $2+\frac{1}{1+}\frac{1}{2+}\frac{1}{1+}\frac{1}{1+}\frac{1}{4+}\frac{1}{1+}\frac{1}{1+}\frac{1}{6}\cdots$; it is possible to show that the $a_i$'s for $i \equiv 2 \ mod\ 3$ are the successive even integers, and all other $a_i$'s are 1.
4. For each $b_i$ you have $b_i^2 - c_i^2 n$ is the least absolute residue of $b_i^2$ modulo $n$. If $p$ divides this least absolute residue, then $b_i^2 \equiv c_i^2 n \ mod\ p$, and this means that $n$ is a quadratic residue modulo $p$.
5. The tables below go through the first value of $i$ such that the least absolute residues of $b_0^2, \ldots, b_i^2$ give a factorization of $n$. In four cases (parts (g), (i), (j), (k)) there is an earlier value of $i$ such that some subset of these residues have corresponding vectors $\vec{\epsilon}_i$ which sum to zero; however, in those cases we end up with $b \equiv \pm c \ mod\ n$.

(a)

| $i$ | 0 | 1 | 2 | 3 |
|---|---|---|---|---|
| $a_i$ | 97 | 1 | 1 | 17 |
| $b_i$ | 97 | 98 | 195 | 3413 |
| $b_i^2 \ mod\ n$ | $-100$ | 95 | $-11$ | 44 |

$B = \{-1, 2, 5, 11\}$, $b = 97\cdot 195\cdot 3413$, $c = 2^2\cdot 5\cdot 11$, $g.c.d.(b+c, n) = 257$.

(b)

| $i$ | 0 | 1 | 2 | 3 |
|---|---|---|---|---|
| $a_i$ | 116 | 2 | 4 | 1 |
| $b_i$ | 116 | 233 | 1048 | 1281 |
| $b_i^2 \ mod\ n$ | $-105$ | 45 | $-137$ | 80 |

$B = \{2, 3, 5\}$, $b = 233 \cdot 1281$, $c = 2^2 \cdot 3 \cdot 5$, $g.c.d.(b+c, n) = 191$.

(c)

| $i$ | 0 | 1 | 2 |
|---|---|---|---|
| $a_i$ | 93 | 1 | 2 |
| $b_i$ | 93 | 94 | 281 |
| $b_i^2 \ mod\ n$ | $-128$ | 59 | $-32$ |

$B = \{-1, 2\}$, $b = 93 \cdot 281$, $c = 2^6$, $g.c.d.(b+c, n) = 67$.

(d)

| $i$ | 0 | 1 | 2 |
|---|---|---|---|
| $a_i$ | 120 | 8 | 3 |
| $b_i$ | 120 | 961 | 3003 |
| $b_i^2$ *mod n* | −29 | 65 | −116 |

$B = \{-1, 2, 29\}$, $b = 120 \cdot 3003$, $c = 2 \cdot 29$, $g.c.d.(b + c, n) = 307$.

(e)

| $i$ | 0 | 1 | 2 | 3 | 4 | 5 | 6 |
|---|---|---|---|---|---|---|---|
| $a_i$ | 111 | 2 | 1 | 2 | 2 | 7 | 1 |
| $b_i$ | 111 | 223 | 334 | 891 | 2116 | 3300 | 5416 |
| $b_i^2$ *mod n* | −82 | 117 | −71 | 89 | −27 | 166 | −39 |

$B = \{-1, 3, 13\}$, $b = 223 \cdot 2116 \cdot 5416$, $c = 3^3 \cdot 13$, $g.c.d.(b + c, n) = 157$.

(f)

| $i$ | 0 | 1 | 2 | 3 | 4 | 5 |
|---|---|---|---|---|---|---|
| $a_i$ | 120 | 1 | 1 | 8 | 2 | 2 |
| $b_i$ | 120 | 121 | 241 | 2049 | 4339 | 10727 |
| $b_i^2$ *mod n* | −127 | 114 | −27 | 98 | −71 | 162 |

$B = \{-1, 2, 3, 7\}$, $b = 2049 \cdot 10727$, $c = 2 \cdot 3^2 \cdot 7$, $g.c.d.(b + c, n) = 199$.

(g)

| $i$ | 0 | 1 | 2 | 3 | 4 | 5 |
|---|---|---|---|---|---|---|
| $a_i$ | 100 | 1 | 1 | 1 | 1 | 2 |
| $b_i$ | 100 | 101 | 201 | 302 | 503 | 1308 |
| $b_i^2$ *mod n* | −123 | 78 | −91 | 97 | −66 | 77 |

$B = \{-1, 2, 3, 7, 11, 13\}$, $b = 101 \cdot 201 \cdot 503 \cdot 1308$, $c = 2 \cdot 3 \cdot 7 \cdot 11 \cdot 13$, $g.c.d.(b + c, n) = 191$.

(h)

| $i$ | 0 | 1 | 2 | 3 | 4 | 5 | 6 | 7 | 8 | 9 |
|---|---|---|---|---|---|---|---|---|---|---|
| $a_i$ | 111 | 1 | 1 | 2 | 1 | 4 | 1 | 6 | 2 | 1 |
| $b_i$ | 111 | 112 | 223 | 558 | 781 | 3682 | 4463 | 5562 | 3138 | 8700 |
| $b_i^2$ *mod n* | −128 | 95 | −67 | 139 | −40 | 163 | −31 | 79 | −115 | 80 |

$B = \{-1, 2, 5\}$, $b = 111 \cdot 781 \cdot 8700$, $c = 2^7 \cdot 5$, $g.c.d.(b + c, n) = 59$.

(i)

| $i$ | 0 | 1 | 2 | 3 | 4 | 5 | 6 | 7 | 8 |
|---|---|---|---|---|---|---|---|---|---|
| $a_i$ | 96 | 1 | 2 | 2 | 5 | 1 | 1 | 1 | 1 |
| $b_i$ | 96 | 97 | 290 | 677 | 3675 | 4352 | 8027 | 3026 | 1700 |
| $b_i^2$ *mod n* | −137 | 56 | −77 | 32 | −107 | 79 | −88 | 89 | −77 |

$B = \{-1, 2, 7, 11\}$, $b = 290 \cdot 1700$, $c = 7 \cdot 11$, $g.c.d.(b + c, n) = 47$.

(j)

| $i$ | 0 | 1 | 2 | 3 | 4 | 5 | 6 | 7 | 8 | 9 |
|---|---|---|---|---|---|---|---|---|---|---|
| $a_i$ | 159 | 1 | 2 | 1 | 1 | 2 | 4 | 1 | 5 | 1 |
| $b_i$ | 159 | 160 | 479 | 639 | 1118 | 2875 | 12618 | 15493 | 13550 | 3532 |
| $b_i^2$ *mod n* | −230 | 89 | −158 | 145 | −115 | 61 | −227 | 50 | −167 | 145 |

$B = \{-1, 2, 5, 23, 29\}$; $b = 639 \cdot 3532$; $c = 5 \cdot 29$; $g.c.d.(b+c, n) = 97$.

(k)

| $i$ | 0 | 1 | 2 | 3 | 4 | 5 | 6 | 7 | 8 |
|---|---|---|---|---|---|---|---|---|---|
| $a_i$ | 133 | 1 | 2 | 4 | 2 | 3 | 1 | 2 | 1 |
| $b_i$ | 133 | 134 | 401 | 1738 | 3877 | 13369 | 17246 | 12115 | 11488 |
| $b_i^2$ *mod n* | −184 | 83 | −56 | 107 | −64 | 161 | −77 | 149 | −88 |

$B = \{-1, 2, 7, 11, 23\}$; $b = 401 \cdot 3877 \cdot 17246 \cdot 11488$; $c = 2^6 \cdot 7 \cdot 11$; $g.c.d.(b + c, n) = 61$.

§ **V.5.**

2. Part 6) is the most time-consuming. Time is bounded by

$$O\left( \sum_{\text{primes } p \leq P} \frac{A}{p} \log p \log n \right) = O(A \log n \log P \log\log P).$$

(The question asked only about steps 1–7; the other time-consuming stage for very large $n$ is finding linearly dependent rows modulo 2 in the matrix of exponents corresponding to the $B$-numbers among the $t^2 - n$.)

3. (a)

| $t$ | $t^2 - n$ | 2 | 13 | 17 | 19 | 29 | 37 | 41 | 47 |
|---|---|---|---|---|---|---|---|---|---|
| 1030 | 14297 | – | – | 1 | – | 2 | – | – | – |
| 1319 | 693158 | 1 | – | 1 | 1 | 1 | 1 | – | – |
| 1370 | 830297 | – | 2 | 3 | – | – | – | – | – |
| 1493 | 1182446 | 1 | – | – | 1 | 2 | 1 | – | – |

Rows 1 and 3 are dependent and lead to the factorization $1879 \cdot 557$.

(b)

| $t$ | $t^2-n$ | 2 | 3 | 5 | 13 | 17 | 19 | 23 | 31 | 37 | 41 |
|---|---|---|---|---|---|---|---|---|---|---|---|
| 1030 | 1209 | – | 1 | – | 1 | – | – | – | 1 | – | – |
| 1043 | 28158 | 1 | 1 | – | 1 | – | 2 | – | – | – | – |
| 1046 | 34425 | – | 4 | 2 | – | 1 | – | – | – | – | – |
| 1047 | 36518 | 1 | – | – | – | – | 1 | – | 2 | – | – |
| 1079 | 104550 | 1 | 1 | 2 | – | 1 | – | – | – | – | 1 |
| 1096 | 141525 | – | 2 | 2 | – | 1 | – | – | – | 1 | – |
| 1123 | 201438 | 1 | 2 | – | – | – | 2 | – | 1 | – | – |
| 1141 | 242190 | 1 | 4 | 1 | 1 | – | – | 1 | – | – | – |
| 1154 | 272025 | – | 3 | 2 | 1 | – | – | – | 1 | – | – |
| 1161 | 288230 | 1 | – | 1 | – | – | 1 | – | – | 1 | 1 |
| 1199 | 377910 | 1 | 2 | 1 | 1 | 1 | 1 | – | – | – | – |
| 1233 | 460598 | 1 | – | – | – | 1 | 1 | 1 | 1 | – | – |
| 1251 | 505310 | 1 | – | 1 | 3 | – | – | 1 | – | – | – |
| 1271 | 555750 | 1 | 2 | 3 | 1 | – | 1 | – | – | – | – |
| 1284 | 588965 | – | – | 1 | 2 | 1 | – | – | – | – | 1 |
| 1309 | 653790 | 1 | 1 | 1 | – | – | 1 | – | 1 | 1 | – |
| 1325 | 695934 | 1 | 2 | – | – | – | – | 1 | – | – | 2 |
| 1366 | 806265 | – | 2 | 1 | – | – | 1 | 1 | – | – | 1 |
| 1371 | 819950 | 1 | – | 2 | – | – | – | 2 | 1 | – | – |
| 1420 | 956709 | – | 2 | – | 2 | 1 | – | – | – | 1 | – |
| 1504 | 1202325 | – | 1 | 2 | – | 1 | – | 1 | – | – | 1 |

Rows 1, 2 and 7 are dependent mod 2, but do not lead to a nontrivial factor. Rows 1 and 9 are dependent and lead to the factorization $1787 \cdot 593$.

(c)

| $t$ | $t^2-n$ | 2 | 5 | 7 | 11 | 17 | 19 | 37 | 43 | 47 |
|---|---|---|---|---|---|---|---|---|---|---|
| 1001 | 3230 | 1 | 1 | – | – | 1 | 1 | – | – | – |
| 1003 | 7238 | 1 | – | 1 | 1 | – | – | – | – | 1 |
| 1004 | 9245 | – | 1 | – | – | – | – | – | 2 | – |
| 1018 | 37553 | – | – | – | – | 1 | – | – | – | 2 |
| 1039 | 80750 | 1 | 3 | – | – | 1 | 1 | – | – | – |
| 1056 | 116365 | – | 1 | – | – | 1 | – | 2 | – | – |
| 1069 | 143990 | 1 | 1 | 1 | 2 | 1 | – | – | – | – |
| 1086 | 180625 | – | 4 | – | – | 2 | – | – | – | – |
| 1090 | 189329 | – | – | 1 | – | 1 | – | 1 | 1 | – |
| 1146 | 314545 | – | 1 | 1 | 1 | – | 1 | – | 1 | – |
| 1164 | 356125 | – | 3 | 1 | 1 | – | – | 1 | – | – |
| 1191 | 419710 | 1 | 1 | – | – | – | 1 | – | – | 2 |
| 1241 | 541310 | 1 | 1 | 1 | 1 | – | 1 | 1 | – | – |
| 1311 | 719950 | 1 | 2 | 1 | 2 | 1 | – | – | – | – |
| 1426 | 1034705 | – | 1 | 1 | – | 1 | – | 1 | – | 1 |

Rows 1 and 5 are dependent and lead to the factorization $661 \cdot 1511$.

## § VI.1.

1. Either the circle group (if the real curve has one connected component) or the product of the circle group and the two-element group (if it has two connected components). An example of the first is $y^2 = x^3 + x$; an example of the second is $y^2 = x^3 - x$ (for an equation of the form (1), this depends on whether the cubic on the right has 1 or 3 real roots).
2. $n^2$ complex points of order $n$; $n$ real points of order $n$ if $n$ is odd, and either $n$ or $2n$ if $n$ is even, depending on whether the real curve has one or two components.
3. Same examples as in Exercise 1.
4. (a) On the $x$-axis; (b) inflection point; (c) a point where a line from an $x$-intercept of the curve is tangent to the curve (in addition to the points in (a)).
5. (a) 3; (b) 4; (c) 7; (d) 5.
6. Characteristic 2: $x_3 = \frac{y_1^2+y_2^2}{x_1^2+x_2^2} + x_1 + x_2$, $y_3 = c + y_1 + \frac{y_1+y_2}{x_1+x_2}(x_1+x_3)$, and when $P = Q$ we have $x_3 = \frac{x_1^4+a^2}{c^2}$, $y_3 = c + y_1 + \frac{x_1^2+a}{c}(x_1+x_3)$; and for equation (2b): $x_3 = \frac{y_1^2+y_2^2}{x_1^2+x_2^2} + \frac{y_1+y_2}{x_1+x_2} + x_1 + x_2 + a$, $y_3 = \left(\frac{y_1+y_2}{x_1+x_2}\right)(x_1 + x_3) + x_3 + y_1$, and when $P = Q$ we have $x_3 = x_1^2 + \frac{b}{x_1^2}$, $y_3 = x_1^2 + (x_1 + \frac{y_1}{x_1})x_3 + x_3$; characteristic 3: $x_3 = \left(\frac{y_2-y_1}{x_2-x_1}\right)^2 - a - x_1 - x_2$, $y_3 = -y_1 + \frac{y_2-y_1}{x_2-x_1}(x_1 - x_3)$, and when $P = Q$ we have $x_3 = \left(\frac{ax_1-b}{y_1}\right)^2 - a + x_1$, $y_3 = -y_1 + \frac{ax_1-b}{y_1}(x_1 - x_3)$.
7. (a) Show that in each pair $\{a, -a\}$ exactly one of the values $x = \pm a$ leads to 2 solutions $(x, y)$ to the equation (treat $x = 0$ and the point at infinity separately). (b)–(c) Use the fact that $x \mapsto x^3$ is a 1-to-1 map of $\mathbf{F}_q$ to itself when $q \equiv 2 \ mod \ 3$.
8. The following table shows the type of the abelian group for each value of $q$ and each of the two elliptic curves:

| $q$ | 3 | 5 | 7 | 9 | 11 | 13 | 17 | 19 | 23 | 25 | 27 |
|---|---|---|---|---|---|---|---|---|---|---|---|
| $y^2 = x^3 - x$ | $(2,2)$ | $(4,2)$ | $(4,2)$ | $(4,4)$ | $(2,2,3)$ | $(4,2)$ | $(4,4)$ | $(2,2,5)$ | $(4,2,3)$ | $(8,4)$ | $(2,2,7)$ |
| $y^2 = x^3 - 1$ | $--$ | $(2,3)$ | $(2,2)$ | $--$ | $(4,3)$ | $(2,2,3)$ | $(2,9)$ | $(2,2,7)$ | $(8,3)$ | $(2,2,3,3)$ | $--$ |

9. (a) Let $P = (x, y)$. Then $-P = (x, y + 1)$, $2P = (x^4, y^4 + 1)$. (b) We have $2(2P) = (x^{16}, y^{16} + 1 + 1) = (x^{16}, y^{16}) = (x, y) = P$. (c) By part (b), $2P = -P$, i.e., $(x^4, y^4+1) = (x, y+1)$; but this means that $x^4 = x$ and $y^4 = y$, so that $x$ and $y$ are in the field of 4 elements. By Hasse's theorem, the number $N$ of points is within $2\sqrt{4} = 4$ of $4+1$ and within $2\sqrt{16} = 8$ of $16 + 1$, i.e., $N = 9$.

10. The denominator of the zeta function is always $(1-T)(1-pT)$; the following table shows the numerator for $p = 5, 7, 11, 13$:

| | | | | |
|---|---|---|---|---|
| $y^2 = x^3 - x$ | $1+2T+5T^2$ | $1+7T^2$ | $1+11T^2$ | $1-6T+13T^2$ |
| $y^2 = x^3 - 1$ | $1+5T^2$ | $1-4T+7T^2$ | $1+11T^2$ | $1-2T+13T^2$ |

11. In both cases there is no solution $(x, y)$ to the equation over $\mathbf{F}_p$, so the only point is the point at infinity. The numerator of the zeta function is $1-2T+2T^2$ and $1-3T+3T^2$, respectively. Then $N_r = \mathbf{N}((1+i)^r - 1)$ and $\mathbf{N}((1+\omega)^r - 1)$, respectively, where $\omega = (-1+i\sqrt{3})/2$.

§ **VI.2.**

1. Pick elements of $\mathbf{F}_q$ at random, and stop when you find $g$ such that $g^{(q-1)/2} = -1$ (rather than $+1$).
2. Let $x \in \mathbf{F}_q$ correspond to $m$. (a) Let $f(x) = x^3 - x$. Note that precisely one of $f(x)$, $f(-x) = -f(x)$ is a square. Let $y = f(x)^{(q+1)/4}$. Then show that either $(x, y)$ or $(-x, y)$ is a point on the curve. (b) Choose any $y$, set $x = (y^2+y)^{(2-q)/3}$ (unless $y = 0$ or $-1$, in which case set $x = 0$), and show that $(x, y)$ is on the curve.
3. (a) The sequence of points $(x, y)$ is:

$$(562, 576),\ (581, 395),\ (484, 214),\ (501, 220),\ (1, 0),\ (1, 0),\ (144, 565).$$

   (b) ICANT (I can't).
4. (a) $E$ *mod* $p$ has a noncyclic subgroup, namely, the group of points of order 2; (b) $E$ *mod* $p$ has a subgroup of order 2 or 4, namely, the points of order 2.
5. Use the formulas in Example 5 of §1. (a) Use congruence modulo 3 to show that in both cases ($r$ odd and $r$ even) one has $3|N_r$. (b) When $4|r$ we have: $N_r = (2^{r/2}-1)^2 = (2^{r/4}+1)^2(2^{r/4}-1)^2$, which is divisible by an $(r/4)$-bit prime if and only if $r/4$ is a prime for which $2^{r/4}-1$ is a Mersenne prime; it is divisible by an $(r/4+1)$-bit prime if and only if $r/4 = 2^k$ with $2^{2^k}+1$ a Fermat prime.
6. (a) The $\mathbf{F}_p$-points then form a proper subgroup of the $\mathbf{F}_{p^r}$-points (by Hasse's theorem), and that subgroup has more than 1 element (also by Hasse's theorem). Thus, $N_r$ has a proper divisor. (b) In both cases let $E$ have equation $y^2 + y = x^3 - x + 1$; one easily checks that over $\mathbf{F}_2$ or $\mathbf{F}_3$ the curve has no points except for the point at infinity $O$. Thus, the argument in part (a) does not apply, and one finds that when $p = 2$ we have $N_2 = 5$, $N_3 = 13$, $N_5 = 41$, $N_7 = 113$, $N_{11} = 2113$ (note that the zeta-function is $(1-2T+2T^2)/(1-T)(1-2T)$; for $r$ prime $N_r$ is prime if and only if the so-called "complex Mersenne number" $(1+i)^r - 1$ is a prime in the Gaussian integers, or equivalently, if and only if $2^r + 1 - (\frac{2}{r})2^{(r+1)/2}$ is a prime, where $(\frac{2}{r})$ is the Legendre symbol); when $p = 3$ we have $N_2 = 7$, $N_5 = 271$, $N_7 = 2269$ (here the zeta-function is $(1-3T+3T^2)/(1-T)(1-3T)$).
7. (a) $y^2 + y = x^3 + \alpha$, where $\alpha$ is either of the elements of $\mathbf{F}_4$ not in $\mathbf{F}_2$. (b) The zeta-function is $(1-4T+4T^2)/(1-T)(1-4T)$, and the two

reciprocal roots of the numerator are both 2; then use the remark at the end of §1. (c) The double of $(x, y)$ is $(x^4, y^4)$ (note that the 4th-power map is the "Frobenius" map, i.e., the generator of the Galois group of $\mathbf{F}_{4^r}$ over $\mathbf{F}_4$). (d) Doubling any point $r$ times gives $(x^{4^r}, y^{4^r}) = (x, y)$, i.e., any $P \in E$ satisfies $2^r P = P$.

8. (a) Use the fact that something is in $\mathbf{F}_2$ if and only if it satisfies $x^2 = x$; and also the fact that $(a+b)^2 = a^2 + b^2$ in a field of characteristic 2. (b) The map $z \mapsto z+1$ gives a 1-to-1 correspondence between the $z$'s with trace 0 and the $z$'s with trace 1. (c) Choose random $x \in \mathbf{F}_{2^r}$, substitute the cubic $x^3 + ax + b$ for $z$ in $g(z)$, and if $z = x^3 + ax + b$ lands in the 50% of elements with trace 0, then the point $(x, g(z))$ is on the curve.
9. When working with $E$ modulo $p$, one uses the same formulas (4)–(5) of §1, and one gets the point at infinity when one adds two smaller multiples $kP = k_1P + k_2P$ which, when reduced modulo $p$, have the same $x$-coordinate and the negative of each other's $y$-coordinate. That is equivalent to conditions (1)–(2) in the exercise.
10. The denominator of $8P$ is divisible by $p = 23$, and so $P$ *mod* 23 has order 8 on $E$ *mod* 23, by Exercise 9. However, Hasse's theorem shows that $E$ *mod* 23 has more than 8 points.
11. $(676, 182)$, $(385, 703)$; $(595, 454)$, $(212, 625)$; $(261, 87)$, $(77, 369)$; $(126, 100)$, $(66, 589)$; $(551, 606)$, $(501, 530)$; $(97, 91)$, $(733, 110)$; $(63, 313)$, $(380, 530)$.

§ **VI.3.**

1. (a) $1 - 1/q$; (b) $1 - 1/q$.
3. (a) If $n = 2^{2^k} + 1$ is prime, then any $a$ with $\left(\frac{a}{n}\right) = -1$ has this property. See Exercise 15 of § II.2 concerning $a = 3, 5, 7$. On the other hand, if $p$ is a proper prime divisor of $n$, and if $a^{2^{2^k-1}} \equiv -1$, then $2^{2^k}$ but not $2^{2^k-1}$ is a multiple of the order of $a$ modulo $p$, i.e., this order is $2^{2^k} = n - 1 > p - 1$, which is impossible. (b) First suppose that $n = 2^p - 1$ is prime. To show that $E$ *mod* $n$ has $2^p$ points, see Exercise 7(a) of § VI.1. To show that the group is cyclic, prove that there are only two points of order 2, because the cubic $x^3 + x$ has only one root modulo $n$. Then any of the 50% of the points which generate $E$ *mod* $n$ (i.e., which are not the double of any point in $E$ *mod* $n$) have the properties (1)–(2). Conversely, suppose that $n$ has a proper prime divisor $\ell$. If $P$ satisfied properties (1)–(2), then on $E$ *mod* $\ell$ the order of $P$ would divide $2^p$ but not $2^{p-1}$, i.e., it would be $2^p$. But then $2^p = n+1$ would divide the number of points on $E$ *mod* $\ell$, and this contradicts Hasse's theorem, which tells us that this number is $< \ell + 2\sqrt{\ell} + 1$. To generate random points on $E$ *mod* $n$, choose $x \in \mathbf{Z}/n\mathbf{Z}$ randomly. If $b = x^3 + x$ happens to be a square modulo $n$, then setting $y = b^{(n+1)/4}$ will give $y^2 \equiv b \cdot b^{(n-1)/2} \equiv x^3 + x$. (See Remark 1 at the end of § II.2.)

§ **VI.4.**

1. $g.c.d.(2^k - 1, n) = n$, but $g.c.d.(3^k - 1, n) = 127$; $n = 127 \cdot 421$.
2. The probability that a random residue $a$ in $(\mathbf{Z}/p\mathbf{Z})^*$ satisfies $p|a^k - 1$ is one out of $(p-1)/g.c.d.(k, p-1)$. Since there is little chance that $a^k - 1$ will be divisible by any other divisor of $n$, this is also an estimate of the probability that $g.c.d.(a^k - 1, n) = p$.
3. (a) 3 out of 41; (b) 22 out of 41; (c) 25 out of 127; (d) 68 out of 127; (e) 105 out of 399.
4. Choose $k = 2^6 \cdot 3^4 \cdot 5^2$. Here are the first value of $a$ for which the method gives a factor, the factor it gives, and the value of $k_1$ for which the algorithm terminates: (a) 1, 37, $2^3$; (b) 2, 71, $2^6 \cdot 3^4 \cdot 5$; (c) 1, 67, $2^6 \cdot 3^4 \cdot 5$; (d) 1, 47, $2^6 \cdot 3$; (e) 2, 79, $2^6 \cdot 3^4 \cdot 5^2$; (f) 1, 73, $2^6 \cdot 3$; (g) 5, 53, $2^2$; (h) 4, 59, $2^6 \cdot 3^2$; (i) 1, 47, $2^6 \cdot 3$; (j) 3, 97, $2^6 \cdot 3$; (k) 1, 61, $2^6 \cdot 3^4 \cdot 5^2$.
5. If the latter possibility occurred, it would mean that $\ell'(k_1/\ell)P \ mod\ p = O \ mod\ p$ for some $\ell' < \ell$, while $(k_1/\ell)P \ mod\ p \neq O \ mod\ p$. But $\ell'$ is a product of primes $\ell^* < \ell$, and our choice of exponents in (2) ensured that for each such $\ell^*$ the highest power of $\ell^*$ that could divide the order of $P \ mod\ p$ in $E \ mod\ p$ already occurred in $(\ell^*)^{\alpha_{\ell^*}}$, i.e., in $k_1/\ell$.
6. (a) If $n$ happens to be divisible only by primes which are $\equiv 3 \ mod\ 4$, then there are always $p+1$ points on $E \ mod\ p$ for $p|n$ (see Exercise 7(a) of §1 for the case $a = -1$; but the same argument applies for any $a$). In that case it won't help to vary $a$ if $p+1$ is divisible by a large prime for each $p|n$. (b) If $n$ happens to be divisible only by primes $p \equiv 2 \ mod\ 3$, then there are always $p+1$ points (see Exercise 7(b) of §1), and so again it won't help to vary $b$ if $p+1$ is divisible by a large prime for each $p|n$.
7. Generate pairs $(E, P)$ where $E$ has equation $y^2 = x(x-a)(x-b)$; then $E$ has four points of order 2, including the point at infinity (see Exercise 4(a) of §VI.1). To do this, choose random $a, x, y_0$; set $y = x(x-a)y_0$ and then $b = x - yy_0$.

# Index